Lecture Notes in Control and Information Sciences 246

Editor: M. Thoma

Springer-Verlag London Ltd.

Dirk Aeyels, Françoise Lamnabhi-Lagarrigue
and Arjan van der Schaft (Eds)

Stability and Stabilization of Nonlinear Systems

Springer

Editors

Dirk Aeyels, Professor
Dept. of Systems Dynamics, Universiteit Gent, Technologiepark-Zwijnaarde 9, 9052 Gent, Belgium

Françoise Lamnabhi-Lagarrigue, Docteur d'état
Laboratoire des Signaux et Systèmes, CNRS, SUPELEC, 91192 GIF-SUR-YVETTE, France

Arjan van der Schaft, Professor
Dept. of Applied Mathematics, University of Twente, PO Box 217, 7500 Enschede, The Netherlands

ISBN 978-1-85233-638-7 **ISBN 978-1-84628-577-6 (eBook)**

DOI 10.1007/978-1-84628-577-6

British Library Cataloguing in Publication Data
Stability and stabilization of nonlinear systems : editors,
D. Aeyels, F. Lamnabhi-Lagarrigue and A. van der Schaft. -
(Lecture notes in control and information sciences)
1.Nonlinear systems 2.Nonlinear control theory
I.Aeyels, D. II.Lamnabhi-Lagarrigue, F. (Françoise)
III.Schaft, A.J. van der
629.8'36
ISBN 978-1-85233-638-7

Library of Congress Cataloging-in-Publication Data
A catalog record for this book is available from the Library of Congress

Originally published by Springer-Verlag London Berlin Heidelberg **in 1999**

Typesetting: Camera ready by contributors

69/3830-543210 Printed on acid-free paper SPIN 10723684

Preface

The chapters of this book have been presented at the 1st Workshop of the Nonlinear Control Network*, which was held in Ghent, March 15,16, 1999. These contributions give an overview of some of the current and emerging trends in nonlinear systems and control theory. As editors of this book we would very much like to thank the speakers at this workshop for their stimulating presentations and for their efforts to bring this material to its current form, which we are sure will provide stimulating reading as well.

Dirk Aeyels
Françoise Lamnabhi-Lagarrigue
Arjan J. van der Schaft

*The Nonlinear Control Network is a four year project within the framework of the European Commission's Training and Mobility of Researchers (TMR) Programme that started on December 1, 1997. There are nine partners involved:

Dirk Aeyels Dirk.Aeyels@rug.ac.be	Universiteit Gent Belgium
Alfonso Baños abanos@dif.um.es	Universidad de Murcia Spain
Fritz Colonius Colonius@math.uni-augsburg.de	Universität Augsburg Germany
Alberto Isidori isidori@giannutri.caspur.it	Università di Roma Italy
Françoise Lamnabhi-Lagarrigue lamnabhi@lss.supelec.fr	Centre National de la Recherche Scientifique France (coordinator)
David H. Owens D.H.Owens@sheffield.ac.uk	University of Sheffield England
Arjan J. van der Schaft a.j.vanderschaft@math.utwente.nl	Universiteit Twente The Netherlands
Fatima Silva Leite fleite@mat.uc.pt	Universidade de Coimbra Portugal
John Tsinias jtsin@math.ntua.gr	National Technical University of Athens Greece

url: Nonlinear Control Network http://www.supelec.fr/lss/NCN

Contents

1. Disturbance attenuation for discrete-time feedforward nonlinear systems

*Tarek Ahmed-Ali**, *Frédéric Mazenc*** and *Françoise Lamnabhi-Lagarrigue**

*Laboratoire des Signaux et Systèmes, C.N.R.S.
SUPELEC
91192 Gif-sur-Yvette, France
ahmed@lss.supelec.fr
lamnabhi@lss.supelec.fr

**Centre for Process Systems Engineering
Imperial College of Science Technology and Medecine
London SW7 2AZ, U.K.
f.mazenc@ic.ac.uk

Summary.

In this paper the inverse optimal stabilization problem is solved for nonlinear nonaffine control discrete-time systems which are globally stable when uncontrolled. Stabilizing feedback laws and nonquadratic cost functionals are constructed. The result is applied to feedforward systems.

1.1 Introduction

Inverse optimal control problems for linear system have been studied more than thirty years ago by Kalman [7] and next by Anderson and Moore [1]. However, inverse optimality for nonlinear systems is a more recent subject studied by Moylan and Anderson [16] and later by Freeman and Kokotović [2], [3]. Other works have followed [5, Section 3.5], [4, 8, 9]. This area of the nonlinear control theory arises from the wish to determine for nonlinear systems stabilizing feedbacks having the good performances of those resulting from the optimization theory without having to solve the Hamilton-Jacobi-Bellman equation which is not always a feasible task. In the inverse optimal approach, a stabilizing feedback is designed first and then it is shown to be optimal for a cost functional of the form

$$\int_0^{\infty} \left(l(x) + u^{\top} R(x)u\right) dt$$

where $l(x)$ is positive and and $R(x)$ is positive definite. In other words, the functions $l(x)$ and $R(x)$ are *a posteriori* determined from the data of a particular stabilizing feedback, rather than *a priori* chosen by the designer i.e. regardless of any stabilizing feedback.

So far, to the best knowledge of our, all the works on this subject are concerned with continuous-time systems. The purpose of the present paper is to address the problem of inverse optimal stabilization for general classes of nonlinear discrete-time systems. More precisely, its aim is twofold. In a first part, we show that the stabilizing feedbacks designed in [13, Appendix A] and in [14, Section 2] for nonlinear discrete-time systems

$$x_{i+1} = f(x_i) + g(x_i, u_i)u_i \tag{1.1}$$

with $f(0) = 0$, $x \in \mathbf{R}^n$ and $u \in \mathbf{R}$ which are globally stable when $u = 0$ also minimize cost functionals of the form

$$J = \sum_{i=0}^{\infty} \left(\alpha(x_i, u_i) + \beta(x_i, u_i)|u_i|^2\right) .$$

The functions $\alpha(x, u)$ and $\beta(x, u)$ of the cost functional we consider depend on u, although in inverse optimal control for continous-time systems, no such cost functional are in general considered. This is due to the specificities of discrete-time systems and in Section 1.3 we show how to infer from these costs (ater a slight modification) disturbance attenuation properties. In a second part, we apply the results of the first to a particular class of discrete-time feedforward systems i.e. systems which admit a representation of the form:

$$\begin{aligned} z_{i+1} &= \mathcal{F}(z_i) + \psi(z_i, \xi_i) + g_2(z_i, \xi_i, u_i)u_i + m_1(z_i, \xi_i, u_i, d_i)d_i \\ \xi_{i+1} &= a(\xi_i) + g_1(z_i, \xi_i, u_i)u_i + m_2(z_i, \xi_i, u_i, d_i)d_i \end{aligned}$$

where $m_1(z_i, \xi_i, u_i, d_i)d_i$ and $m_2(z_i, \xi_i, u_i, d_i)d_i$ are disturbances, where $z_{i+1} = \mathcal{F}(z_i)$ is globally stable, $\xi_{i+1} = a(\xi_i)$ is globally asymptotically stable and locally exponentially stable $\psi(z, 0) = 0$ for all z. Observe that the study of this class of systems is, from a practical point of view, appealing. On the one hand, many physical systems are described by feedforward equations (Cart-pendulum system, Ball and Beam with friction term, PVTOL) and, on the other hand the technique provides us with bounded feedbacks for null-controllable linear systems. These systems, in continuous-time, have been studied for the first time by A. Teel in [19] where a family of stabilizing feedbacks is displayed. Many extentions this pioneer work have followed: see [13, 5, 15, 12, 18] where new family of controls, control Lyapunov functions and output feedback results are given. A discrete-time version of the main result of [13] is proved in [14]. In continuous time, it is shown in [17, Section 6.2.2] that the forwarding design applied to affine systems has stability margins. This result is proved via inverse optimal results. This result has no direct equivalent in discrete-time because, on the one hand, even affine discrete-time feedforward systems cannot be rendered passive in the classical sense and, on the other hand, only bounded stabilizing feedbacks are available: see [14]. However, our work owes a great deal to the technique of proof of this result as long as to those of [8], where is stressed the link there is between inverse optimal control and disturbance attenuation, and [20] where a disturbance attenuation result for continuous-time feedforward system is given. To consider a slightly larger class of systems than the one studied in [14], we propose a discrete-time version of the main result of [5] to construct a Lyapunov function enabling us to prove disturbance attenuation properties of the closed-loop systems.

Preliminaries and definitions

1. Throughout the paper we assume that the functions encountered are sufficiently smooth.

2. We denote by χ_i the solution of the discrete-time system:

$$\chi_{i+1} = \mathcal{H}(\chi_i) \tag{1.2}$$

with the initial condition $\chi_0 = \chi$.

3. For a sequence χ_i solution of (1.2) and a function $\mathcal{V}(\chi)$ we denote by $\Delta\mathcal{V}$ the term $\mathcal{V}(\chi_{i+1}) - \mathcal{V}(\chi_i)$.

4. A function $\mathcal{V}(\chi_i)$ is positive definite if

$$\mathcal{V}(\chi) > 0 , \forall\chi \neq 0 .$$

5. The *inverse optimal stabilization problem* for discrete-time systems (1.1) is solvable if there exist positive real-valued functions $\alpha(x,u)$ and $\beta(x,u)$ such that there exists a feedback law $u(x)$ which globally asymptotically stabilizes (1.1) and at the same time minimizes the cost functional

$$J = \int_0^{\infty} (\alpha(x,u) + \beta(x,u)u^2)dt.$$

1.2 Inverse optimal control

In this section, we study the inverse optimal problem for the discrete-time nonlinear systems (1.1) and introduce the following assumptions:

H1. There exists a proper, positive definite function $V(x)$ and such that $V(f(x)) - V(x) \leq 0$.

H1'. There exists a positive definite function $V(x)$ such that $V(f(x)) - V(x) \leq 0$.

H2. The sets

$$\Omega = \{x \in \mathbf{R}^n : V(f^{i+1}(x)) = V(f^i(x)) \ , \ i = 0,1,2,\ldots\}$$

$$S = \{x \in \mathbf{R}^n : \frac{\partial V}{\partial x}(f^{i+1}(x))g(f^i(x),0) = 0 \ , \ i = 0,1,2,\ldots\}$$

are such that

$$\Omega \bigcap S = \{0\}.$$

Remark. The system (1.1) is a single-input system. Generalizations of our results to the case of multi-input systems can be carried out but it turns out that the proofs are then much more intricate.

1.2.1 Globally asymptotically stabilizing feedback

Let us recall a result which is an immediate consequence of [13, Lemma II.4] or of the feedback design of [14, Section2] and is an extension of [11, Corollary 3.1].

Theorem 1.2.1. *Consider the discrete-time systems (1.1). Assume that Assumptions H1 and H2 are satisfied. Then for all function* $\mu(x) > 0$*, there exists a smooth function* $\phi(x)$ *such that the following feedback control*

$$\overline{u}(x) = -\phi(x)h(x,0), \qquad 0 < \phi(x) \leq \mu(x) \tag{1.3}$$

$$h(x,u) = \int_0^1 \frac{\partial V}{\partial x}(f(x) + g(x,u)u\theta)g(x,u)d\theta \tag{1.4}$$

globally asymptotically stabilizes the system (1.1).

Theorem 1.2.2. *Consider the discrete-time systems (1.1). Assume that Assumptions H1' and H2 are satisfied. Then, for all function* $\mu(x) > 0$ *such that there exists a smooth function* $\phi(x)$ *such that all the solutions of (1.1) in closed-loop with the following feedback control*

$$\overline{u}(x) = -\phi(x)h(x,0), \qquad 0 < \phi(x) \leq \mu(x) \tag{1.5}$$

$$h(x,u) = \int_0^1 \frac{\partial V}{\partial x}(f(x) + g(x,u)u\theta)g(x,u)d\theta \tag{1.6}$$

are bounded, the system (1.1) is globally asymptotically stabilizes by the feedbacks (1.5).

Discussion of Theorem 1.2.1 and Theorem 1.2.2.

i) Observe that the feedbacks (1.3), (1.5) are given by explicit formulas and not as the implicit solutions of nonlinear algebraic equations which do not necessarily admit a solution as those proposed in [10] are.

ii) Theorem 1.2.1 is a discrete-time nonaffine version of the Jurdjevic-Quinn theorem [6]: Assumption H1 guarantees the global stability of (1.1) with $u = 0$ and the technical Assumption H2 guarantees that a detectability property which allows to conclude by invoking the LaSalle invariance principle with arbitrarily small feedbacks is satisfied.

iii) The main difference there is between Theorem 1.2.1 and Theorem 1.2.2 is clear: the first theorem requires the knowledge of a proper function $V(x)$ but not the second.

1.2.2 Optimal criterion design for discrete-time systems

Let us state the main result.

Theorem 1.2.3. *Consider the system (1.1). Assume that the assumptions of Theorem 1.2.1 or Theorem 1.2.2 are satisfied. Then, for all function $\mu(x) > 0$ the inverse optimal stabilization problem is solved by the control law*

$$\overline{u}(x) = -\phi(x)h(x,0), \qquad 0 < \phi(x) \leq \mu(x) \tag{1.7}$$

with $h(x,u)$ given in (1.4) and the cost functional

$$J = \sum_{i=0}^{\infty}(\alpha(x_i,u_i) + \beta(x_i,u_i)u_i^2) \tag{1.8}$$

with

$$\alpha(x,u) = V(x) - V(f(x)) + \frac{1}{2}\phi(x)\rho(u)h(x,0)^2 \tag{1.9}$$

$$\beta(x,u) = k(x,u) + \frac{\rho(u)}{2\phi(x)} - h(x,0)\frac{1-\rho(u)}{u} \tag{1.10}$$

where $\rho(u)$ is a strictly positive function given by an explicit formula and such that $\rho(0) = 1$ and $k(x,u)$ is defined by

$$-h(x,u) = -h(x,0) + k(x,u)u \; . \tag{1.11}$$

Proof. Theorem 1.2.1 or Theorem 1.2.2 provide us with a globally asymptotically stabilizing feedback $\overline{u}(x)$ of the form (1.3) or (1.5). Let us consider the following criterion

$$S = -\sum_{i=0}^{\infty}(V(x_{i+1}) - V(x_i)) + \sum_{i=0}^{\infty}\frac{\rho(u_i)}{2\phi(x_i)}(u_i - \overline{u}(x_i))^2 .$$

Observe that, since $\rho(u)$ is a positive function, $u = \overline{u}(x)$ minimizes S because, when a globally asymptotically feedback is applied,

$$S \; = \; V(x_0) + \sum_{i=0}^{\infty}\frac{\rho(u_i)}{2\phi(x_i)}(u_i - \overline{u}(x_i))^2 .$$

The function S rewrites as:

$$\begin{aligned} S = &-\sum_{i=0}^{\infty}(V(f(x_i)) - V(x_i)) - \sum_{i=0}^{\infty} h(x_i, u_i)u_i \\ &+\sum_{i=0}^{\infty}\left[\frac{\rho(u_i)}{2\phi(x_i)}u_i^2 - \frac{\rho(u_i)}{\phi(x_i)}\overline{u}(x_i)u_i + \frac{\rho(u_i)}{2\phi(x_i)}\overline{u}(x_i)^2\right] . \end{aligned}$$

Using (1.11), we get:

$$\begin{aligned} S = &-\sum_{i=0}^{\infty}(V(f(x_i)) - V(x_i)) \\ &+\sum_{i=0}^{\infty} k(x_i, u_i)u_i^2 - \sum_{i=0}^{\infty} h(x_i, 0)u_i \\ &+\sum_{i=0}^{\infty}\left[\frac{\rho(u_i)}{2\phi(x_i)}u_i^2 + h(x_i, 0)\rho(u_i)u_i + \frac{\rho(u_i)}{2\phi(x_i)}\overline{u}(x_i)^2\right] . \end{aligned}$$

Regrouping the terms differently, we obtain:

$$\begin{aligned} S = &\sum_{i=0}^{\infty}\left[(V(x_i) - V(f(x_i))) + \frac{1}{2}\phi(x_i)\rho(u_i)h(x_i, 0)^2\right] \\ &+\sum_{i=0}^{\infty}\left[k(x_i, u_i) + \frac{\rho(u_i)}{2\phi(x_i)} - h(x_i, 0)\frac{1-\rho(u_i)}{u_i}\right] u_i^2 . \end{aligned}$$

Therefore $S = J$ where J is the function defined in (1.8). Assumption **H1** guarantees that $\alpha(x, u)$ is positive. The proof is completed by using the following lemma

Lemma 1.2.1. *Functions $\phi(x)$ and $\rho(u)$ such that $\beta(x, u)$ is a strictly positive function can be determined.*

Proof. Since $\rho(u)$ is smooth and such that $\rho(0) = 1$, the function $\left|\frac{1-\rho(u)}{u}\right|$ is bounded on a neighborhood of the origin. It follows readily that the expression of a function $\phi(x)$ such that, for all x, u

$$\frac{\rho(u)}{4\phi(x)} - h(x,0)\frac{1-\rho(u)}{u} \geq 0$$

can be given. Next, let us prove that we may determine $\phi(x)$ and $\rho(u)$ so that

$$k(x,u) + \frac{\rho(u)}{4\phi(x)} \geq 0. \tag{1.12}$$

Determining explicit formulas of positive functions $\lambda(x), \mu(u)$ such that $|k(x,u)| \leq \lambda(x)\mu(u)$ and $\mu(0) \leq 1$ is always a feasible task. When this inequality holds, (1.12) is satisfied if

$$\frac{1}{\phi(x)} \geq 4\lambda(x)\frac{1}{\rho(u)}\mu(u).$$

which is met with $\phi(x) \leq \frac{1}{4\lambda(x)+1}$ and $\rho(u) \geq \mu(u)$. This allows us to conclude our proof.

1.3 Disturbance attenuation for discrete-time systems

1.3.1 Inverse optimal $\mathcal{H}_\infty$ problem

In this part we consider the system

$$x_{i+1} = f(x_i) + g(x_i, u_i)u_i + m(x_i, u_i, d_i)d_i \tag{1.13}$$

with $f(0) = 0$ and $x \in \mathbf{R}^n$, $u \in \mathbf{R}$, $d \in \mathbf{R}$. We assume that $m(x,u,d)$ is known and that d_i is an unknown sequence of class L^2.

Define a function $h_d(x,u,d)$ as:

$$\begin{aligned} h_d(x,u,d)d = & \int_0^1 \left[\frac{\partial V}{\partial x}(f(x) + g(x,u)u\theta + m(x,u,d)d\theta) \right. \\ & \left. \times\, g(x,u)\right] d\theta \\ & - \int_0^1 \left[\frac{\partial V}{\partial x}(f(x) + g(x,u)u\theta)g(x,u)\right] d\theta \\ & + \int_0^1 \left[\frac{\partial V}{\partial x}(f(x) + g(x,u)u\theta + m(x,u,d)d\theta) \right. \\ & \left. \times m(x,u,d)d\right] d\theta \end{aligned} \tag{1.14}$$

and $k_d(x,u,d)$ by

$$-h_d(x,u,d) = -h_d(x,u,0) + k_d(x,u,d)d \tag{1.15}$$

Lemma 1.3.1. *Consider the system (1.13). Assume that the assumptions of Theorem 1.2.1 or Theorem 1.2.2 are satisfied. Then, for all function $\mu(x) > 0$ the following problem is solved:*

The control law

$$\overline{u}(x) = -\phi(x)h(x,0)\,, \qquad 0 < \phi(x) \leq \mu(x) \tag{1.16}$$

with $h(x,u)$ given in (1.4) minimizes the cost functional

$$J_d = \sup_{\{d_i \in \mathcal{D}\}} \sum_{i=0}^{\infty} (\alpha(x_i,u_i) + \beta(x_i,u_i)u_i^2 - \beta_d(x_i,u_i,d_i)d_i^2) \tag{1.17}$$

with $\mathcal{D}$ the set such that all the solutions of the system (1.1) in closed-loop with $\overline{u}(x)$ with distubances $(d_i) \in \mathcal{D}$ go to the origin with

$$\begin{aligned} \alpha(x,u) = V(x) - V(f(x)) + \tfrac{1}{2}\phi(x)\rho(u)h(x,0)^2 \\ + \tfrac{1}{2}\phi_d(x,u)h_d(x,u,0)^2 \end{aligned} \tag{1.18}$$

$$\beta(x,u) = k(x,u) + \frac{\rho(u)}{2\phi(x)} - h(x,0)\frac{1-\rho(u)}{u} \tag{1.19}$$

where $\rho(u)$ is a strictly positive function given by an explicit formula and such that $\rho(0) = 1$, $k(x,u)$ is the function defined in (1.11),

$$\begin{aligned} \beta_d(x,u,d) = -k_d(x,u,d) + \tfrac{\rho_d(d)}{2\phi_d(x,u)} \\ -h_d(x,u,0)\tfrac{1-\rho_d(d)}{d} + \tfrac{\rho_d(d)-1}{2d^2\phi_d(x,u)}\overline{d}(x,u)^2 \end{aligned} \tag{1.20}$$

where $\rho_d(d)$ is a strictly positive function given by an explicit formula and such that $\rho(0) = 1$, $k_d(x,u,d)$ and $h_d(x,u,d)$ are the functions given in respectively (1.15), (1.14) and $\overline{d}(x,u) = \phi_d(x,u)h_d(x,u,0)$ where $\phi_d(x,u)$ is a stricly positive function given by an explicit formula.

Proof. Theorem 1.2.1 or Theorem 1.2.2 provide us with a globally asymptotically stabilizing feedback $\overline{u}(x)$ of the form (1.3) or (1.5) when $d_i = 0$ for all i. Similarly, let

$$\overline{d}(x,u) = \phi_d(x,u)h_d(x,u,0) \tag{1.21}$$

where $\phi_d(x,u)$ is a stricly positive function to be chosen later. Let us consider the following criterion

$$\begin{aligned} S_d = -\sum_{i=0}^{\infty}(V(x_{i+1}) - V(x_i)) + \sum_{i=0}^{\infty}\frac{\rho(u_i)}{2\phi(x_i)}(u_i - \overline{u}(x_i,u_i))^2 \\ -\sum_{i=0}^{\infty}\frac{\rho_d(d_i)}{2\phi_d(x_i,u_i)}(d_i - \overline{d}(x_i,u_i))^2 \,. \end{aligned}$$

Observe that, since $\rho(u)$ is a positive function, $u = \overline{u}(x)$ minimizes S when $(d_i) \in \mathcal{D}$ because, when a globally asymptotically stabilizing feedback is applied,

$$S_d = V(x_0) + \sum_{i=0}^{\infty} \frac{\rho(u_i)}{2\phi(x_i)}(u_i - \overline{u}(x_i, u_i))^2 - \sum_{i=0}^{\infty} \frac{\rho_d(d_i)}{2\phi_d(x_i, u_i)}(d_i - \overline{d}(x_i, u_i))^2 .$$

The function S_d rewrites as:

$$\begin{aligned} S_d = & -\sum_{i=0}^{\infty}(V(f(x_i)) - V(x_i)) - \sum_{i=0}^{\infty} h(x_i, u_i)u_i - \sum_{i=0}^{\infty} h_d(x_i, u_i, d_i)d_i \\ & + \sum_{i=0}^{\infty}\left[\frac{\rho(u_i)}{2\phi(x_i)}u_i^2 - \frac{\rho(u_i)}{\phi(x_i)}\overline{u}(x_i)u_i + \frac{\rho(u_i)}{2\phi(x_i)}\overline{u}(x_i)^2\right] \\ & - \sum_{i=0}^{\infty}\left[\frac{\rho_d(d_i)}{2\phi_d(x_i, u_i)}d_i^2 - \frac{\rho_d(d_i)}{\phi_d(x_i, u_i)}\overline{d}(x_i, u_i)d_i + \frac{\rho_d(d_i)}{2\phi_d(x_i, u_i)}\overline{d}(x_i, u_i)^2\right] . \end{aligned}$$

According to (1.11) and (1.15), we have:

$$\begin{aligned} S_d = & -\sum_{i=0}^{\infty}(V(f(x_i)) - V(x_i)) + \sum_{i=0}^{\infty} k(x_i, u_i)u_i^2 - \sum_{i=0}^{\infty} h(x_i, 0)u_i \\ & + \sum_{i=0}^{\infty}\left[\frac{\rho(u_i)}{2\phi(x_i)}u_i^2 + h(x_i, 0)\rho(u_i)u_i + \frac{\rho(u_i)}{2\phi(x_i)}\overline{u}(x_i)^2\right] \\ & + \sum_{i=0}^{\infty} k_d(x_i, u_i, d_i)d_i^2 - \sum_{i=0}^{\infty} h_d(x_i, u_i, 0)d_i \\ & - \sum_{i=0}^{\infty}\left[\frac{\rho_d(d_i)}{2\phi_d(x_i, u_i)}d_i^2 - \frac{\rho_d(d_i)}{\phi_d(x_i, u_i)}\overline{d}(x_i, u_i)d_i + \frac{\rho_d(d_i)}{2\phi_d(x_i, u_i)}\overline{d}(x_i, u_i)^2\right] . \end{aligned}$$

Regrouping the terms differently, we obtain:

$$\begin{aligned} S_d = & \sum_{i=0}^{\infty}\left[(V(x_i) - V(f(x_i))) + \frac{1}{2}\phi(x_i)\rho(u_i)h(x_i, 0)^2 \right. \\ & \left. + \frac{\rho_d(0)}{2\phi_d(x_i, u_i)}\overline{d}(x_i, u_i)^2\right] \\ & + \sum_{i=0}^{\infty}\left[k(x_i, u_i) + \frac{\rho(u_i)}{2\phi(x_i)} - h(x_i, 0)\frac{1 - \rho(u_i)}{u_i}\right] u_i^2 \\ & - \sum_{i=0}^{\infty}\left[-k_d(x_i, u_i, d_i) + \frac{\rho_d(d_i)}{2\phi_d(x_i, u_i)} \right. \\ & \left. - h_d(x_i, u_i, 0)\frac{1 - \rho_d(d_i)}{d_i} + \frac{\rho_d(d_i) - 1}{2d_i^2\phi_d(x_i, u_i)}\overline{d}(x_i, u_i)^2\right] d_i^2 . \end{aligned}$$

Therefore $S = J$ where J is the function defined in (1.8). Assumption **H1** guarantees that $\alpha(x,u)$ is positive. The proof is completed by using the following lemma

Lemma 1.3.2. *Functions $\phi_d(x,u)$ and $\rho_d(d)$ such that $\beta_d(x,u,d)$ is a strictly positive function can be determined.*

Proof. The proof is similar to the proof of Lemma 1.2.1.

1.3.2 An $\mathcal{L}_2$ disturbance attenuation result

In this section, we introduce assumptions ensuring that for a system (1.13) all the sequences $d_i \in L^2$ belong to $\mathcal{D}$.

H3. There exists a function $B(x,u)$ such that

$$|\beta_d(x,u,d)| \leq B(x,u)$$

H4. There exists $c > 0$ such that,

$$\alpha(x,\overline{u}(x)) + \beta(x,\overline{u}(x))\overline{u}(x)^2 \geq c\frac{|x|^2}{1+|x|^2} \tag{1.22}$$

We state the main result of the section.

Theorem 1.3.1. *Assume that the system (1.13) satisfies the assumption of Theorem 1.2.3 when $d_i = 0$ for all i and the assumptions H3 and H4. Then there exists $C > 0$ such that for all sequence $d_i \in L^2$, the solution of (1.13) with $x_0 = 0$ satisfies:*

$$\sum_{i=0}^{\infty} |x_i|^2 \leq C \sum_{i=0}^{\infty} d_i^2$$

Proof. When $u = \overline{u}(x)$,

$$\begin{aligned} J_d = & \sup_{\{d_i \in \mathcal{D}\}} \sum_{i=0}^{\infty} \left[\alpha(x_i,\overline{u}(x_i)) + \beta(x_i,\overline{u}(x_i))\overline{u}(x_i)^2 \right. \\ & \qquad \left. -\beta_d(x_i,\overline{u}(x_i),d_i)d_i^2\right] \\ = & V(x_0) \end{aligned} \tag{1.23}$$

So when the initial condition is at the origin, we deduce that for all $(d_i) \in \mathcal{D}$,

$$\sum_{i=0}^{\infty} \left[\alpha(x_i,\overline{u}(x_i)) + \beta(x_i,\overline{u}(x_i))\overline{u}(x_i)^2\right] \leq \sum_{i=0}^{\infty} \beta_d(x_i,\overline{u}(x_i),d_i)d_i^2 \tag{1.24}$$

Using Assumption H3 and Assumption H4, we deduce that there exists $C > 0$ (which can be explicitly determined) such that

$$\sup_{\{d_i \in \mathcal{D}\}} \sum_{i=0}^{\infty} |x_i|^2 \leq C \sup_{\{d_i \in \mathcal{D}\}} \sum_{i=0}^{\infty} d_i^2$$

Moreover, according to Assumption H4 and the fact that $|\beta_d(x, \overline{u}(x), d)|$ is smaller than a function independent from d, all the sequences $d_i \in L_2$ belong to $\mathcal{D}$.

This concludes our proof.

1.4 Feedforward discrete-time nonlinear systems

In this section, we particularize the results of Section 1.3 to the class of the discrete-time feedforward systems i.e. systems having the following representation:

$$\begin{aligned} z_{i+1} &= \mathcal{F}(z_i) + \psi(z_i, \xi_i) + g_1(z_i, \xi_i, u_i)u_i \\ \xi_{i+1} &= a(\xi_i) + g_2(z_i, \xi_i, u_i)u_i \end{aligned} \tag{1.25}$$

with $z_i \in \mathbf{R}^{n_z}$, $\xi_i \in \mathbf{R}^{n_\xi}$, $u \in \mathbf{R}$. We introduce the following assumptions

A1. There exists a proper, positive definite function $W_1(\cdot)$ which is zero at the origin and such that for all z,

$$W_1(\mathcal{F}(z)) - W_1(z) \leq 0 \ .$$

A2. There exist a function $W_2(\cdot)$ positive definite radially unbounded zero at the origin and a function $\nu(\cdot)$ positive definite and zero at the origin such that

$$W_2(a(\xi)) - W_2(\xi) \leq -\nu(\xi) \ .$$

Moreover both $W_2(\cdot)$ and $\nu(\cdot)$ are lower bounded on a neighborhood of the origin by a positive definite quadratic function.

A3. There exist two differentiable positive functions $\gamma_0(\xi)$ and $\gamma_1(\xi)$ zero at the origin and such that

$$|\psi(z, \xi)| \leq \gamma_0(\xi) + \gamma_1(\xi) W_1(z)$$

A4. The following inequality is satisfied:

$$\left| \frac{\partial W_1}{\partial z}(z) \right| \leq 1 \ .$$

Discussion of the assumptions.

• The assumptions A1 to A4 are the standard assumptions of the forwarding approach. We conjecture that they can be relaxed in the time-varying context as they are relaxed in [15] in the continuous-time context.

• The family of systems (1.25) is slightly larger than the one studied in [14] since it is not required on $z_{i+1} = \mathcal{F}(z_i)$ to be linear.

• As an immediate consequence of A2, we have that the system $\xi_{i+1} = a(\xi_i)$ is globally asymptotically stable and locally exponentially stable.

• If is known a function $\mathcal{W}_1(z)$ such that

$$\left|\frac{\partial \mathcal{W}_1}{\partial z}(z)\right| \leq \mathcal{L}(\mathcal{W}_1(z))$$

where $\mathcal{L}(\cdot)$ is a positive function such that $\frac{1}{1+\mathcal{L}(\cdot)} \notin L^1$, then Assumption A4 is satisfied by

$$W_1(z) = \int_0^{\mathcal{W}_1(z)} \frac{1}{\mathcal{L}(s)+1} ds .$$

If $\mathcal{W}_1(z)$ is a quadratic form, the corresponding function $W_1(z)$ satisfies a linear growth property and Assumption A3 is a linear growth assumption imposed on the coupling term.

In order to derive inverse optimal controls from the result of the previous part, we first prove that the assumptions A1 to A4 ensure the Lyapunov stability of the free system associated with (1.25). We construct a candidate Lyapunov function depending on the given functions $W_1(\cdot)$ and $W_2(\cdot)$.

The construction we adopt mimics the one proposed in [5] (see also [15]) for continuous-time feedforward systems. An alternative construction is given in [14] in a slightly more restrictive context.

1.4.1 Stability of the uncontrolled system

Let us introduce the notations: $x = (z, \xi)^\top$,

$$f(z,\xi) = \begin{pmatrix} \mathcal{F}(z) + \psi(z,\xi) \\ a(\xi) \end{pmatrix},$$

$$g(z,\xi,u) = \begin{pmatrix} g_1(z,\xi,u) \\ g_2(z,\xi,u) \end{pmatrix}.$$

Theorem 1.4.1. *Assume that the system (1.25) satisfies the Assumptions* **A1** *to* **A4**. *Then this system is Lyapunov stable when* $u = 0$ *and there exists a function zero at the origin, positive definite and of the form*

$$V(x) = W_1(z) + \Phi(z,\xi) + W_2(\xi) \tag{1.26}$$

such that $\Delta V \leq 0$ *when* $u = 0$.

Remark. One may easily deduce from the forthcoming proof a family of Lyapunov functions for (1.25): for all real-valued functions $k(\cdot), l(\cdot)$ of class $\mathcal{K}^\infty$ there exists a cross-term $\Phi_{kl}(z,\xi)$ such that $V_{kl}(x) = l(W_1(z)) + \Phi_{kl}(z,\xi) + k(W_2(\xi))$ is proper, positive definite zero at zero and such that $\Delta V_{kl} \leq 0$.

Proof. First, let us determine a cross term function $\Phi(z,\xi)$ such that the candidate Lyapunov function given in (1.26) satisfies $\Delta V \leq 0$. The expression of the variation ΔV is

$$\begin{aligned}\Delta V = {} & W_1(\mathcal{F}(z_i) + \psi(z_i,\xi_i)) - W_1(z_i) \\ & + \Phi(z_{i+1},\xi_{i+1}) - \Phi(z_i,\xi_i) + W_2(a(\xi_i)) - W_2(\xi_i)\ .\end{aligned}$$

Since for all z,ξ

$$\begin{aligned}W_1(\mathcal{F}(z) + \psi(z,\xi)) - W_1(\mathcal{F}(z)) = {} & \\ & \int_0^1 \left(\frac{\partial W_1}{\partial z}(\mathcal{F}(z) + \psi(z,\xi)\theta)\psi(z,\xi)\right) d\theta\end{aligned}$$

Assumption A1 implies that ΔV is negative if

$$\begin{aligned}\Phi(z_{i+1},\xi_{i+1}) - \Phi(z_i,\xi_i) = {} & \\ & - \int_0^1 \left(\frac{\partial W_1}{\partial z}(\mathcal{F}(z_i) + \psi(z_i,\xi_i)\theta)\psi(z_i,\xi_i)\right) d\theta\ .\end{aligned}$$

Let us denote the right hand side of this expression by $-q(z_i,\xi_i)$. It straightforwardly follows from (1.27) that

$$\Phi(z,\xi) = \sum_{i=0}^{\infty} q(z_i,\xi_i) \tag{1.27}$$

provided that the right hand side of (1.27) is a well-defined function.

The next part of the proof consists of showing that this power series converges. From the definition of $q(z,\xi)$ and Assumptions A3 and A4 we successively obtain:

$$\begin{aligned}|q(z,\xi)| & \leq \left|\int_0^1 \frac{\partial W_1}{\partial z}(\mathcal{F}(z) + \psi(z,\xi)\theta)\psi(z,\xi)d\theta\right| \\ & \leq |\psi(z,\xi)| \leq \gamma_0(\xi) + \gamma_1(\xi)W_1(z)\ .\end{aligned}$$

From Assumption A2, we deduce that there exists a positive and increasing function $\tilde{\gamma}(\xi)$ zero at the origin and a strictly positive real number r such that:

$$|q(z_i,\xi_i)| \leq \tilde{\gamma}(\xi)e^{-ri}[W_1(z_i)+1] . \tag{1.28}$$

To conclude, we prove that the sequence $W(z_i)$ is bounded.

Using Assumptions A3 and A4, we deduce that

$$\begin{aligned} W_1(z_{i+1}) - W_1(z_i) &\leq W_1(\mathcal{F}(z_i) + \psi(z_i,\xi_i)) - W_1(\mathcal{F}(z_i)) \\ &\leq \gamma_0(\xi_i) + \gamma_1(\xi_i)W_1(z_i) . \end{aligned}$$

Using Assumption A2, we deduce that there exist a function $\Gamma_1(\xi)$ smooth, positive, zero at zero and $r > 0$ such that:

$$\frac{W_1(z_{i+1})+1}{W_1(z_i)+1} \leq 1 + \Gamma_1(\xi)e^{-ri} .$$

It follows that

$$\ln(W_1(z_{i+1})+1) \leq \ln(W_1(z)+1) + \sum_{j=0}^{i} \ln\left[1+\Gamma_1(\xi)e^{-rj}\right]$$

which implies that there exists $\Gamma_2(\xi)$ such that for all integer l,

$$W_1(z_l) \leq (W_1(z)+1)\Gamma_2(\xi) . \tag{1.29}$$

To show that $V(z,\xi)$ defined in (1.26) is a positive function, consider

$$\begin{aligned} W_1(z) + \Phi(z,\xi) &= W_1(z) + \sum_{i=0}^{\infty}[W_1(\mathcal{F}(z_i) + \psi(z_i,\xi_i)) \\ &\quad -W_1(\mathcal{F}(z_i))] \\ &= W_1(z) + \sum_{i=0}^{\infty}[W_1(z_{i+1}) - W_1(\mathcal{F}(z_i))] . \end{aligned}$$

For all integer $J > 0$, we have

$$\begin{aligned} W_1(z) + \Phi(z,\xi) &= \sum_{i=0}^{J}[W_1(z_i) - W_1(\mathcal{F}(z_i))] + W_1(z_{J+1}) \\ &\quad + \sum_{i=J+1}^{\infty}[W_1(z_{i+1}) - W_1(\mathcal{F}(z_i))] . \end{aligned}$$

According to Assumption A1, the term $\sum_{i=0}^{J}[W_1(z_i) - W_1(\mathcal{F}(z_i))]$ is positive.

On the other hand, using Assumptions A3 and A4, one can prove easily that:

$$\lim_{J \to +\infty} \sum_{i=J+1}^{\infty} [W_1(z_{i+1}) - W_1(\mathcal{F}(z_i))] = 0 .$$

It follows that $W_1(z) + \Phi(z, \xi)$ is positive. Since $\Phi(z, 0) = 0$ for all z, it straightforwardly follows that $V(z, \xi)$ is a positive definite function.

Remark. It is worth noting that surprisingly, Assumptions A1 to A4 do not guarantee that $V(x)$ is radially unbounded whereas similar assumptions in the continuous-time context ensure that the corresponding function $V(x)$ is radially unbounded.

1.4.2 Disturbance attenuation property of feedforward systems

Consider the system

$$\begin{aligned} z_{i+1} &= \mathcal{F}(z_i) + \psi(z_i, \xi_i) + g_2(z_i, \xi_i, u_i)u_i \\ &\quad + m_1(z_i, \xi_i, u_i, d_i)d_i \\ \xi_{i+1} &= a(\xi_i) + g_1(z_i, \xi_i, u_i)u_i + m_2(z_i, \xi_i, u_i, d_i)d_i \end{aligned} \tag{1.30}$$

with $z_i \in \mathbf{R}^{n_z}$, $\xi_i \in \mathbf{R}^{n_\xi}$, $u \in \mathbf{R}$, $d \in \mathbf{R}$. Let us state a disturbance attenuation result for feedforward systems.

Corollary 1.4.1. *Assume that the system (1.30) satisfies the Assumptions A1 to A4. Assume that the cross term of the function $V(x)$ provided by Theorem 1.4.1 is continuously differentiable and that Assumption H2 is satisfied. Then the inverse optimal stabilization problem is solvable. Moreover, if the system (1.30) satisfies the the assumptions of Theorem 1.2.3 when $d_i = 0$ for all i and the assumptions H3 and H4. Then there exists $C > 0$ such that for all sequence $d_i \in L^2$, the solution of (1.13) with $x_0 = 0$ satisfies:*

$$\sum_{i=0}^{\infty} |x_i|^2 \leq C \sum_{i=0}^{\infty} d_i^2$$

Remark. Observe that Corollary 1.4.1 can be applied repeatedly.

Proof. The solvability of the inverse optimal stabilization problem is an immediate consequence of Theorem 1.2.3 and Theorem 1.4.1 when $V(x)$ is a proper function. When $V(x)$ is not a proper function then Assumption H1' and not Assumption H1 is satisfied. To apply Theorem 1.2.2, we have to prove that there exist feedbacks of the form (1.3) which do not destabilize the system. This can be done by using the arguments similar to those invoked above in the proof of Theorem 1.4.1. The disturbance attenuation result is a consequence of Theorem 1.3.1. This concludes our proof.

References

1. B.D.O. Anderson and J.B. Moore: *Linear Optimal Control*, Prentice-Hall, Englewood Cliffs, NJ, 1971.

2. R.A. Freeman and P.V. Kokotovic: *Inverse optimality in robust stabilization*, SIAM Journal on Control and Optimization, vol. 34, pp. 1365–1391, 1996.

3. R.A. Freeman and P.V. Kokotovic: *Robust Nonlinear Control Design*, Birkhauser, Boston, 1996.

4. W.M. Haddad, V. Chellaboina and J.L. Fausz: *Optimal Nonlinear Disturbance Rejection Control for Nonlinear Cascade Systems*, Int. J. Contr., 68 (1997) 997-1018.

5. M. Jankovic, R. Sepulchre and P.V. Kokotovic: *Global stabilization of an enlarged class of cascade nonlinear systems.* IEEE Trans. on Automatic Control, vol.41, no.12, pp.1723-1735, 1996.

6. V. Jurdjevic, J.P. Quinn : *Controllability and stability.* J. Differential Equations, vol.4 , pp. 381-389, 1978.

7. R.E. Kalman: *When is a linear control systtem optimal ?* Trans. ASME Ser. D: J. Basic En., vol. 86, pp. 1-10, 1964.

8. M. Krstić and Zhong-Hua Li: *Inverse optimal design of input-to-state stabilizing nonlinear controllers*, Preprint 1997.

9. Zhong-Hua Li and M. Krstić: *Optimal design of adaptive tracking controllers for nonlinear systems*, Proc. ACC, Albuquerque, New Mexico, 1997.

10. Wei Lin: *Further results on global stabilization of discrete-time nonlinear systems.* Systems & Control Letters, 29, 51–59, 1996.

11. Wei Lin and C. Byrnes: *Passivity and Absolute Stabilization of a Class of Discrete-time Nonlinear Systems.* Automatica, Vol. 31. No.2, pp. 263-267, 1995.

12. F. Mazenc : *Stabilization of Feedforward Systems Approximated by a Nonlinear Chain of Integrators.* Systems & Control Letters 32 (1997) 223-229.

13. F. Mazenc and L. Praly: *Adding an integration and Global asymptotic stabilization of feedforward systems.* IEEE Trans. on Automatic Control, vol.41, no.11, pp.1559-1578, 1996.

14. F. Mazenc and H. Nijmeijer: *Forwarding in discrete-time nonlinear systems.* Int. J. Control, 1998, vol. 71, No.5, pp. 823-835.

15. F. Mazenc, R. Sepulchre and M. Jankovic: *Lyapunov functions for stable cascades and applications to global stabilization.* 36th CDC conference, San Diego, 1997 and to appear in IEEE Trans. Automatic Control.

16. P.J. Moylan and B.D. Anderson: *Nonlinear regulator theory and an inverse optimal control problem.* IEEE Trans. Automatic Control, vol. 18, pp. 460-465, 1973.

17. R. Sepulchre, M. Jankovic and P.V. Kokotovic, *Constructive Nonlinear Control,* Springer-Verlag, 1997.

18. H.J. Susmann, E.D. Sontag and Y. Yang: *A general result on the stabilization of linear systems using bounded control.* IEEE Trans. on Automatic Control, vol. 39, pp. 2411-2425, 1994.

19. A. Teel: *Feedback Stabilization: Nonlinear Solutions to Inherently Nonlinear Problems.* PhD Dissertation, University of California, Berkeley, 1992.

20. A. Teel: *On L_2 performance induced by feedbacks with multiple saturations.* ESAIM: Control, Optimisation and Clculus of Variations, vol. 1, pp. 225-240, Spetembre 1996.

2. Further results on decoupling with stability for Hamiltonian systems

Alessandro Astolfi and Laura Menini***

*Dip. di Elettronica e Informazione
Politecnico di Milano
Piazza Leonardo Da Vinci 32
20133, Milano, Italy
and
Dept. Electrical and Electronic Engineering
Imperial College
Exhibition Road
London SW7 2BT, England.
a.astolfi@ic.ac.uk

**Dip. Informatica, Sistemi e Produzione
Univ. Roma Tor Vergata
via di Tor Vergata 110
00133 Roma, Italy.
menini@disp.uniroma2.it

Summary.

The problem of input-output decoupling with stability by, possibly dynamic, state-feedback is addressed for Hamiltonian systems. As well known, to decide if the problem is solvable, and which class of state-feedback has to be used, the stability properties of the $P^{\perp}$, P^{*} and Δ_{mix} dynamics are to be investigated. For this reason, on the way to the main result, it is shown that, for general Hamiltonian systems, such dynamics are not necessarily Hamiltonian. On the other hand, it is shown that, for linear simple Hamiltonian systems, both the $P^{\perp}$ and the P^{*} dynamics are Hamiltonian (whereas, as well known, the Δ_{mix} dynamics are empty). Moreover, for a class of nonlinear simple Hamiltonian systems, a simple to check necessary and sufficient condition for the solvability of the problem via dynamic state-feedback is proposed. Several examples, clarifying the role of different classes of state-feedback control laws (either static or dynamic) in the solution of the problem, are proposed.

2.1 Introduction and motivations

The problem of input-output decoupling (or, equivalently, noninteraction) with stability for nonlinear systems has been studied by several authors, see [5, 10, 8, 12, 15, 7, 3]. Necessary and sufficient conditions for the existence of either static or dynamic state-feedback control laws yielding stable noninteractive closed-loop control systems have been proposed, and systematic procedures for the design of such control laws have been given. Despite the elegant characterization of the problem, which is based on geometric control theory, the applicability of the theory to physical systems has not received (to the best of the authors knowledge) enough interest. A notable exception is the paper [6]. Therein, the problem of decoupling with stability is addressed for the class of (nonlinear) Hamiltonian systems by means of a particular class of state-feedback control laws. It has been shown in [6] that Hamiltonian systems can be put in a particular canonical form, which will be exploited in this chapter too, since it highly facilitates the analysis of the problem. As a matter of fact, by using such a canonical form, it is easier to take into account the well known fact, proven in [11, Chapter 12], that the zero dynamics of Hamiltonian systems are Hamiltonian. In [6] it is shown that, if a particular class of static state-feedback control laws is considered, decoupling and asymptotic stability are not jointly achievable for Hamiltonian systems whose zero dynamics are non-trivial. Nevertheless, under suitable hypotheses, decoupling with simple stability can be obtained, as shown in [6], by means of static state-feedback control laws in the mentioned class.

The work reported in this chapter stems from the consideration that, by using more general state-feedback control laws, either static or dynamic, the problem can be solved for larger classes of systems. It is well known (see [14, 9, 13]) that, in the case of linear systems, the problem can be solved by means of dynamic state-feedback for all those systems for which the two problems of stabilization and of input-output decoupling are separately solvable. This implies that, if dynamic state-feedback is allowed, the problem of input-output decoupling with asymptotic stability is generically solvable for linear controllable Hamiltonian systems.

The importance of Hamiltonian systems in the modeling of practical situations is well known (see [11, Chapter 12] and the concise exposition in [1]), hence it seems of interest to investigate if the general results concerning the problem of noninteracting control with stability, reported in [3, 7], assume special characterization when applied to general, nonlinear, Hamiltonian systems, in view of their special properties.

Hamiltonian systems are not asymptotically stable at any equilibrium, although they can be stable. As shown in [6], a sufficient condition for stability of a Hamiltonian system is that the Hamiltonian function has an isolated local minimum at the equilibrium. Hence it is of interest to know whether the dynamics of certain subsystems are Hamiltonian, since this fact can highly facilitate the tests for stability, needed, as will be recalled in the following, to decide if the considered problem of decoupling with stability is solvable.

In Section 2.3, it is shown that, contrarily to what holds true for the zero dynamics, the $P^{\perp}$, P^* and $\Delta_{\rm mix}$ dynamics[1] of general Hamiltonian systems are not necessarily Hamiltonian. Such facts are shown, by means of a simple low-order example. In particular, since all these dynamics are contained in the zero dynamics, they may not be Hamiltonian if their dimension is smaller than that of the zero dynamics. Since the mentioned dynamics are responsible for obstructions to the solvability of the problem (if either the $P^{\perp}$ or the $\Delta_{\rm mix}$ dynamics are not "stable", then the problem is not solvable, whereas, if the P^* dynamics are not "stable", then the problem is not solvable by means of static state-feedback, but may be solvable by means of dynamic state-feedback), it follows that their "stability" has to be checked (without using the special results on stability of Hamiltonian systems) in order to decide if the problem is indeed solvable, and, in case it does, which class of control laws can be adopted.

The case of simple Hamiltonian systems is considered in detail in Subsection 2.3.1. As a matter of fact, simple Hamiltonian systems have special geometric properties which are exploited to prove that, under suitable assumptions, the submanifolds on which the P^* and $\Delta_{\rm mix}$ dynamics are defined are symplectic. For the special case of linear simple Hamiltonian systems, whose $\Delta_{\rm mix}$ dynamics are trivial, it is shown that both the $P^{\perp}$ and P^* dynamics are Hamiltonian.

The main result of this chapter, reported in Section 2.4, consists of a simple to check condition for the solvability of the problem of decoupling with stability, which is necessary and sufficient for the class of systems considered and can be used with respect to different stability requirements. In what follows, when referred to nonlinear systems, all the proposed results are local, *i.e.*, valid in a suitable neighborhood of the equilibrium configuration of the system.

Finally, in Section 2.6, two mechanical systems are studied. In the first one the problem of decoupling with asymptotic stability in the first approximation is solved by means of static state-feedback for a system having non-trivial zero dynamics; in the second one, using the results in Section 2.4, it is shown that the problem of decoupling with (simple) stability can be solved by means of dynamic state-feedback, for a system having unstable zero dynamics.

A preliminary version of the main results reported in this chapter has been published in [2].

[1] Definitions are reported in Section 2.3

Notation In what follows, e_i denotes the i-th column of the identity matrix of proper dimension, 0_i denotes the zero vector in $\mathbb{R}^i$, $\mathrm{Im}(M)$ denotes the range of matrix M, $\emptyset$ denotes the zero distribution and $\{\emptyset\}$ denotes the empty set.

2.2 Background on Hamiltonian systems

In this chapter, the problem of input-output decoupling with stability will be tackled for a class of 2 inputs-2 outputs Hamiltonian systems, namely, systems described, in Hamiltonian form, by means of the following equations:

$$\begin{aligned}
\dot{q}_i &= \frac{\partial H(q,\, p,\, u)}{\partial p_i}, & i &= 1, \ldots, n, \\
\dot{p}_i &= -\frac{\partial H(q,\, p,\, u)}{\partial q_i}, & i &= 1, \ldots, n, \\
y_1 &= q_1, \\
y_2 &= q_2,
\end{aligned} \tag{2.1}$$

where the Hamiltonian function $H(q,\, p,\, u)$ has the form:

$$H(q,\, p,\, u) = H_0(q,\, p) - q_1\, u_1 - q_2\, u_2, \tag{2.2}$$

and is assumed to be sufficiently smooth.

The components q_i, $i = 1, 2, \ldots, n$, $n \geq 2$, of the vector q are suitable configuration coordinates, whereas the components p_i, $i = 1, 2, \ldots, n$, of the vector p are the corresponding generalized momenta; the first two degrees of freedom q_1 and q_2 are actuated by means of the external inputs u_1, u_2. Note that y_1 and y_2 are the so-called "natural outputs" (see [11, Chapter 12]). A subclass of Hamiltonian systems of special importance is the class of simple Hamiltonian systems, in which the function H_0 has the form:

$$H_0(q,\, p) = \frac{1}{2} p^T G(q)\, p + V(q). \tag{2.3}$$

The square n-dimensional symmetric matrix $G(q)$ is assumed to be positive definite for every q in its domain: this condition is satisfied by many Hamiltonian systems of practical interest, *e.g.*, by those representing mechanical systems, in which the term $\frac{1}{2} p^T G(q)\, p$ corresponds to the kinetic energy. Moreover it is also assumed that $\frac{\partial V}{\partial q}(0) = 0$, so that the point $q = 0$, $p = 0$, is an equilibrium point for system (2.1).

Let $x := \left[q^T \; p^T \right]^T$, and let the vector fields $f(x)$, g_1, g_2, $h_1(x)$, $h_2(x)$, be given by:

$$
\begin{aligned}
f_i(x) &:= \frac{\partial H_0(q, p)}{\partial p_i}, \quad i = 1, 2, \ldots, n, \\
f_{n+i}(x) &:= -\frac{\partial H_0(q, p)}{\partial q_i}, \quad i = 1, 2, \ldots, n, \\
g_1 &= e_{n+1}, \\
g_2 &= e_{n+2}, \\
h_1(x) &:= q_1, \\
h_2(x) &:= q_2,
\end{aligned} \tag{2.4}
$$

so that equations (2.1) can be rewritten as:

$$
\begin{aligned}
\dot{x} &= f(x) + g_1 u_1 + g_2 u_2, \\
y_1 &= h_1(x), \\
y_2 &= h_2(x).
\end{aligned}
$$

Let the characteristic numbers ρ_1, ρ_2, of system (2.1) at the point $q = 0$, $p = 0$, be defined as in [11, Definition 8.7], *i.e.*

$$
\begin{aligned}
\rho_i = \min_{k \in \mathbf{Z}^+} \{ & L_{g_j} L_f^l h_i(x) = 0, \forall l = 0, 1, \ldots, k-1, \forall j = 1, 2, \\
& \forall x \text{ in a neighborhood of } 0, \exists j : L_{g_j} L_f^k h_i(0) \neq 0 \}, \; i = 1, 2
\end{aligned} \tag{2.5}
$$

and let the decoupling matrix $A(\cdot)$ be defined, in a neighborhood of the origin, as the 2-dimensional square matrix

$$
A(x) := \begin{bmatrix} L_{g_1} L_f^{\rho_1} h_1(x) & L_{g_2} L_f^{\rho_1} h_1(x) \\ L_{g_1} L_f^{\rho_2} h_2(x) & L_{g_2} L_f^{\rho_2} h_2(x) \end{bmatrix}.
$$

In the case of simple Hamiltonian systems, it is well known [11, Chapter 12] that $\rho_1 = \rho_2 = 1$, and that $A(x) = G_{11}(q)$, where $G_{11}(q)$ is the 2×2 leading submatrix of $G(q)$. Therefore, the matrix $A(x)$ is nonsingular everywhere. Consider a feedback control law of the form:

$$
u = \alpha(x) + \beta(x)\, v, \tag{2.6}
$$

where $\alpha(x)$ and $\beta(x)$ are defined as:

$$
\begin{aligned}
\alpha(x) &:= -\left(A(x)\right)^{-1} b(x), \quad \beta(x) := \left(A(x)\right)^{-1}, \\
b(x) &:= \begin{bmatrix} L_f^2 h_1(x) \\ L_f^2 h_2(x) \end{bmatrix},
\end{aligned}
$$

and $v = [v_1 \; v_2]^T$ is the vector of the new inputs. If the vectors ξ_i are given by

$$
\xi_1 = \begin{bmatrix} h_1(x) \\ L_f h_1(x) \end{bmatrix} = \begin{bmatrix} q_1 \\ \dot{q}_1 \end{bmatrix}, \qquad \xi_2 = \begin{bmatrix} h_2(x) \\ L_f h_2(x) \end{bmatrix} = \begin{bmatrix} q_2 \\ \dot{q}_2 \end{bmatrix}, \tag{2.7}
$$

and the vectors $\tilde{q}$ and $\tilde{p}$ are given by $\tilde{q} = \left[q_3 \cdots q_n\right]^T$, $\tilde{p} = \left[p_3 \cdots p_n\right]^T$, the closed-loop system can be written as

$$\dot{\xi}_1 = A\xi_1 + B\,v_1, \tag{2.8a}$$
$$\dot{\xi}_2 = A\xi_2 + B\,v_2, \tag{2.8b}$$
$$\dot{\tilde{q}} = \frac{\partial \hat{H}}{\partial \tilde{p}}\left(\tilde{q}, \tilde{p}, \xi_1, \xi_2\right) + R_1\left(\tilde{q}, \tilde{p}, \xi_1, \xi_2\right), \tag{2.8c}$$
$$\dot{\tilde{p}} = -\frac{\partial \hat{H}}{\partial \tilde{q}}\left(\tilde{q}, \tilde{p}, \xi_1, \xi_2\right) + R_2\left(\tilde{q}, \tilde{p}, \xi_1, \xi_2\right), \tag{2.8d}$$
$$y_1 = [1\;0]\,\xi_1, \tag{2.8e}$$
$$y_2 = [1\;0]\,\xi_2, \tag{2.8f}$$

where

$$A = \begin{bmatrix} 0 & 1 \\ 0 & 0 \end{bmatrix}, \quad B = \begin{bmatrix} 0 \\ 1 \end{bmatrix},$$

$\hat{H}\left(\tilde{q}, \tilde{p}, \xi_1, \xi_2\right)$ is the function $H_0(q, p)$, written in the new coordinates, and $R_1\left(\tilde{q}, \tilde{p}, \xi_1, \xi_2\right)$, $R_2\left(\tilde{q}, \tilde{p}, \xi_1, \xi_2\right)$ are suitable functions with the property that $R_i\left(\tilde{q}, \tilde{p}, 0, 0\right) = 0$, $i = 1, 2$, for all $\tilde{q}$, $\tilde{p}$ in a neighborhood of the origin.

For simple Hamiltonian systems, the results in [6], [11, Chapter 12] and [7, Chapter 6] imply that, if Δ^* denotes the largest locally controlled invariant distribution contained in $\mathrm{Ker}(dh_1) \cap \mathrm{Ker}(dh_2)$ then

$$\Delta^* = \mathrm{span}\left\{\frac{\partial}{\partial \tilde{q}_1}, \ldots, \frac{\partial}{\partial \tilde{q}_{n-2}}, \frac{\partial}{\partial \tilde{p}_1}, \ldots, \frac{\partial}{\partial \tilde{p}_{n-2}}\right\},$$

hence $\tilde{q}$ and $\tilde{p}$ can be used to describe the zero dynamics of the system (2.1), (2.3), *i.e.*

$$\begin{aligned} \dot{\tilde{q}} &= \frac{\partial \tilde{H}}{\partial \tilde{p}}\left(\tilde{q}, \tilde{p}\right), \\ \dot{\tilde{p}} &= -\frac{\partial \tilde{H}}{\partial \tilde{q}}\left(\tilde{q}, \tilde{p}\right), \end{aligned} \tag{2.9}$$

where the restricted Hamiltonian $\tilde{H}\left(\tilde{q}, \tilde{p}\right) := \hat{H}\left(\tilde{q}, \tilde{p}, 0, 0\right)$ can be computed explicitly by means of the following equation (see [11, Chapter 12]):

$$\tilde{H}\left(\tilde{q}, \tilde{p}\right) = \frac{1}{2}\tilde{p}^T\left(G_{22} - G_{12}^T G_{11}^{-1} G_{12}\right)\left(\begin{bmatrix} 0 \\ \tilde{q} \end{bmatrix}\right)\tilde{p} + V\left(\begin{bmatrix} 0 \\ \tilde{q} \end{bmatrix}\right).$$

Note that equations (2.9) coincide with equations (2.8c) and (2.8d) for $\xi_1 = \xi_2 = 0$.

As observed in [6], the zero dynamics of general Hamiltonian systems of the form (2.1) are Hamiltonian; for simple Hamiltonian systems this implies that, if $n > 2$, any "decentralized" feedback control law described by the equations

$$\begin{aligned} v_1 &= \epsilon_{11} q_1 + \epsilon_{12} \dot{q}_1, \\ v_2 &= \epsilon_{21} q_2 + \epsilon_{22} \dot{q}_2, \end{aligned} \tag{2.10}$$

with $\epsilon_{ij} \in \mathbb{R}$, cannot achieve asymptotic stability for the closed-loop system. Nevertheless, if (0, 0) is a stable equilibrium point for (2.9), (simple) stability can be achieved by a proper choice of the gains ϵ_{11}, ϵ_{12}, ϵ_{21}, ϵ_{22}, as discussed in [6]. In this chapter, it is shown that, allowing more general state-feedback control laws, static or dynamic, the problem of input-output decoupling with stability (simple or asymptotic, depending on the properties of the given system) can be solved for a wider class of Hamiltonian systems.

2.3 Some geometric properties

Given a general nonlinear system of the form

$$\begin{aligned} \dot{\overline{x}} &= \overline{f}(\overline{x}) + \overline{g}(\overline{x})\overline{u}, \\ \overline{y} &= \overline{h}(\overline{x}), \end{aligned} \tag{2.11}$$

with $\overline{x} \in \mathbb{R}^n$, $\overline{u}$, $\overline{y} \in \mathbb{R}^m$, $\overline{f}(0) = 0$, $\overline{h}(0) = 0$, satisfying suitable regularity assumptions (see [3] and [7, Chapter 7]), several approaches can be adopted in order to solve the problem of noninteraction with *stability*, depending on the geometric properties of (2.11). The main results of the general theory will be now summarized; to this purpose some notations and well-known properties are recalled. For the sake of simplicity, in the first half of the present section, the exposition will be limited to the case in which the stability requirement is that of asymptotic stability in the first approximation. For the objectives to be pursued in this chapter, it is sufficient to restrict the attention to the class of systems for which the characteristic numbers ρ_i, $i = 1, 2, \ldots, m$, can be defined, similarly to equation (2.5), and the decoupling matrix $\overline{A}(\overline{x})$, also defined similarly to what has been done for 2 inputs-2 outputs systems, is non-singular at $\overline{x} = 0$. Consider any regular static state-feedback control law $\overline{u}$, of the form:

$$\overline{u} = \overline{\alpha}(\overline{x}) + \overline{\beta}(\overline{x})\overline{v}, \tag{2.12}$$

such that the closed-loop system is noninteractive (such a control law exists by virtue of the assumptions on the matrix $\overline{A}(\overline{x})$) and rewrite the closed-loop system (2.11), (2.12) as follows:

$$\begin{aligned} \dot{\overline{x}} &= \tilde{f}(\overline{x}) + \tilde{g}(\overline{x})\overline{v}, \\ \overline{y} &= \overline{h}(\overline{x}), \end{aligned}$$

i.e. define $\tilde{f}(\overline{x}) := \overline{f}(\overline{x}) + \overline{g}(\overline{x})\overline{\alpha}(\overline{x})$ and $\tilde{g}(\overline{x}) := \overline{g}(\overline{x})\overline{\beta}(\overline{x})$. Let the distributions P and P^* be defined as follows:

$$P = \langle \overline{f}, \overline{g}_1, \overline{g}_2, \dots, \overline{g}_m | \operatorname{span}\{\overline{g}_j, j = 1, 2, \dots, m\}\rangle,$$
$$P^* = \bigcap_{i=1}^{m} P_i^*,$$

where

$$P_i^* = \langle \tilde{f}, \tilde{g}_1, \tilde{g}_2, \dots, \tilde{g}_m | \operatorname{span}\{\tilde{g}_j, j = 1, 2, \dots, m, j \neq i\}\rangle,$$

$\tilde{g}_i(\overline{x})$ and $\overline{g}_i(\overline{x})$ denote the i-th column of $\tilde{g}(\overline{x})$ and $\overline{g}(\overline{x})$, respectively, and, as usual, $\langle \tau_1, \tau_2, \dots, \tau_q | \Delta\rangle$ denotes the smallest distribution which contains Δ and is invariant under the vector fields $\tau_1, \tau_2, \dots, \tau_q$. Assume that the origin $\overline{x} = 0$, is a regular point for P, P_i^*, $i = 1, 2, \dots, m$. It is stressed that the distribution P does not change after a regular state-feedback, *i.e.*,

$$P = \langle \tilde{f}, \tilde{g}_1, \tilde{g}_2, \dots, \tilde{g}_m | \operatorname{span}\{\tilde{g}_j, j = 1, 2, \dots, m\}\rangle;$$

moreover, if a suitable set of coordinates $\phi = [\phi_1 \ \phi_2 \ \dots \ \phi_\nu]^T$ is chosen so that $P^\perp = \operatorname{span}\{d\phi\}$, independently of the choice of the state-feedback (2.12), the subsystem associated with $P^\perp$ is described by equations of the form

$$\dot{\phi} = f_\phi(\phi), \tag{2.13}$$

i.e., it is not affected at all by the inputs. In the following, the improper notation "$P^\perp$ dynamics" will be used to refer to (2.13), even if, in general, such a subsystem is not associated to any invariant distribution.

Moreover, if S^* is the integral submanifold of P^* containing the origin $\overline{x} = 0$, it has been proven [7, Lemma 7.3.4] that S^* is locally invariant under $\tilde{f}(\overline{x})$ and the restriction of $\tilde{f}(\overline{x})$ to S^* (P^* dynamics), *i.e.*

$$\dot{\overline{x}} = \tilde{f}(\overline{x})\Big|_{S^*}, \tag{2.14}$$

does not depend on the choice of the particular static state-feedback control law (2.12), provided that input-output decoupling is achieved.

Finally, let Δ_{mix} be the distribution generated by the vector fields τ defined as in [7, Section 7.4]:

$$\tau = \left[ad_{\tilde{f}}^{k_q}\tilde{g}_{i_q}, \left[\dots, \left[ad_{\tilde{f}}^{k_2}\tilde{g}_{i_2}, ad_{\tilde{f}}^{k_1}\tilde{g}_{i_1}\right]\right]\right], \ q \geq 2, \ k_i \geq 0,$$
$$i_r \neq i_s \text{ for some pair } (r, s).$$

Let L^* denote the integral submanifold of Δ_{mix} containing the origin $\overline{x} = 0$. The restriction of the closed-loop system to L^* (Δ_{mix} dynamics), *i.e.*

$$\dot{\overline{x}} = \tilde{f}(\overline{x})\Big|_{L^*} \tag{2.15}$$

does not depend on the choice of the particular static or dynamic state-feedback control law, provided that input-output decoupling is achieved (see [7, Proposition 7.4.1]).

To decide which class of control laws has to be used to solve the problem of decoupling with stability for system (2.11), and to decide whether the problem is indeed solvable, the illustrative diagram reported in Figure 2.3 can be considered. It clarifies the relations existing between the invariant dynamics and subsystems described above, which are all contained in the zero dynamics of the system (see [7, Chapter 7] for detail).

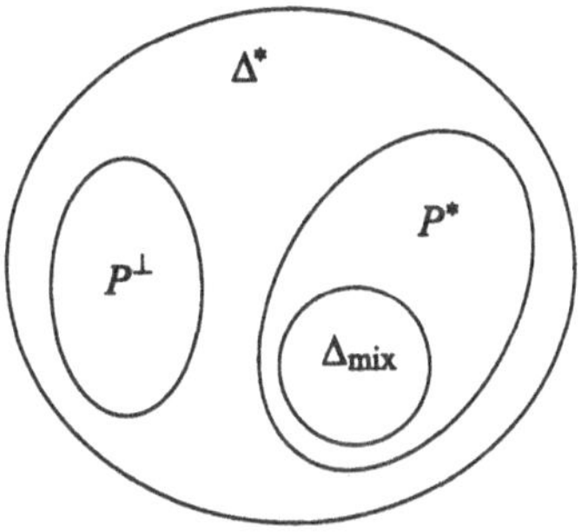

Fig. 2.1. Qualitative diagram illustrating the relevant dynamics concerned with the problem of decoupling with stability.

It is clear that, if the subsystem (2.13) associated with $P^\perp$ is not asymptotically stable in the first approximation, a control law yielding a closed loop system which is stable and noninteractive does not exist. On the other hand, if the zero dynamics of the system are stable in the first approximation, the problem can be solved easily by means of the standard decoupling feedback of the form (2.12), composed with a linear "decentralized" state-feedback, similar to the one considered in (2.10) for Hamiltonian systems. With an abuse of notation, such a state-feedback will be called in the following "decentralized", or, in the case of simple Hamiltonian systems, in which, as previously stated, $\rho_i = 1$, "PD-like decentralized". Such a class of state-feedback has been used in [6] to solve the problem of input-output decoupling with simple stability in the case of Hamiltonian systems having stable zero dynamics. As well known, stability of the zero dynamics is not necessary to solve the noninteracting control problem with stability. As a matter of fact, as shown in [7, Chapter 7] and [3], if general static state-feedback control laws are allowed, the problem is solvable if and only if the system itself is stabilizable in the first approximation and the P^* dynamics (2.14) are asymptotically stable in the first approximation. If this last condition is not satisfied, the problem might still be solvable by means of dynamic state-feedback. Under some regularity assumptions, in [7, 3] it is shown that a necessary and sufficient condition for the existence of a dynamic state-feedback control law which solves the problem of input-output decoupling with asymptotic stability in the first approximation is that the system itself is stabilizable in the first approximation and the linearization about the origin of the Δ_{mix} dynamics (2.15) is asymptotically stable. Observe, finally, that for linear systems the distribution $\Delta_{\text{mix}} = \emptyset$. Hence, a linear system can be rendered noninteractive and stable, by means of dynamic state-feedback, if and only if the two problems of stabilization and noninteracting control are separately solvable [14, 9, 13]. In particular, the problem of decoupling with stability is always solvable for simple linear controllable Hamiltonian systems, with respect to the natural outputs.

Hamiltonian systems are not asymptotically stable at any equilibrium, although they can be stable. As mentioned above, it is of interest to know whether the $P^{\perp}$, P^{*} and Δ_{mix} dynamics are Hamiltonian, since this fact can highly facilitate the tests for stability, needed, as described above, to decide if the considered problem of decoupling with stability is solvable and what class of control laws has to be considered in order to find a solution.

Despite the fact that the zero dynamics of Hamiltonian systems are Hamiltonian, it will be now shown that, for Hamiltonian systems of the form (2.1), (2.2), neither one of the three subsystems (2.13), (2.14) and (2.15) is Hamiltonian, in general. This will be done by means of simple counterexamples.

Example 2.3.1. Consider the following nonlinear system:

$$\begin{aligned} \dot{q}_1 &= p_1, \\ \dot{p}_1 &= v_1, \\ \dot{q}_2 &= p_2, \\ \dot{p}_2 &= v_2, \\ \dot{q}_3 &= p_3 + a_1 q_1 + a_2 q_2 + \alpha\, q_1 q_2, \\ \dot{p}_3 &= q_3 - d_1 q_1 - d_2 q_2 - \delta\, q_1 q_2, \end{aligned} \tag{2.16}$$

with outputs

$$\begin{aligned} y_1 &= q_1, \\ y_2 &= q_2 \end{aligned}$$

obtained by means of the static state-feedback control law:

$$\begin{aligned} u_1 &= a_1 p_3 + \alpha\, q_2 p_3 + d_1 q_3 + \delta\, q_2 q_3 + v_1, \\ u_2 &= a_2 p_3 + \alpha\, q_1 p_3 + d_2 q_3 + \delta\, q_1 q_3 + v_2, \end{aligned}$$

applied to the Hamiltonian system of the form (2.1), with Hamiltonian function:

$$\begin{aligned} H(q, p, u) = \frac{1}{2}\,(p_1^2 + p_2^2 + p_3^2) + (a_1 q_1 + a_2 q_2 + \alpha\, q_1 q_2)\, p_3 - \\ \frac{1}{2}\, q_3^2 + (d_1 q_1 + d_2 q_2 + \delta\, q_1 q_2)\, q_3 - q_1 u_1 - q_2 u_2, \end{aligned}$$

where a_i, d_i, α, δ, are real parameters of the system. Note that the zero dynamics of such a system can be simply written as

$$\begin{aligned} \dot{q}_3 &= p_3, \\ \dot{p}_3 &= q_3. \end{aligned} \tag{2.17}$$

We are now ready to prove the following facts.

Fact 1 *The $P^{\perp}$ dynamics of Hamiltonian systems of the form* (2.1) *need not be Hamiltonian.*

Consider system (2.16). Let $a_1 = a_2 = d_1 = d_2 = 1$, $\alpha = \delta = 0$. Simple calculations show that

$$P = \text{span}\,\{e_1, e_2, e_3, e_4, e_5 - e_6\}.$$

Hence, $\phi := q_3 + p_3$ is such that $P^{\perp} = \text{span}\,\{\mathrm{d}\phi\}$; the resulting subsystem (2.13) is

$$\dot{\phi} = \phi, \tag{2.18}$$

and it is clearly non-Hamiltonian. Since (2.18) is unstable, the problem of decoupling with stability is not solvable.

Fact 2 *The P^* dynamics of Hamiltonian systems of the form* (2.1) *need not be Hamiltonian.*

Consider system (2.16). Let $a_1 = d_1 = a_2 = \alpha = \delta = 1$, $d_2 = 0$ and observe that (locally around the origin)

$$\begin{aligned} P_1^* &= \text{span}\,\{e_3, e_4, e_5, e_6\}, \\ P_2^* &= \text{span}\,\{e_1, e_2, e_5 - c_6\}, \end{aligned}$$

whence

$$P^* = P_1^* \cap P_2^* = \text{span}\,\{e_5 - e_6\}.$$

As a consequence, $S^* = \{q_1 = p_1 = q_2 = p_2 = 0,\, p_3 = -q_3\}$ and the dynamics (2.14) can be written as

$$\dot{q}_3 = -q_3,$$

which is asymptotically stable (and, obviously, non-Hamiltonian). Moreover, as $P^{\perp} = \emptyset$, the problem of input-output decoupling with asymptotic stability in the first approximation can be solved by means of a static state-feedback. On the contrary, since the zero dynamics (2.17) are clearly unstable, the system cannot be rendered noninteractive and stable with a PD-type decentralized control law.

Fact 3 *The Δ_{mix} dynamics of Hamiltonian systems of the form* (2.1) *need not be Hamiltonian.*

Consider again system (2.16), with $a_1 = d_1 = a_2 = \alpha = \delta = 1$, $d_2 = 0$. The vector field $\left[\tilde{g}_1, ad_{\tilde{f}}^3 \tilde{g}_2\right] = e_5 - e_6$, belongs to Δ_{mix}. As $\Delta_{\text{mix}} \subset P^*$, it is evident that $\Delta_{\text{mix}} \equiv P^*$, hence the Δ_{mix} dynamics are not Hamiltonian.

2.3.1 Simple Hamiltonian Systems

It must be noted that the Hamiltonian system considered in Example 2.3.1 is not simple. The problem of determining if the stronger structure of simple Hamiltonian systems implies that their $P^{\perp}$, P^* or Δ_{mix} dynamics are Hamiltonian is investigated in this subsection. A positive answer is given, in Propositions 2.3.1 and 2.3.2, for the special case of linear simple Hamiltonian systems (limited to the $P^{\perp}$ and P^* dynamics since the Δ_{mix} dynamics are defined on a zero dimensional submanifold). For nonlinear simple Hamiltonian systems, relatively to the P^* and the Δ_{mix} dynamics, which can be seen as the dynamics of a Hamiltonian system (the zero dynamics) restricted to the integral submanifold of suitable nonsingular, involutive and invariant distributions, the following considerations can be carried out, in order to prove that, under some assumptions, the submanifolds on which they are defined are symplectic.

Assume that a simple Hamiltonian system is given as in (2.9), with the Hessian matrix with respect to $\tilde{q}$ of the function $V(\begin{bmatrix} 0 \\ \tilde{q} \end{bmatrix})$ nonsingular at the origin, which is assumed to be an equilibrium point. Let Δ be a nonsingular and involutive distribution of dimension $2r$, $r < \tilde{n}$, with $\tilde{n}$ being the dimension of $\tilde{q}$, which is f-invariant. Let $\chi_1(\tilde{q}, \tilde{p}), \ldots, \chi_s(\tilde{q}, \tilde{p})$ be s smooth functions such that $\{\chi_1(\tilde{q}, \tilde{p}), \ldots, \chi_s(\tilde{q}, \tilde{p}), \dot{\chi}_1(\tilde{q}, \tilde{p}), \ldots, \dot{\chi}_s(\tilde{q}, \tilde{p})\}$ is a set of $2s$ independent functions, where, as usual

$$\dot{\chi}_i = \frac{\partial \chi_i}{\partial \tilde{q}} \dot{\tilde{q}} + \frac{\partial \chi_i}{\partial \tilde{p}} \dot{\tilde{p}}.$$

Assume, further, that span$\{d\chi, d\dot{\chi}\} = \Delta^{\perp}$, *i.e.*, let $s = \tilde{n} - r$. Observe that this is true if $\chi(\tilde{q}, \tilde{p}) = \chi(\tilde{q})$. In such a case, if $d\chi \in \Delta^{\perp}$, it is easy to show that $d\dot{\chi} \in \Delta^{\perp}$, and also the independence of the set of functions $\{\chi_1(\tilde{q}), \ldots, \chi_s(\tilde{q}), \dot{\chi}_1(\tilde{q}, \tilde{p}), \ldots, \dot{\chi}_s(\tilde{q}, \tilde{p})\}$ is easily proven, by virtue of the fact that the matrix $G(\tilde{q})$ is nonsingular. Under such assumptions, it can be proven that the $2s \times 2s$-dimensional matrix $C(x)$ defined by

$$C(x) := \left[\begin{array}{c|c} \{\chi_i, \chi_j\}(x) & \{\chi_i, \dot{\chi}_j\}(x) \\ \hline \{\dot{\chi}_i, \chi_j\}(x) & \{\dot{\chi}_i, \dot{\chi}_j\}(x) \end{array}\right],$$

where $\{a, b\}$ denotes the standard Poisson bracket of the functions a, b, is nonsingular at the origin. Hence, by virtue of [11, Remark 12.37], the integral submanifold of Δ containing $x = 0$, is symplectic, for all x in a some neighborhood of the origin. This condition is necessary, but not sufficient, for the dynamics of the given system restricted to Δ to be Hamiltonian.

The following result, which can be proved without assuming that the Hessian matrix with respect to $\tilde{q}$ of the function $V(\begin{bmatrix} 0 \\ \tilde{q} \end{bmatrix})$ is nonsingular, concerns linear simple Hamiltonian systems, which are obtained from the general case (2.1)–(2.3) by letting $G(q)$ be a constant symmetric and positive definite matrix, $G \in \mathbb{R}^{n \times n}$, and $V(q) = q^T U q$, where $U \in \mathbb{R}^{n \times n}$ is also symmetric.

Proposition 2.3.1. *The $P^{\perp}$ dynamics of linear simple Hamiltonian systems are Hamiltonian.*

Proof. The state space equation of the system are given by:

$$\begin{aligned} \dot{q} &= Gp \\ \dot{p} &= -Uq + e_1 u_1 + e_2 u_2, \end{aligned}$$

¿from which it is clear that the dynamic matrix is the Hamiltonian matrix $\Theta := \begin{bmatrix} 0 & G \\ -U & 0 \end{bmatrix}$. A direct computation shows that

$$P^{\perp} = \left\{ \begin{bmatrix} Gw_1 \\ w_2 \end{bmatrix}, \quad w_1, w_2 \in \mathrm{Im}\,(\mathcal{R}) \right\}, \tag{2.19}$$

where $\mathcal{R}$ is the reachability matrix of a linear system with two inputs, having as dynamic matrix UG and as input matrix $\begin{bmatrix} e_1 & e_2 \end{bmatrix}$.

The following properties of the Hamiltonian matrix Θ can be easily proven, by taking into account that the eigenvalues of the matrix GU are all real, and the matrix GU is diagonalizable:

(P1) if $\lambda \in \mathbb{R}$, $\lambda \neq 0$ is an eigenvalue of Θ, with eigenvector $\begin{bmatrix} v_q \\ v_p \end{bmatrix} \in \mathbb{R}^{2n}$, then $-\lambda$ is also an eigenvalue, with eigenvector $\begin{bmatrix} v_q \\ -v_p \end{bmatrix}$;

(P2) if $\lambda \in \mathbb{R}$, $\lambda \neq 0$ is an eigenvalue of Θ, with algebraic multiplicity m, then Θ has m linearly independent eigenvectors relative to the eigenvalue λ;

(P3) Θ has null eigenvalues if and only if U has null eigenvalues; moreover, if m is the algebraic multiplicity of the null eigenvalue for U, and $v_1, v_2, \dots,$ $v_m \in \mathbb{R}^n$ are corresponding m linearly independent eigenvectors, then the Jordan structure of Θ relative to the null eigenvalue is constituted by m 2×2 Jordan blocks and the generalized eigenspace of Θ relative to the null eigenvalue is the following:

$$V_0 = \text{span}\left\{\begin{bmatrix} v_1 \\ 0 \end{bmatrix}, \begin{bmatrix} v_1 \\ G^{-1}v_1 \end{bmatrix}, \dots, \begin{bmatrix} v_m \\ 0 \end{bmatrix}, \begin{bmatrix} v_m \\ G^{-1}v_m \end{bmatrix}\right\};$$

(P4) the eigenvalues of Θ which are not real lie on the imaginary axis; moreover, $\jmath\omega$, $\omega \neq 0$, is a complex eigenvalue of Θ if and only if ω^2 is an eigenvalue of GU; if m is the algebraic multiplicity of the eigenvalue ω^2 for GU, and $v_1, v_2, \dots, v_m \in \mathbb{R}^n$ are corresponding m linearly independent eigenvectors, then Θ admits m linearly independent complex eigenvectors relative to the eigenvalue $\jmath\omega$ and m linearly independent complex eigenvectors relative to the eigenvalue $-\jmath\omega$, and a real basis for the sum $V_{\jmath\omega} + V_{-\jmath\omega}$ of the eigenspaces relative to the eigenvalues $\jmath\omega$ and $-\jmath\omega$ is composed of

$$\left\{\begin{bmatrix} v_1 \\ 0 \end{bmatrix}, \begin{bmatrix} 0 \\ G^{-1}v_1 \end{bmatrix}, \dots, \begin{bmatrix} v_m \\ 0 \end{bmatrix}, \begin{bmatrix} 0 \\ G^{-1}v_m \end{bmatrix}\right\}.$$

Since P is Θ-invariant (this is a consequence of the properties of the distribution P in the general case, and, in the case of linear systems can be easily proven from (2.19)), then, in view of [13, Proposition 0.4], P is a direct sum of its intersections with eigenspaces (possibly, generalized). Moreover, in view of (2.19), it is easy to see that

$$\begin{bmatrix} v_q \\ v_p \end{bmatrix} \in P \Rightarrow \begin{bmatrix} v_q \\ -v_p \end{bmatrix} \in P, \quad \forall v_q, v_p \in \mathbb{R}^n, \tag{2.20a}$$

$$\begin{bmatrix} \alpha\, v_q \\ G^{-1}v_q \end{bmatrix} \in P \Rightarrow \begin{bmatrix} v_q \\ 0 \end{bmatrix}, \begin{bmatrix} 0 \\ G^{-1}v_q \end{bmatrix} \in P, \quad \forall v_q \in \mathbb{R}^n, \forall \alpha \in \mathbb{R}. \tag{2.20b}$$

Properties (P1)-(P4) and (2.20a)-(2.20b) imply that a basis for $\mathbb{R}^{2n}$ can be taken as follows:

$$
\mathcal{B} = \left\{ \underbrace{\begin{bmatrix} v_{q1} \\ 0 \end{bmatrix}, \begin{bmatrix} 0 \\ G^{-1}v_{q1} \end{bmatrix}, \ldots, \begin{bmatrix} v_{qi} \\ 0 \end{bmatrix}, \begin{bmatrix} 0 \\ G^{-1}v_{qi} \end{bmatrix}}_{\text{basis } \mathcal{B}_1 \text{ for } (V_{\jmath\omega_1} + V_{-\jmath\omega_1}) \cap P}, \ldots, \right.
$$

$$
\underbrace{\begin{bmatrix} v_{q(i+1)} \\ v_{p(i+1)} \end{bmatrix}, \begin{bmatrix} v_{q(i+1)} \\ -v_{p(i+1)} \end{bmatrix}, \ldots, \begin{bmatrix} v_{qj} \\ v_{pj} \end{bmatrix}, \begin{bmatrix} v_{qj} \\ -v_{pj} \end{bmatrix}}_{\text{basis for } (V_{\lambda_1} + V_{-\lambda_1}) \cap P}, \ldots,
$$

$$
\underbrace{\begin{bmatrix} v_{q(j+1)} \\ 0 \end{bmatrix}, \begin{bmatrix} v_{q(j+1)} \\ G^{-1}v_{q(j+1)} \end{bmatrix}, \ldots, \begin{bmatrix} v_{qk} \\ 0 \end{bmatrix}, \begin{bmatrix} v_{qk} \\ G^{-1}v_{qk} \end{bmatrix}}_{\text{basis for } V_0 \cap P},
$$

$$
\left. \underbrace{\begin{bmatrix} v_{q(k+1)} \\ 0 \end{bmatrix}, \begin{bmatrix} 0 \\ G^{-1}v_{q(k+1)} \end{bmatrix}, \ldots, \begin{bmatrix} v_{qs} \\ 0 \end{bmatrix}, \begin{bmatrix} 0 \\ G^{-1}v_{qs} \end{bmatrix}}_{\text{complement of } \mathcal{B}_1 \text{ to a basis for } V_{\jmath\omega_1} + V_{-\jmath\omega_1}}, \ldots \right\}
$$

where $2k$ is the dimension of P and $v_{q1}, \ldots v_{qn}$ are suitable orthogonal eigenvectors of GU. When rewritten in the basis $\mathcal{B}$, matrix Θ becomes block diagonal, with diagonal blocks $\Theta_1 \in \mathbb{R}^{2k \times 2k}$ and $\Theta_2 \in \mathbb{R}^{2(n-k) \times 2(n-k)}$. Matrices Θ_1 and Θ_2 are, in turn, block diagonal, and their diagonal blocks, all of dimension 2×2, are of the following kinds:

$$
\begin{bmatrix} 0 & -\omega^2 \\ 1 & 0 \end{bmatrix}, \quad \begin{bmatrix} \lambda & 0 \\ 0 & -\lambda \end{bmatrix}, \quad \begin{bmatrix} 0 & 1 \\ 0 & 0 \end{bmatrix}, \quad \omega^2, -\lambda^2 \text{ eigenvalues of } U.
$$

Hence, it is easily seen that the block diagonal skew symmetric matrix $J \in \mathbb{R}^{2(n-k) \times 2(n-k)}$, having its diagonal blocks all equal to $\begin{bmatrix} 0 & 1 \\ -1 & 0 \end{bmatrix}$ is such that $\Theta_2^T J + J\Theta_2 = 0$. Now, notice that matrix Θ_2 is related through a suitable similarity transformation $\widetilde{\Theta}_2 = T^{-1}\Theta_2 T$, to any matrix $\widetilde{\Theta}_2$ describing the $P^\perp$ dynamics, hence, by defining $\widetilde{J} = T^T JT$, it follows that $\widetilde{\Theta}_2^T \widetilde{J} + \widetilde{J}\widetilde{\Theta}_2 = 0$.

This last equation proves that the $P^\perp$ dynamics are Hamiltonian. In fact, as is well known, the dynamics $\dot{\tilde{x}} = \widetilde{\Theta}_2 \tilde{x}$ can be seen as generated by the Hamiltonian function $\widetilde{H} = \frac{1}{2}\tilde{x}^T M \tilde{x}$, with $M := \widetilde{J}\widetilde{\Theta}_2$, and the Poisson structure corresponding to the skew symmetric matrix $\widetilde{J}^{-1}$, in the same coordinates $\tilde{x}$.
□

Proposition 2.3.2. *The P^* dynamics of linear simple Hamiltonian systems are Hamiltonian.*

Proof. Partitioning matrices G and U as follows:

$$
G = \begin{bmatrix} G_{11} & G_{21}^T \\ G_{21} & G_{22} \end{bmatrix}, \quad U = \begin{bmatrix} U_{11} & U_{21}^T \\ U_{21} & U_{22} \end{bmatrix},
$$

where $G_{11}, U_{11} \in \mathbb{R}^{2\times 2}$, letting $\tilde{G} = G_{22} - G_{21}G_{11}^{-1}G_{21}^T$ and using the notations in Section 2.2, after a first state-feedback of the form (2.6), the closed-loop dynamics are described by

$$\dot{\xi}_1 = A\,\xi_1 + B\,v_1, \tag{2.21a}$$
$$\dot{\xi}_2 = A\,\xi_2 + B\,v_2, \tag{2.21b}$$
$$\dot{\tilde{q}} = G_{21}G_{11}^{-1}\begin{bmatrix}\xi_{12}\\ \xi_{22}\end{bmatrix} + \tilde{G}\tilde{p}, \tag{2.21c}$$
$$\dot{\tilde{p}} = -U_{21}\begin{bmatrix}\xi_{11}\\ \xi_{21}\end{bmatrix} - U_{22}\tilde{q}. \tag{2.21d}$$

Now, call γ_1 and γ_2 the two columns of $G_{21}G_{11}^{-1}$, and U_{21}^1 and U_{21}^2 the two columns of U_{21} and define

$$\overline{\gamma}_1 := U_{21}^1 + U_{22}\gamma_1,$$
$$\overline{\gamma}_2 := U_{21}^2 + U_{22}\gamma_2.$$

A direct computation shows that

$$P^* = P_1^* \cap P_2^* = \left\{\begin{bmatrix}0_4\\ \tilde{G}w_1\\ w_2\end{bmatrix}, \quad w_1,\, w_2 \in \mathrm{Im}\,(\mathcal{R}_1) \cap \mathrm{Im}\,(\mathcal{R}_2)\right\}, \tag{2.22}$$

where $\mathcal{R}_1$ and $\mathcal{R}_2$ are the reachability matrices of two linear single input systems of order $\tilde{n} = n - 2$ having the same dynamic matrix $U_{22}\tilde{G}$ and as input vectors $\overline{\gamma}_1$ and $\overline{\gamma}_2$, respectively. This clearly shows that P^* has even dimension: let $\dim(P^*) = 2r$. Moreover, the special structure of P^* implies that, if X is defined as follows:

$$X := \left\{ v_q \in \mathbb{R}^{\tilde{n}} \; : \; \begin{bmatrix}0_4\\ v_q\\ 0\end{bmatrix} \in P^* \right\},$$

then X is a r-dimensional vector subspace of $\mathbb{R}^{\tilde{n}}$. Now, consider the Hamiltonian system constituted by the zero dynamics of the given n-dimensional system, which is an Hamiltonian system with no inputs, characterized by the Hamiltonian function $\overline{H}(\tilde{q}, \tilde{p}) = \frac{1}{2}\tilde{p}^T\tilde{G}\tilde{p} + \frac{1}{2}\tilde{q}^T U_{22}\tilde{q}$. Let $\{\ell_1, \ell_2, \ldots, \ell_s\}$, $s = \tilde{n} - r$, be a basis for $X^\perp$, and define the s functions $\chi_1 := \ell_1^T\tilde{q}$, $\ldots$, $\chi_s := \ell_s^T\tilde{q}$. A direct computation shows that $\dot{\chi}_i = \ell_i^T\tilde{G}\tilde{p}$, so that, if $L = [\ell_1\,\ell_2\,\ldots\,\ell_s]$, the matrix C defined above (which is constant in view of the fact that the Hamiltonian system under consideration is linear) is given by:

$$C = \begin{bmatrix} 0 & L^T\tilde{G}L\\ L^T\tilde{G}L & 0\end{bmatrix},$$

which is clearly nonsingular, thus proving that the integral submanifold of P^* containing the origin is symplectic. Now, since $\{\overline{H}, \chi_i\} = \ell_i^T \tilde{G}\tilde{p} = 0$ and $\{\overline{H}, \dot{\chi}_i\} = -\ell_i^T \tilde{G} U_{22}\tilde{q} = 0$, for all $\begin{bmatrix} \tilde{q} \\ \tilde{p} \end{bmatrix}$ such that $\begin{bmatrix} 0_4 \\ \tilde{q} \\ \tilde{p} \end{bmatrix} \in P^*$, in view of [11, Lemma 12.39] the P^* dynamics are Hamiltonian. □

2.4 Decoupling with stability by dynamic state-feedback

In this section, the problem of input-output decoupling with stability will be dealt with for the class of simple Hamiltonian systems given by (2.1), (2.2), (2.3), using dynamic state-feedback control laws.

In order to tackle jointly several problems related with different stability requirements, let the symbol $\mathbb{C}_g$ denote the region of the complex plane, symmetric about the real axis, where the eigenvalues of the linear approximation of the closed-loop system are desired to lie: in particular, let $\mathbb{C}_g$ denote the closed left half-plane, or the open left half-plane, or the half-plane $\{s \in \mathbb{C} : \mathrm{Re}(s) < -\alpha\}$, if either stability, or asymptotic stability in the first approximation, or asymptotic stability with a prescribed rate of convergence α, being α a positive real number, is required, respectively. Moreover, let $\mathbb{C}_b := \mathbb{C} - \mathbb{C}_g$.

It is assumed, without loss of generality, because of the considerations in Section 2.2, that a suitable static state-feedback control law, of the form (2.6), has already been applied to the given Hamiltonian system to achieve noninteraction. Hence, one can start from the equations (2.8a)–(2.8f) for the closed-loop system, *i.e.*

$$\begin{aligned} \dot{\xi}_1 &= A\,\xi_1 + B\,v_1, \\ \dot{\xi}_2 &= A\,\xi_2 + B\,v_2, \\ \dot{z} &= F\,z + L\,\xi_1 + M\,\xi_2 + \vartheta(\xi_1, \xi_2, z), \\ y_1 &= [1\ 0]\,\xi_1, \\ y_2 &= [1\ 0]\,\xi_2, \end{aligned} \tag{2.23}$$

where the vector $z \in \mathbb{R}^{2n-4}$ is given by $z := [\tilde{q}^T\ \tilde{p}^T]^T$, F, L, M are real matrices of suitable dimensions, and the vector valued function $\vartheta(\xi_1, \xi_2, z)$ is such that

$$\frac{\partial\vartheta}{\partial\xi_1}(0, 0, 0) = 0, \quad \frac{\partial\vartheta}{\partial\xi_2}(0, 0, 0) = 0, \quad \frac{\partial\vartheta}{\partial z}(0, 0, 0) = 0.$$

In order to restrict the attention to a class of systems which require dynamic state-feedback control laws in order to be rendered stable and decoupled, the following two assumptions (a) and (b) are made.

(a) the two pairs:

$$\left(\begin{bmatrix} A & 0 \\ L & F \end{bmatrix}, \begin{bmatrix} B \\ 0 \end{bmatrix}\right) \quad \text{and} \quad \left(\begin{bmatrix} A & 0 \\ M & F \end{bmatrix}, \begin{bmatrix} B \\ 0 \end{bmatrix}\right)$$

are controllable,

(b) $\sigma(F) \cap \mathbb{C}_b \neq \{\emptyset\}$, where the symbol $\sigma(\cdot)$ denotes the spectrum of the matrix at argument.

Notice that assumption (a) implies that $P^* \equiv \Delta^* \equiv \text{span}\{\frac{\partial}{\partial z}\}$, hence assumption (b) implies that the problem of decoupling with stability is not solvable by means of static state-feedback.

Now, let V_g^F, V_b^F denote the two F-invariant subspaces of $\mathbb{R}^{2n-4}$ such that

$$\mathbb{R}^{2n-4} = V_g^F \oplus V_b^F,$$
$$\sigma\left(F|_{V_g^F}\right) \subset \mathbb{C}_g,$$
$$\sigma\left(F|_{V_b^F}\right) \subset \mathbb{C}_b.$$

Let a linear coordinate transformation be defined on P^* such that, if the new coordinates $\tilde{z}$ are given by $\tilde{z} = T\,z$, then one has

$$\tilde{F} := T\,F\,T^{-1} = \begin{bmatrix} F_g & 0 \\ 0 & F_b \end{bmatrix},$$

with $\sigma(F_g) = \sigma\left(F|_{V_g^F}\right)$ and $\sigma(F_b) = \sigma\left(F|_{V_b^F}\right)$.

Let $\tilde{z} =: \left[z_g^T \; z_b^T\right]^T$ be the partition of $\tilde{z}$ corresponding to the block partition of $\tilde{F}$, and, finally, let the vector $\tilde{\vartheta}(\xi_1, \xi_2, \tilde{z})$, defined by

$$\tilde{\vartheta}(\xi_1, \xi_2, \tilde{z}) := T\,\vartheta(\xi_1, \xi_2, T^{-1}\tilde{z}),$$

be partitioned according to the partition of $\tilde{z}$:

$$\tilde{\vartheta}(\xi_1, \xi_2, \tilde{z}) = [\vartheta_g^T(\xi_1\,\xi_2, \tilde{z}) \;\; \vartheta_b^T(\xi_1, \xi_2, \tilde{z})]^T.$$

The following assumption (c) considerably simplifies the problem.

(c) The vector $\vartheta_b(\xi_1, \xi_2, \tilde{z})$ is a function of the variables ξ_1, ξ_2 only:

$$\vartheta_b(\xi_1, \xi_2, \tilde{z}) =: \psi(\xi_1, \xi_2), \quad \forall(\xi_1, \xi_2, \tilde{z}) \text{ in a neighborhood of } (0, 0, 0).$$

The following result provides a condition to solve the problem of decoupling with stability by means of dynamic state-feedback, which is, in general, much easier to check than the necessary and sufficient conditions based on the explicit computation of the distribution Δ_{mix}. In order to apply the results recalled in Section 2.3, valid for general nonlinear systems, giving the necessary conditions for the existence of a solution, the following technical assumption is introduced:

(d) The origin $(\xi_1, \xi_2, z) = (0, 0, 0)$ is a regular point of the distribution Δ_{mix} of system (2.23).

We are now ready to state the main result of this section.

Proposition 2.4.1. *Under assumptions (a), (b), (c) and (d), a dynamic state-feedback control law which solves the problem of input-output decoupling with either*

(A) (simple) stability, or

(B) asymptotic stability in the first approximation, or

(C) asymptotic stability with a prescribed convergence rate,

exists only if the following conditions hold in a neighborhood of $\xi_1 = 0$, $\xi_2 = 0$ *(it is recalled that* $\xi_1 = [q_1\ \dot{q}_1]^T$, $\xi_2 = [q_2\ \dot{q}_2]^T$*):*

(i) $$\psi_{\dot{q}_1 \dot{q}_2}(\xi_1, \xi_2) = 0,$$

(ii) $$\psi_{q_1 \dot{q}_2}(\xi_1, \xi_2) = \psi_{q_2 \dot{q}_1}(\xi_1, \xi_2),$$

(iii) $$\frac{1}{2} F_b\Big(\psi_{q_1 \dot{q}_2}(\xi_1, \xi_2) + \psi_{q_2 \dot{q}_1}(\xi_1, \xi_2)\Big) + \psi_{q_1 q_2}(\xi_1, \xi_2) - \dot{q}_1\, \psi_{q_1 q_2 \dot{q}_1}(\xi_1, \xi_2) - \dot{q}_2\, \psi_{q_1 q_2 \dot{q}_2}(\xi_1, \xi_2) = 0.$$

In cases (B) and (C), conditions (i), (ii) and (iii) are also sufficient for the existence of a solution, whereas, in case (A), a set of sufficient conditions is given by (i), (ii), (iii) and the following condition:

(iv) the equilibrium of the dynamical system

$$\dot{z}_g = F_g\, z_g + \vartheta_g(0, 0, \begin{bmatrix} z_g \\ 0 \end{bmatrix}) \tag{2.24}$$

is stable.

□

2.5 Proof of proposition 2.4.1

In view of the proposed notations, and of assumption (c), system (2.23) can be rewritten as follows:

$$\dot{\xi}_1 = A\,\xi_1 + B\,v_1, \tag{2.25a}$$
$$\dot{\xi}_2 = A\,\xi_2 + B\,v_2, \tag{2.25b}$$
$$\dot{z}_g = F_g\,z_g + L_g\,\xi_1 + M_g\,\xi_2 + \vartheta_g(\xi_1,\,\xi_2,\,\begin{bmatrix} z_g \\ z_b \end{bmatrix}), \tag{2.25c}$$
$$\dot{z}_b = F_b\,z_b + L_b\,\xi_1 + M_b\,\xi_2 + \psi(\xi_1,\,\xi_2), \tag{2.25d}$$

where L_g, L_b, M_g and M_b are real matrices of suitable dimensions.

The proof of Proposition 2.4.1 is organized as follows. First, it is shown that hypotheses (i), (ii) and (iii) are necessary for the existence of a solution of the given control problem. Secondly, the design procedure of a dynamic state-feedback compensator is outlined, on the basis of the algorithm proposed in [7, Chapter 7]. In a third step, it is shown that, under hypotheses (i), (ii) and (iii), such a compensator solves the problem of decoupling with stability in cases (B) and (C), *i.e.* when the stability requirement can be checked on the basis of the properties of the linearized system. Lastly, by means of some results from the Center Manifold Theory [4], it is shown that, under hypotheses (i), (ii), (iii) and (iv), the proposed compensator solves the problem in case (A).

Necessity of (i), (ii) and (iii). In order to see that each of conditions (i), (ii) and (iii) is necessary for the existence of a solution, rewrite system (2.25a)–(2.25d) in the general form

$$\dot{\overline{x}} = \tilde{f}(\overline{x}) + \tilde{g}_1 v_1 + \tilde{g}_2 v_2,$$

where the state vector $\overline{x} \in \mathbb{R}^{2n}$ is given by $\overline{x} = [\xi_1^T\ \xi_2^T\ z_g^T\ z_b^T]^T$, and compute the following vectors, which certainly belong to Δ_{mix}

$$\left[ad_{\tilde{f}}\tilde{g}_1,\,\tilde{g}_2\right] = \left[0\ 0\ 0\ 0\ l_{g10}(\overline{x})\ l_{b10}(\overline{x})\right]^T,$$
$$\left[ad_{\tilde{f}}\tilde{g}_1,\,ad_{\tilde{f}}\tilde{g}_2\right] = \left[0\ 0\ 0\ 0\ l_{g11}(\overline{x})\ l_{b11}(\overline{x})\right]^T,$$
$$\left[ad^2_{\tilde{f}}\tilde{g}_1,\,ad_{\tilde{f}}\tilde{g}_2\right] = \left[0\ 0\ 0\ 0\ l_{g21}(\overline{x})\ l_{b21}(\overline{x})\right]^T,$$

with

$$l_{b10}(\overline{x}) = \psi_{\dot{q}_1\dot{q}_2}(\xi_1,\,\xi_2),$$
$$l_{b11}(\overline{x}) = -\psi_{\dot{q}_1 q_2}(\xi_1,\,\xi_2) + \psi_{q_1\dot{q}_2}(\xi_1,\,\xi_2),$$
$$l_{b21}(\overline{x}) = F_b\psi_{\dot{q}_1 q_2}(\xi_1,\,\xi_2) - \dot{q}_2\psi_{\dot{q}_1 q_2 q_2}(\xi_1,\,\xi_2) - \dot{q}_1\psi_{q_1\dot{q}_1 q_2}(\xi_1,\,\xi_2)$$
$$+\psi_{q_1 q_2}(\xi_1,\,\xi_2),$$

and with $l_{g10}(\bar{x})$, $l_{g11}(\bar{x})$ and $l_{g21}(\bar{x})$ being suitable vectors of dimension n_g. It is clear that, if any of conditions (i), (ii) and (iii) is not satisfied, then $\Delta_{\mathrm{mix}} \cap \mathrm{span}\left\{\frac{\partial}{\partial z_b}\right\} \neq 0$, hence the Δ_{mix} dynamics of system (2.25a)–(2.25d) cannot be stable, with respect to the given stability requirement.

Structure of the overall control system. In order to design a dynamic state-feedback compensator, solving the problem of decoupling with stability, consider the following nonsingular coordinates transformation:

$$\tilde{z}_b = z_b + \tau(q_1, q_2),$$

in which the vector $\tau(q_1, q_2)$ is defined as

$$\tau(q_1, q_2) := \frac{1}{2}\int_0^{q_2}\int_0^{q_1}\left(\psi_{q_1\dot{q}_2}\left(\begin{bmatrix}\eta_1\\ \dot{q}_1\end{bmatrix}, \begin{bmatrix}\eta_2\\ \dot{q}_2\end{bmatrix}\right) + \psi_{q_2\dot{q}_1}\left(\begin{bmatrix}\eta_1\\ \dot{q}_1\end{bmatrix}, \begin{bmatrix}\eta_2\\ \dot{q}_2\end{bmatrix}\right)\right) \mathrm{d}\eta_1\, \mathrm{d}\eta_2. \quad (2.26)$$

In the new coordinates, system (2.25a)–(2.25d) is described by equations (2.25a), (2.25b), (2.25c) and

$$\dot{\tilde{z}}_b = F_b\tilde{z}_b + L_b\xi_1 + M_b\xi_2 + \tilde{\psi}(\xi_1, \xi_2),$$

where, by virtue of hypotheses (i), (ii) and (iii) and of equation (2.26), the vector $\tilde{\psi}(\xi_1, \xi_2)$ can be seen to satisfy the following four identities

$$\tilde{\psi}_{q_1q_2}(\xi_1, \xi_2) = 0, \quad \tilde{\psi}_{q_1\dot{q}_2}(\xi_1, \xi_2) = 0,$$
$$\tilde{\psi}_{\dot{q}_1q_2}(\xi_1, \xi_2) = 0, \quad \tilde{\psi}_{\dot{q}_1\dot{q}_2}(\xi_1, \xi_2) = 0,$$

in a neighborhood of $\xi_1 = 0$, $\xi_2 = 0$. This implies that the vector $\tilde{\psi}(\xi_1, \xi_2)$ can be written as follows:

$$\tilde{\psi}(\xi_1, \xi_2) = \tilde{\psi}_1(\xi_1) + \tilde{\psi}_2(\xi_2);$$

whence, it is easy to verify that $\Delta_{\mathrm{mix}} \subset \mathrm{span}\left\{\frac{\partial}{\partial z_g}\right\}$.

Therefore, on the basis of the synthesis procedure reported in [7, Section 7.5], valid for general nonlinear systems, it is possible to design a dynamic state-feedback compensator for the subsystem

$$\begin{aligned}\dot{\xi}_1 &= A\,\xi_1 + B\,v_1,\\ \dot{\xi}_2 &= A\,\xi_2 + B\,v_2,\\ \dot{\tilde{z}}_b &= F_b\tilde{z}_b + L_b\xi_1 + M_b\xi_2 + \tilde{\psi}_1(\xi_1) + \tilde{\psi}_2(\xi_2),\end{aligned} \quad (2.27)$$

which solves the problem of noninteraction with asymptotic stability in the first approximation (in both cases (A) and (B)), or with the desired convergence rate (in case (C)). Such a compensator is of the form

$$\begin{aligned} \dot{x}_c &= \gamma(\xi_1, \xi_2, \tilde{z}_b, x_c) + G_c w, \\ v &= \eta(\xi_1, \xi_2, \tilde{z}_b, x_c) + w, \end{aligned} \tag{2.28}$$

where x_c is the state vector, $x_c \in \mathbb{R}^{4+2n_b}$, with $n_b := \dim\left(V_b^F\right)$ and w is the vector of the new inputs, $w = \left[\, w_1 \; w_2 \,\right]^T$, where w_1 does not affect y_2 and w_2 does not affect y_1.

Now, letting $x_e := [\xi_1^T \; \xi_2^T \; \tilde{z}_b^T \; x_c^T]^T$, the closed-loop system (2.27), (2.28) can be written as

$$\begin{aligned} \dot{x}_e &= f_e(x_e) + g_{1\,e}(x_e) w_1 + g_{2\,e}(x_e) w_2, \\ y_1 &= h_{1\,e}(x_e), \\ y_2 &= h_{2\,e}(x_e), \end{aligned} \tag{2.29}$$

and, in view of the design procedure adopted, it is decoupled and asymptotically stable in the first approximation (with the desired convergence rate, in case (C)).

Sufficiency of (i), (ii) and (iii) in cases (B) and (C). Simple considerations relative to the linearization about the origin of the overall control system, constituted by (2.29) and by

$$\dot{z}_g = F_g z_g + L_g \xi_1 + M_g \xi_2 + \vartheta_g \left(\xi_1, \xi_2, \begin{bmatrix} z_g \\ \tilde{z}_b - \tau(q_1, q_2) \end{bmatrix} \right), \tag{2.30}$$

suffice to prove the proposition with respect to the stability requirements (B) and (C), in view of the fact that the outputs y_1 and y_2 are not affected by z_g.

Sufficiency of (i), (ii), (iii) and (iv) in case (A). The stability of the origin $x_e = 0$, $z_g = 0$, for the overall control system, composed of (2.29) and (2.30), can be proven by means of well known results from the Center Manifold Theory (see [4] and [7, Appendix B]). To this end, consider a change of coordinates on the state space of system (2.24) such that, if the new coordinates $\tilde{z}_g$ are $\tilde{z}_g = \tilde{T} z_g$, then one has

$$\tilde{F}_g := \tilde{T} F_g \tilde{T}^{-1} = \begin{bmatrix} F_{g-} & 0 \\ 0 & F_{g0} \end{bmatrix},$$

with $\sigma(F_{g-}) \subset \{\lambda \in \mathbb{C}, \operatorname{Re}(\lambda) < 0\}$ and $\sigma(F_{g0}) \subset \{\lambda \in \mathbb{C}, \operatorname{Re}(\lambda) = 0\}$.

Let $\tilde{z}_g = \left[z_{g-}^T \; z_{g0}^T\right]^T$ be the partition of $\tilde{z}_g$ corresponding to the block partition of $\tilde{F}_g$. In the new coordinates, system (2.24) can be rewritten as

$$\begin{aligned} \dot{z}_{g-} &= F_{g-} z_{g-} + \vartheta_{g-}(z_{g-}, z_{g0}), \\ \dot{z}_{g0} &= F_{g0} z_{g0} + \vartheta_{g0}(z_{g-}, z_{g0}), \end{aligned} \tag{2.31}$$

with

$$\vartheta_{g-}\left(z_{g-},\, z_{g0}\right) := \tilde{T}_{-}\vartheta_g \left(0,\, 0,\, \begin{bmatrix} \tilde{T}^{-1}\begin{bmatrix} z_{g-} \\ z_{g0} \end{bmatrix} \\ 0 \end{bmatrix}\right),$$

$$\vartheta_{g0}\left(z_{g-},\, z_{g0}\right) := \tilde{T}_{0}\vartheta_g \left(0,\, 0,\, \begin{bmatrix} \tilde{T}^{-1}\begin{bmatrix} z_{g-} \\ z_{g0} \end{bmatrix} \\ 0 \end{bmatrix}\right),$$

where $\tilde{T}_{-}$, $\tilde{T}_{0}$ are the two row blocks of the partition $\tilde{T} =: \begin{bmatrix} \tilde{T}_{-} \\ \tilde{T}_{0} \end{bmatrix}$ of matrix $\tilde{T}$, corresponding to the block partition of $\tilde{F}_g$. Since the origin $z_g = 0$ is an equilibrium point of system (2.24), then it is clear that functions $\vartheta_{g-}(\cdot,\, \cdot)$ and $\vartheta_{g0}(\cdot,\, \cdot)$ vanish at $\begin{bmatrix} z_{g-} \\ z_{g0} \end{bmatrix} = \begin{bmatrix} 0 \\ 0 \end{bmatrix}$, hence a mapping $z_{g-} = \pi(z_{g0})$, defined on a neighborhood U of $z_{g0} = 0$, such that the set

$$S := \left\{ \begin{bmatrix} z_{g-} \\ z_{g0} \end{bmatrix} \in \mathbb{R}^{n_g} :\ z_{g0} \in U,\, z_{g-} = \pi(z_{g0}) \right\}$$

is a center manifold for system (2.31), exists. Moreover, by hypothesis (iv), it follows that the dynamics of system (2.31) restricted to S, described by the equation

$$\dot{z}_{g0} = F_{g0} z_{g0} + \vartheta_{g0}\left(\pi\left(z_{g0}\right), z_{g0}\right), \quad z_{g0} \in U, \tag{2.32}$$

are necessarily stable. Now, in order to see that this implies stability for the overall control system, which can be written as

$$\begin{aligned} \dot{x}_e &= f_e(x_e) + g_{1e} v_1 + g_{2e} v_2, \\ \dot{z}_{g-} &= F_{g-} z_{g-} + L_{-} x_e + \tilde{\vartheta}_{g-}\left(x_e,\, z_{g-},\, z_{g0}\right), \\ \dot{z}_{g0} &= F_{g0} z_{g0} + L_0 x_e + \tilde{\vartheta}_{g0}\left(x_e,\, z_{g-},\, z_{g0}\right), \end{aligned} \tag{2.33}$$

where the matrices L_{-}, L_0 take into account the terms linear in ξ_1, ξ_2 appearing in equation (2.25c), and

$$\tilde{\vartheta}_{g-}\left(x_e,\, z_{g-},\, z_{g0}\right) := \tilde{T}_{-}\vartheta_g \left(\xi_1,\, \xi_2,\, \begin{bmatrix} \tilde{T}^{-1}\begin{bmatrix} z_{g-} \\ z_{g0} \end{bmatrix} \\ \tilde{z}_b - \tau(q_1,\, q_2) \end{bmatrix}\right),$$

$$\tilde{\vartheta}_{g0}\left(x_e,\, z_{g-},\, z_{g0}\right) := \tilde{T}_{0}\vartheta_g \left(\xi_1,\, \xi_2,\, \begin{bmatrix} \tilde{T}^{-1}\begin{bmatrix} z_{g-} \\ z_{g0} \end{bmatrix} \\ \tilde{z}_b - \tau(q_1,\, q_2) \end{bmatrix}\right),$$

a further linear coordinates transformation is needed. Define

$$F_e := \left.\frac{\partial f_e(x_e)}{\partial x_e}\right|_{x_e=0},$$

and let the last vector component of the new coordinates vector $\left[x_e^T \; z_{g-}^T \; \tilde{z}_{g0}^T\right]^T$ be given by

$$\tilde{z}_{g0} = P_0 x_e + z_{g0},$$

where the matrix $P_0 \in \mathbb{R}^{(2n-4-2n_b)\times(8+3n_b)}$ is such that

$$F_{g0}P_0 - P_0 F_e = L.$$

The existence of such a matrix P_0 is guaranteed by the fact that

$$\sigma(F_{g0}) \cap \sigma(F_e) = \{\emptyset\}.$$

System (2.33), with $v_1 = v_2 = 0$, can be rewritten as follows:

$$\begin{aligned} \dot{x}_e &= F_e x_e + \Delta f_e(x_e), \\ \dot{z}_{g-} &= F_{g-} z_{g-} + L_- x_e + \tilde{\vartheta}_{g-}\left(x_e, z_{g-}, \tilde{z}_{g0} - P_0 x_e\right), \\ \dot{\tilde{z}}_{g0} &= \Delta f_e^0(x_e) + F_{g0}\tilde{z}_{g0} + \tilde{\vartheta}_{g0}\left(x_e, z_{g-}, \tilde{z}_{g0} - P_0 x_e\right), \end{aligned} \tag{2.34}$$

where the functions $\Delta f_e(x_e) := f_e(x_e) - F_e x_e$ and $\Delta f_e^0(x_e) := P_0 f_e(x_e) - F_{g0}P_0 x_e + L_0 x_e$, vanish, together with their Jacobian matrices with respect to x_e, at $x_e = 0$.

It is easy to see that the set

$$S_e := \left\{ \begin{bmatrix} x_e \\ z_{g-} \\ \tilde{z}_{g0} \end{bmatrix} \in \mathbb{R}^{(8+3n_b+n_g)} : \; \tilde{z}_{g0} \in U, \, x_e = 0, z_{g-} = \pi(\tilde{z}_{g0}) \right\}$$

is a center manifold for system (2.34), and that the dynamics of system (2.34) restricted to S_e coincide with (2.32) if z_{g0} is replaced by $\tilde{z}_{g0}$. The claim follows from the Reduction Principle [4]. □

2.6 Examples

In this section, two examples, stemming from simple mechanical systems, are presented and the application of the theory reported in Sections 2.2, 2.3 and 2.4 is discussed.

The first example is concerned with a simple physical system which can be rendered input-output decoupled and asymptotically stable in the first approximation by means of a suitable static state-feedback control law, although its zero dynamics are not trivial and, being Hamiltonian, obviously not asymptotically stable.

Example 2.6.1. Consider the system represented in Figure 2.6.1, which is composed of three equal bodies having mass m each, which slide along an horizontal axis, namely the x axis of some inertial reference frame. Any kind of friction is neglected in the proposed model of the system. The first body, whose position at time t is denoted by $q_1(t)$, is connected to a fixed point at $x = 0$ by means of a nonlinear elastic spring (so-called hardening spring), having length equal to zero, when undeformed, and exerting a force equal in modulus to

$$F(\ell) = k\,\ell + k'\ell^3,$$

on the bodies at its extremities, when deformed up to length ℓ. The second body, whose position is denoted by $q_2(t)$, is connected to the first one by means of a nonlinear elastic spring equal to the one described previously, whereas the third body, whose position is denoted by $q_3(t)$, is connected in the same way to the second one. The two control inputs of the system are two external forces, $u_1(t)$ and $u_2(t)$, applied to the first and to the second body, respectively. The natural outputs of the system are $y_1 = q_1$, $y_2 = q_2$.

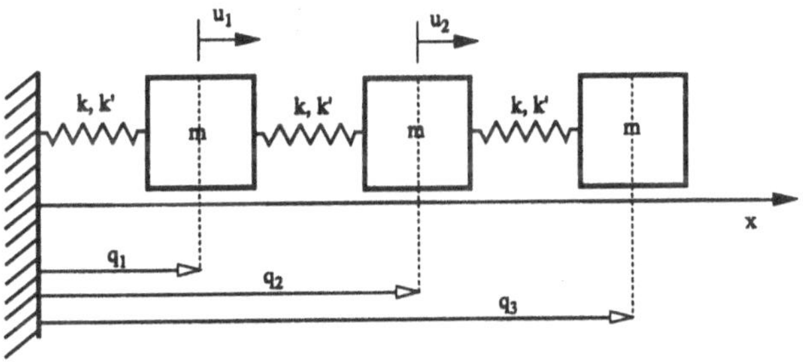

Fig. 2.2. The mechanical system considered in Example 2.6.1.

Since the kinetic and potential energies of the system are given by

$$T(\dot q) = \frac{1}{2}m\left(\dot q_1^2 + \dot q_2^2 + \dot q_3^2\right),$$

$$V(q) = \frac{1}{2}k\left(q_1^2 + (q_2 - q_1)^2 + (q_3 - q_2)^2\right) + \frac{1}{4}k'\left(q_1^4 + (q_2 - q_1)^4 + (q_3 - q_2)^4\right),$$

the equations of motion can be written as

$$\ddot{q}_1 = \frac{1}{m}\left(k(-2\,q_1 + q_2) + k'\left(-q_1^3 + (q_2 - q_1)^3\right) + u_1\right),$$
$$\ddot{q}_2 = \frac{1}{m}\left(k(q_1 - 2\,q_2 + q_3) + k'\left(-(q_2 - q_1)^3 + (q_3 - q_2)^3\right) + u_2\right),$$
$$\dot{q}_3 = \frac{1}{m}p_3,$$
$$\dot{p}_3 = k(q_2 - q_3) + k'(q_2 - q_3)^3,$$

where $p_3 := m\,\dot{q}_3$.

After a first state-feedback (2.6) the closed-loop system is decoupled and has the form:

$$\ddot{q}_1 = v_1, \tag{2.35a}$$
$$\ddot{q}_2 = v_2, \tag{2.35b}$$
$$\dot{q}_3 = \frac{1}{m}p_3, \tag{2.35c}$$
$$\dot{p}_3 = k(x_2 - q_3) + k'(x_2 - q_3)^3. \tag{2.35d}$$

The zero dynamics of system (2.35a)-(2.35d) can be written as

$$\dot{q}_3 = \frac{1}{m}p_3,$$
$$\dot{p}_3 = -k\,q_3 - k'q_3^3,$$

since they are not asymptotically stable (they are those of an undamped Duffing oscillator), the approach of [6] cannot succeed in obtaining decoupling with asymptotic stability. However, it is clear that the distribution P^* has zero dimension for system (2.35a)-(2.35d), hence the system (which is controllable in the first approximation) can be stabilized asymptotically by means of a static state-feedback which preserves decoupling. As a matter of fact, any feedback control law described by equations of the form

$$v_1 = -\epsilon_{11}x_1 - \epsilon_{12}\dot{x}_1 + w_1, \quad \epsilon_{11} > 0,\ \epsilon_{12} > 0,$$
$$v_2 = \alpha(x_2,\ \dot{x}_2,\ q_3, p_3) + w_2,$$

which asymptotically stabilizes the first approximation of subsystem (2.35b) -(2.35d), succeeds in obtaining a closed-loop system which is decoupled (the input w_i does not affect the output y_j, $i \neq j$) and asymptotically stable in the first approximation. In particular, $\alpha(x_2,\ \dot{x}_2,\ q_3, p_3)$ can be chosen as a static state-feedback assigning the eigenvalues of the linear approximation of subsystem (2.35b)-(2.35d). This has been done in the case of $m = 1$, $k = 20$, $k' = 20$, assigning the eigenvalues in $\{-5, -10, -15, -20\}$, whereas, to stabilise subsystem (2.35a), $\epsilon_{11} = 25$ and $\epsilon_{12} = 10$ have been taken. The results of a significant simulation of the behavior of the overall control system are reported in Figure 2.6.1. Starting from initial conditions equal to zero for all the state variables, two piece-wise constant input functions $w_1(\cdot)$ and $w_2(\cdot)$

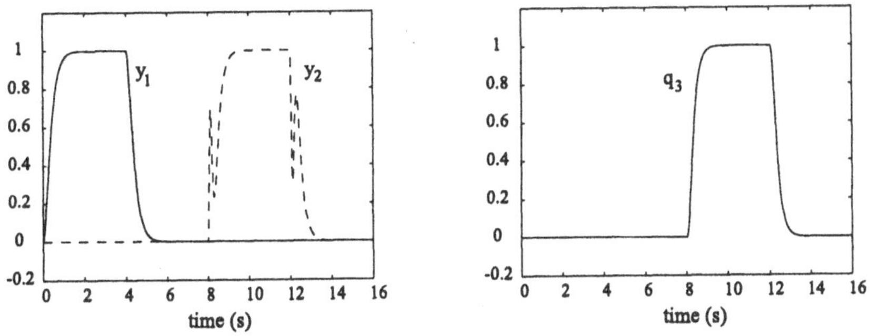

Fig. 2.3. Simulation results for the noninteracting control system obtained in Example 2.6.1: time behavior of the two outputs y_1 and y_2 (left) and of the non-actuated configuration variable q_3 (right).

have been applied, one at each of the two decoupled channels: $w_1(t)$ different ¿from zero for $t \in [0, 4]$ and $w_2(t)$ different from zero for $t \in [8, 12]$. In the left plot one can see that each output y_i is not affected by the values of the input function w_j, $j \neq i$, whereas in the right plot it is possible to appreciate the time behavior of the position of the non-actuated mass. □

The second example illustrates the results in Section 2.4: it consists of an unstable system, for which decoupling and simple stability are jointly achievable, if the use of dynamic state-feedback control is allowed.

Example 2.6.2. Consider the system represented in Figure 2.6.2, which is composed of four heavy dimensionless carts, denoted by C_1, C_2, C_3 and C_4, which are subject to the gravitational field, of magnitude g, and are constrained to move along specified curves lying on a vertical plane. The four carts, C_1 and C_2 having mass m, C_3 and C_4 having mass M, interact between them through mechanical couplings involving other massless objects, as described in the following. Any kind of friction is neglected in the proposed model of the system.

On the vertical plane an inertial reference frame xOy is defined, whose y axis is parallel to the gravity acceleration vector and has opposite direction.

The two carts C_3 and C_4 are constrained to slide along two curves, Γ_a and Γ_b, each of them parameterized through its curvilinear abscissa $s \in [-2, 2]$:

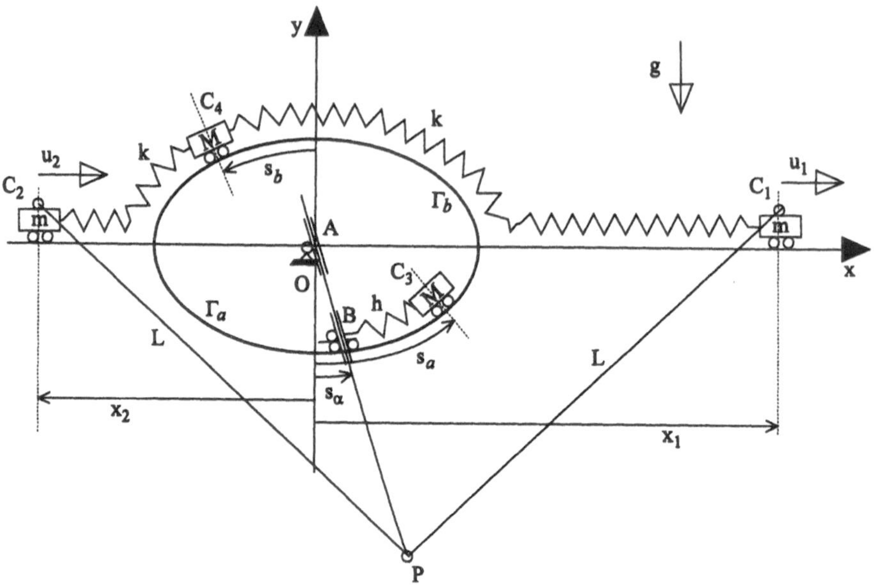

Fig. 2.4. The mechanical system considered in Example 2.6.2.

$$\Gamma_a : \begin{cases} x_a(s) = \dfrac{s}{2}\sqrt{1-\dfrac{s^2}{4}} + \arcsin\left(\dfrac{s}{2}\right), \\ y_a(s) = -1 + \dfrac{s^2}{4}, \end{cases}$$

$$\Gamma_b : \begin{cases} x_b(s) = -\dfrac{s}{2}\sqrt{1-\dfrac{s^2}{4}} - \arcsin\left(\dfrac{s}{2}\right), \\ y_b(s) = 1 - \dfrac{s^2}{4}. \end{cases}$$

The cart C_3 slides along Γ_a, hence its position at time t is given by $\big(x_a(s_a(t)),\, y_a(s_a(t))\big)$ and $s_a(t)$ can be taken as its configuration coordinate; similarly, the position of the cart C_4, which slides along Γ_b, is $(x_b(s_b(t)),\, y_b(s_b(t)))$ and $s_b(t)$ can be taken as its configuration coordinate.

The two carts C_1 and C_2 slide along the x axis, so that their configuration coordinates can be simply taken as x_1 and x_2. Carts C_1 and C_2 are subject to two external forces having direction parallel to the x axis and intensity u_1 and u_2, respectively (the only control inputs). Two linear, elastic, massless springs, having length $L_0 > 2$, when undeformed, and elastic constant k, connect the cart C_4 with the carts C_1 and C_2; as shown in Figure 2.6.2, such springs lie on the same curves along which C_1, C_2 and C_4 are constrained to slide.

A massless cylinder, denoted with B in Figure 2.6.2, whose axis belongs to the plane xOy, is free to rotate about a hinge, whose axis is perpendicular to the plane xOy, placed on a massless cart, which is constrained to slide along Γ_a; the dimensions of B, of the hinge and of the cart are all negligible, so that those three objects can be considered as a single point, whose position is $(x_a(s_\alpha(t)), y_a(s_\alpha(t)))$, with $s_\alpha(t)$ being the value of the curvilinear abscissa of such a point on Γ_a. A linear, elastic, massless spring, having length equal to zero, when undeformed, and elastic constant h, connects C_3 with the cart on which B is hinged; such a spring is also constrained to lie on Γ_a. A further mechanical coupling is established between C_1 and C_2 and the cart supporting B, by means of three massless rigid bars, also lying on the plane xOy, which are hinged at one extremity with the hinge P. Two of the bars, having length $L > \sqrt{1+(L_0-2+\pi/2)^2}$, are hinged at the other extremity, one at each of the carts C_1 and C_2. The third bar, whose length is not relevant, provided that it is greater than the maximum distance between the hinge P and the origin O, is constrained by means of two prismatic one degree of freedom couplings, the first with the cylinder B, and the other with a second dimensionless and massless cylinder, denoted by A. The axis A also belongs to the plane xOy, and, by means of a dimensionless hinge placed at O, it is assured that the central point of A coincides with the origin O.

The described interaction between the carts C_1 and C_2, the cylinders A and B, and the three bars hinged at P, guarantees that the curvilinear abscissa $s_\alpha(t)$ of B along Γ_a, is a function of the position coordinates $x_1(t)$ and $x_2(t)$ of C_1 and C_2. The function $s_\alpha = s_\alpha(x_1, x_2)$ is defined implicitly by means of the equation $F(x_1, x_2, s_\alpha) = 0$, where

$$F(x_1, x_2, s_\alpha) := x_a(s_\alpha)\sqrt{L^2 - \left(\frac{x_1 - x_2}{2}\right)^2} + y_a(s_\alpha)\frac{x_1 + x_2}{2},$$

in the domain of interest: $x_1 > \pi/2$, $x_2 < -\pi/2$, $x_1 - x_2 < 2L$, $-2 < s_\alpha < 2$.

The system is in equilibrium if $x_1 = x_e$, $x_2 = -x_e$, $s_a = s_b = 0$, where $x_e := L_0 - 2 + \pi/2$. Therefore a suitable vector of configuration coordinates is $q = [x_1 - x_e \quad x_2 + x_e \quad s_a \quad s_b]^T$; in the following the motion of the system around the origin $q = 0$ is considered.

The kinetic and potential energies of the system can be written as follows:

$$T(\dot{q}) = \frac{1}{2}\left(m\left(\dot{q}_1^2 + \dot{q}_2^2\right) + M\left(\dot{q}_3^2 + \dot{q}_4^2\right)\right), \tag{2.36}$$

$$\begin{aligned} V(q) &= M g\left(y_a(q_3) + y_b(q_4)\right) + \\ &\frac{1}{2}\left(k(q_1+q_4)^2 + k(q_2+q_4)^2 + h\left(q_3 - s_\alpha(q_1+x_e, q_2-x_e)\right)^2\right). \end{aligned} \tag{2.37}$$

Using (2.36) and (2.37) and the discussion in [11, Chapter 12], it follows that

$$H_0(q, p) = T\left(\left[p_1/m \; p_2/m \; p_3/M \; p_4/M\right]^T\right) + V(q),$$

in which $p = [m\,\dot{q}_1 \quad m\,\dot{q}_2 \quad M\,\dot{q}_3 \quad M\,\dot{q}_4]^T$. As a result, the state space equations describing the system can be obtained as in Section 2.2:

$$\ddot{q}_1 = \frac{1}{m}(\overline{f}_2(q) + u_1), \tag{2.38a}$$

$$\ddot{q}_2 = \frac{1}{m}(\overline{f}_4(q) + u_2), \tag{2.38b}$$

$$\dot{q}_3 = \frac{1}{M}p_3, \tag{2.38c}$$

$$\dot{q}_4 = \frac{1}{M}p_4, \tag{2.38d}$$

$$\dot{p}_3 = h\,(s_\alpha(q_1 + x_e,\, q_2 - x_e) - q_3) - \frac{M\,g}{2}q_3, \tag{2.38e}$$

$$\dot{p}_4 = -k\,(q_1 + q_2 + 2\,q_4) + \frac{M\,g}{2}q_4, \tag{2.38f}$$

where $\overline{f}_2(\cdot)$ and $\overline{f}_4(\cdot)$ are suitable functions of q; the natural outputs are

$$y_1 = q_1,$$
$$y_2 = q_2.$$

After a first state-feedback of the form

$$\begin{aligned} u_1 &= -\overline{f}_2(q) + m\,v_1, \\ u_2 &= -\overline{f}_4(q) + m\,v_2, \end{aligned} \tag{2.39}$$

if ξ_1 and ξ_2 are given by (2.7), the system (2.38a)–(2.38f), (2.39) can be written in the form (2.23), with $z = [q_3\; q_4\; p_3\; p_4]^T$. It is easy to see that assumption (a) holds. In order to study the stability properties, in the first approximation, of the zero dynamics, the matrix F has to be considered:

$$F = \begin{bmatrix} 0 & 0 & 1/M & 0 \\ 0 & 0 & 0 & 1/M \\ -h - g\,M/2 & 0 & 0 & 0 \\ 0 & -2\,k + g\,M/2 & 0 & 0 \end{bmatrix}.$$

If $k < \frac{g\,M}{4}$, the matrix F has a real eigenvalue λ with positive real part, hence stability of the closed-loop system cannot be achieved by any static state-feedback control law which guarantees decoupling. Since the eigenvalues of F are $\{\lambda,\, -\lambda,\, \jmath\omega,\, -\jmath\omega\}$, with $\lambda = \frac{1}{\sqrt{2}}\sqrt{g - \frac{4\,k}{M}}$ and $\omega = \frac{1}{\sqrt{2}}\sqrt{g + \frac{2\,h}{M}}$, and $\jmath$ being the imaginary unit, it makes sense to check for the existence of a dynamic state-feedback control law guaranteeing decoupling and simple stability. The closed-loop system (2.38a)–(2.38f), (2.39) can be put in the form (2.25a)–(2.25d) by means of a coordinates transformation

$$\tilde{z} = T\, z, \tag{2.40}$$

where the matrix T is such that

$$\tilde{F} = T\, F\, T^{-1} = \begin{bmatrix} \Lambda_1 & \begin{matrix} 0\,0 \\ 0\,0 \end{matrix} \\ \begin{matrix} 0\,0 \\ 0\,0 \end{matrix} & \begin{matrix} -\lambda & 0 \\ 0 & \lambda \end{matrix} \end{bmatrix},$$

with $\sigma(\Lambda_1) = \{-\jmath\omega, \jmath\omega\}$. After such a transformation it turns out that equation (2.25d) is given by:

$$\dot{z}_b = k_1\, q_1 + k_2\, q_2 + \lambda\, z_b, \quad k_1,\, k_2 \in \mathbb{R}, \tag{2.41}$$

hence $\psi(\xi_1, \xi_2) = 0$ and hypotheses (i), (ii) and (iii) are satisfied. In order to check that (iv) also holds, it is sufficient to notice that, by letting $\xi_1 = \xi_2 = 0$, the zero dynamics of the system are given by two decoupled subsystems, the first describing the dynamics of the cart C_3 and the second the ones of C_4. As for cart C_3, when $\xi_1 = \xi_2 = 0$, its motion is described by equations (2.38c) and (2.38e) with $s_\alpha(x_e, -x_e) = 0$:

$$\begin{aligned} \dot{q}_3 &= p_3/M, \\ \dot{p}_3 &= -\,(h + M\, g/2)\, q_3, \end{aligned}$$

¿from which it is evident that such a subsystem is simply stable. As for cart C_4, its constrained dynamics

$$\begin{aligned} \dot{q}_4 &= p_4/M, \\ \dot{p}_4 &= (-2\, k + M\, g/2)\, q_4, \end{aligned}$$

are clearly split by the coordinates transformation (2.40) into the unstable part, given by (2.41) with $q_1 = q_2 = 0$ and an analogous equation relative to an asymptotically stable subsystem. It follows that hypothesis (iv) is satisfied, with the components of z_g being q_3, p_3 and the state variable of the asymptotically stable subsystem just mentioned.

Therefore, simple stability and input-output decoupling can be obtained jointly for this system, if a suitable dynamic state-feedback is used. □

2.7 Conclusions

In this chapter the problem of input-output decoupling with stability has been tackled for a class of nonlinear Hamiltonian systems, by means of (possibly) dynamic state-feedback control laws. Using well known results from nonlinear geometric control, it has been possible to enlarge the class of systems proposed in [6], for which the problem of decoupling with simple stability is solvable by means of static state-feedback. A wider class of systems has been determined for which the problem is solvable if dynamic state-feedback is allowed. Mathematical and physical examples have been presented and discussed.

The $P^{\perp}$, P^* and Δ_{mix} dynamics of Hamiltonian systems have been studied, since they characterize the key properties of the system with respect to the problem of decoupling with stability. In particular, it has been shown that these dynamics need not be Hamiltonian for general Hamiltonian systems, whereas, for simple and linear Hamiltonian systems, the $P^{\perp}$ and P^* dynamics, which are the only ones of interest, are Hamiltonian.

Further work will be devoted to the subject, to enlarge further the class of systems for which the problem of decoupling with stability is solvable and to complete the study of the $P^{\perp}$, P^* and Δ_{mix} dynamics of nonlinear simple Hamiltonian systems.

References

1. A. Astolfi. *Control of Hamiltonian Systems*, pages 81–113. World Scientific, Singapore, 1996.

2. A. Astolfi and L. Menini. Noninteracting control with stability for Hamiltonian systems. In *IFAC Symposium on Nonlinear Control Systems Design (NOLCOS'98)*, volume 2, pages 405–410. Enschede (The Netherlands), July 1998.

3. S. Battilotti. *Noninteracting control with stability for nonlinear systems.* Springer Verlag, 1994.

4. J. Carr. *Applications of Centre Manifold Theory.* Springer Verlag, 1981.

5. I. J. Ha and E. G. Gilbert. A complete characterization of decoupling control laws for a general class of nonlinear control systems. *IEEE Trans. Aut. Contr.*, AC-31:823–830, 1986.

6. H. J. C. Huijberts and A. J. van der Schaft. Input-output decoupling with stability for Hamiltonian systems. *Math. Control Signal Systems*, pages 125–138, 1990.

7. A. Isidori. *Nonlinear control systems.* Springer Verlag, 1994. Third edition.

8. A. Isidori and J. W. Grizzle. Fixed modes and nonlinear noninteractive control with stability. *IEEE Trans. Aut. Contr.*, AC-33:907–914, 1988.

9. A. S. Morse and W. M. Wonham. Decoupling and pole assignment by dynamic compensation. *SIAM J. Control*, 8(3):317–337, 1970.

10. H. Nijmeijer and W. Respondek. Decoupling via dynamic compensation for nonlinear control systems. In *IEEE Conference on Decision and Control*, 1986.

11. H. Nijmeijer and A. J. van der Schaft. *Nonlinear Dynamical Control Systems*. Springer Verlag, 1990.

12. K. G. Wagner. On nonlinear noninteraction with stability. In *Proceedings of the 28-th Conference on Decision and Control*, Tampa, FL, December 1989.

13. W. M. Wonham. *Linear multivariable control: a geometric approach*. Springer Verlag, 1979. Second edition.

14. W. M. Wonham and A. S. Morse. Decoupling and pole assignment in linear multivariable systems: a geometric approach. *SIAM J. Control*, 8(1):1–18, 1970.

15. W. Zhan, A. Isidori, and T. J. Tarn. A canonical dynamic extension algorithm for noninteraction with stability for affine nonlinear systems. *Syst. Contr Lett.*, 17:177–184, 1991.

3. Issues in modelling and control of mass balance systems

Georges Bastin

Centre for Systems Engineering and Applied Mechanics (CESAME)
Université Catholique de Louvain
Bâtiment Euler, 4-6, avenue G.Lemaitre
1348 Louvain la Neuve, Belgium
Fax : +32 10472380
bastin@auto.ucl.ac.be

3.1 Introduction

This paper devoted to mass balance systems is written in a tutorial spirit. The aim is to give a self content presentation of the modelling of engineering systems that are governed by a law of mass conservation and to briefly discuss a fundamental feedback control problem regarding these systems.

Modelling issues are first addressed in Sections 2 to 11. The general state-space model of mass balance systems is presented. The equations of the model are shown to satisfy physical constraints of positivity and mass conservation. These conditions have strong structural implications that lead to particular Hamiltonian and Compartmental representations. The modelling of mass balance systems is illustrated with two simple industrial examples : a biochemical process and a grinding process. Some open loop stability properties are briefly presented and illustrated with these examples.

The control issue is then addressed in Sections 12 to 14. In general, mass balance systems have multiple equilibria, one of them being the operating point of interest which is locally asymptotically stable. However if big enough disturbances occur, the process may be lead by accident to a behaviour which may be undesirable or even catastrophic. The control challenge is then to design a feedback controller which is able to prevent the process from such undesirable behaviours. Two solutions of this problem are briefly described namely (i) robust output feedback control of minimum phase mass balance systems and (ii) robust state feedback stabilisation of the total mass.

3.2 Mass balance systems

In mass balance systems, each state variable x_i ($i = 1, \dots, n$) represents an amount of some material (or some matter) inside the system, while each state equation describes a balance of flows as illustrated in Fig. 3.1 :

$$\dot{x}_i = r_i - q_i + p_i \tag{3.1}$$

where p_i represents the inflow rate, q_i the outflow rate and r_i an internal transformation rate. The flows p_i, q_i and r_i can be function of the state variables $x_1, \dots x_n$ and possibly of control inputs $u_1, \dots, u_m$. The state space model which is the natural behavioural representation of the system is therefore written in vector form :

$$\dot{x} = r(x, u) - q(x, u) + p(x, u) \tag{3.2}$$

As a matter of illustration, some concrete examples of the phenomena that can be represented by the (p, q, r) flow rates in engineering applications are given in Table 1.

Transformations	
	Physical : grinding, evaporation, condensation
	Chemical : reaction, catalysis, inhibition
	Biological : infection, predation, parasitism
Outflows	
	Withdrawals, extraction
	Excretion, decanting, adsorption
	Emigration, mortality
Inflows	
	Supply of raw material
	Feeding of nutrients
	Birth, immigration
	etc...etc...

Table 1.

In this paper, we shall assume that the functions $p(x, u), q(x, u), r(x, u)$ are differentiable with respect to their arguments. The physical meaning of the model (3.2) implies that these functions must satisfy two kinds of conditions : positivity conditions and mass conservation conditions which are explicited hereafter.

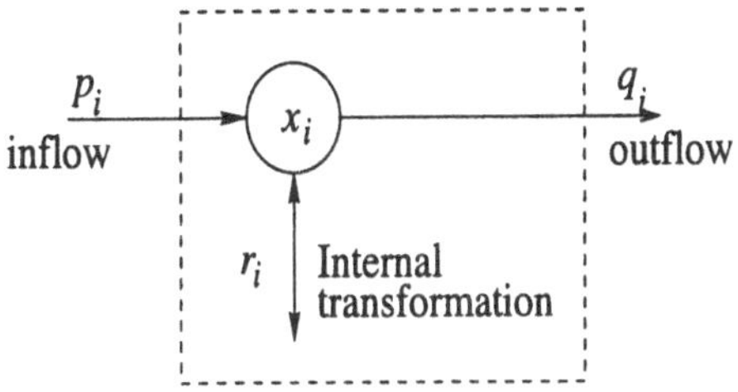

Fig. 3.1. Balance of flows

3.3 Positivity

Since there cannot be negative masses, the model (3.2) makes sense only if the state variables $x_i(t)$ remain *non-negative* for all t :

$$x_i(t) \in R_+$$

where R_+ denotes the set of real non-negative numbers. It follows that :

$$x_i = 0 \implies \dot{x}_i \geq 0 \tag{3.3}$$

whatever the values of $x_j \in R_+$, $j \neq i$ and u_k. This requirement is satisfied if the functions $p(x, u), q(x, u), r(x, u)$ have the following properties :

1. The inflow and outflow functions are defined to be non-negative :

$$\left.\begin{array}{l} p(x,u) \\ q(x,u) \end{array}\right\} : R_+^n \times R^m \rightarrow R_+^n$$

2. There cannot be an outflow if there is no material inside the system :

$$x_i = 0 \Longrightarrow q_i(x, u) = 0 \tag{3.4}$$

3. The transformation rate $r_i(x, u)$: $R_+^n \times R^m \rightarrow R$ may be positive or negative but it must be defined to be positive when x_i is zero :

$$x_i = 0 \Longrightarrow r_i(x, u) \geq 0 \tag{3.5}$$

3.4 Conservation of mass

Provided the quantities x_i are expressed in appropriate normalized units, the total mass contained in the system may be expressed as[1] :

[1] To simplify the notations, it will be assumed throughout the paper that the summation $\sum_i$ is taken over all possible values of i (here $i = 1, \ldots, n$) and $\sum_{i \neq j}$ over all possible values of i except j.

$$M = \sum_i x_i$$

When the system is *closed* (neither inflows nor outflows), the dynamics of M are written :

$$\dot{M} = \sum_i r_i(x, u)$$

It is obvious that the total mass inside a closed system must be conserved ($\dot{M} = 0$), which implies that the transformation functions $r_i(x, u)$ satisfy the condition :

$$\sum_i r_i(x, u) = 0 \tag{3.6}$$

The positivity conditions (3.4)- (3.5) and the mass conservation condition (3.6) have strong structural implications that are now presented.

3.5 Hamiltonian representation

A necessary consequence of the mass conservation condition (3.6) is that $n(n-1)$ functions $r_{ij}(x, u)$ $(i = 1, \ldots, n \; ; \; j = 1, \ldots, n \; ; \; i \neq j)$ may be selected such that :

$$r_i(x, u) = \sum_{j \neq i} r_{ji}(x, u) - \sum_{j \neq i} r_{ij}(x, u) \tag{3.7}$$

(note the indices !). Indeed, the summation over i of the right hand sides of (3.7) equals zero. It follows that any mass balance system (3.2) can be written under the form of a so-called *port-controlled Hamiltonian representation* (see [8], [9]) :

$$\dot{x} = [F(x, u) - D(x, u)] \left(\frac{\partial M}{\partial x} \right)^T + p(x, u) \tag{3.8}$$

where the storage function is the total mass $M(x) = \sum_i x_i$. The matrix $F(x, u)$ is skew-symmetric :

$$F(x, u) = -F^T(x, u)$$

with off-diagonal entries $f_{ij}(x, u) = r_{ji}(x, u) - r_{ij}(x, u)$. The matrix $D(x, u)$ represents the natural damping or dissipation provided by the outflows. It is diagonal and positive :

$$D(x, u) = \text{diag}\ (q_i(x, u)) \geq 0$$

The last term $p(x, u)$ in (3.8) obviously represents a supply of mass to the system from the outside.

3.6 Compartmental representation

There is obviously an infinity of ways of defining the r_{ij} functions in (3.7). We may assume that they are selected to be non-negative :

$$r_{ij}(x,u) : R^n_+ \times R^m \to R_+$$

and differentiable since $r_i(x,u)$ is required to be differentiable.

Then condition (3.5) implies that :

$$x_i = 0 \Rightarrow r_{ij}(x,u) = 0 \tag{3.9}$$

Now, it is a well known fact (see e.g. [3], page 67) that if $r_{ij}(x,u)$ is differentiable and if condition (3.9) holds, then $r_{ij}(x,u)$ may be written as :

$$r_{ij} = x_i \bar{r}_{ij}(x,u)$$

for some appropriate function $\bar{r}_{ij}(x,u)$ which is defined on $R^n_+ \times R^m$, non-negative and at least continuous. Obviously, the same is true for $q_i(x,u)$ due to condition (3.4) :

$$q_i(x,u) = x_i \bar{q}_i(x,u)$$

The functions $\bar{r}_{ij}$ and $\bar{q}_i$ are called fractional rates. It follows that any mass balance system (3.2) can be written under the following alternative representation :

$$\dot{x} = G(x,u)x + p(x,u) \tag{3.10}$$

where $G(x,u)$ is a so-called *compartmental matrix* with the following properties :

1. $G(x,u)$ is a Metzler matrix with non-negative off-diagonal entries :

   $$g_{ij}(x,u) = \bar{r}_{ji}(x,u) \geq 0 \quad i \neq j$$

   (note the inversion of indices !)

2. The diagonal entries of $G(x,u)$ are non-positive :

   $$g_{ii}(x,u) = -\bar{q}_i(x,u) - \sum_{j \neq i} \bar{r}_{ij}(x,u) \leq 0$$

3. The matrix $G(x,u)$ is diagonally dominant :

   $$|g_{ii}(x,u)| \geq \sum_{j \neq i} g_{ji}(x,u)$$

The term *compartmental* is motivated by the fact that a mass balance system may be represented by a network of conceptual reservoirs called compartments. Each quantity (state variable) x_i is supposed to be contained in a compartment which is represented by a box in the network (see Fig. 3.2). The internal transformation rates are represented by directed arcs : there is an arc from compartment i to compartment j when there is a non-zero entry $g_{ji} = \bar{r}_{ij}$ in the compartmental matrix G. These arcs are labeled with the fractional rates $\bar{r}_{ij}$. Additional arcs, labeled respectively with fractional outflow rates $\bar{q}_i$ and inflow rates p_i are used to represent inflows and outflows. Concrete examples of compartmental networks will be given in Fig.3.4 and Fig.3.6.

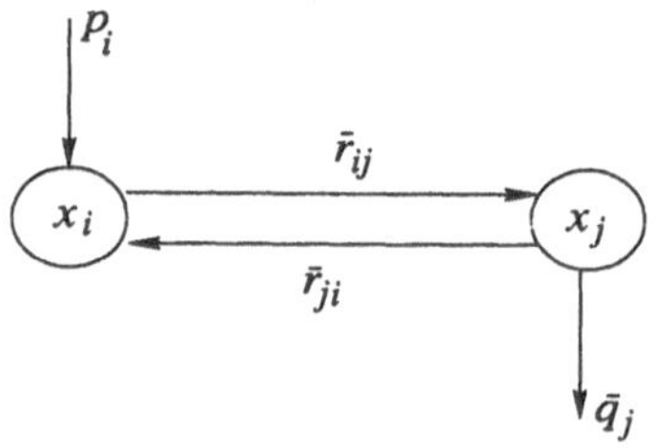

Fig. 3.2. Network of compartments

A compartment is said to be *outflow connected* if there is a path from that compartment to a compartment from which there is an outflow arc. The system is said to be *fully outflow connected* if all compartments are outflow connected. As stated in the following property, the non singularity of a compartmental matrix can be checked directly on the network.

Property 1. For a given value of $(x, u) \in R^n_+ \times R^m$, the compartmental matrix $G(x, u)$ of a mass balance system (3.10) is non singular if and only if the system is fully outflow connected. ■

A proof of this property can be found e.g. in [3].

3.7 Special case : inflow controlled systems

From now on, we will focus on the special case of *inflow-controlled systems* where the inflow rates $p_i(x, u)$ do not depend on the state x and are linear with respect to the control inputs u_k :

$$p_i(x, u) = \sum_k b_{ik} u_k \quad b_{ik} \geq 0 \quad u_k \geq 0$$

while the transformation rates $r_i(x, u)$ and the outflow rates $q_i(x, u)$ are independent of u. The model (3.2) is thus written as :

$$\dot{x} = r(x) - q(x) + Bu \tag{3.11}$$

with B the $n \times m$ matrix with entries b_{ik}.

The Hamiltonian representation specializes as :

$$\dot{x} = [F(x) - D(x)] \left(\frac{\partial M}{\partial x}\right)^T + Bu \tag{3.12}$$

and the compartmental representation as :

$$\dot{x} = G(x)x + Bu \tag{3.13}$$

with appropriate definitions of the matrices $F(x), D(x)$ and $G(x)$.

Two practical examples of single-input inflow-controlled systems are given hereafter.

3.8 Example 1 : a biochemical process

A continuous stirred tank reactor is represented in Fig.3.3. The following biochemical reactions take place in the reactor :

$$A \xrightarrow[X]{} B$$
$$B \xrightarrow[X]{} X$$

where X represents a microbial population and A, B organic matters. The first reaction represents the hydrolysis of species A into species B, catalysed by cellular enzymes. The second reaction represents the growth of microorganisms on substrate B. It is obviously an auto-catalytic reaction. Assuming mass action kinetics, the dynamics of the reactor may be described by the model :

$$\dot{x}_1 = +k_1 x_1 x_2 - d x_1$$
$$\dot{x}_2 = -k_1 x_1 x_2 + k_2 x_1 x_3 - d x_2$$
$$\dot{x}_3 = -k_2 x_1 x_3 - d x_3 + du$$

with the following notations and definitions :

x_1 = concentration of species X in the reactor
x_2 = concentration of species B in the reactor

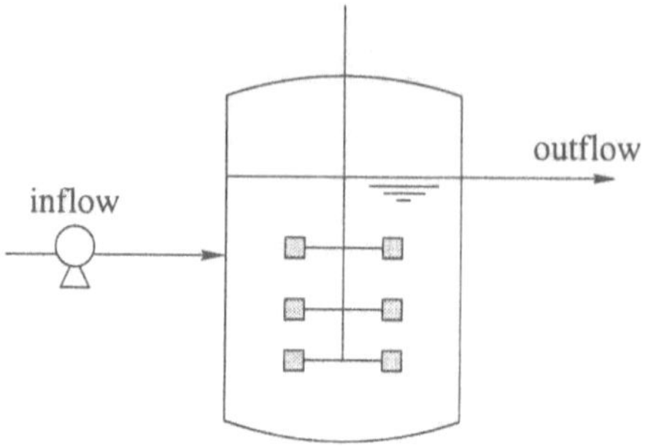

Fig. 3.3. Stirred tank reactor

x_3 = concentration of species A in the reactor
d = dilution rate
u = concentration of species A in the influent
k_1, k_2 = rate constants.

This could be for instance the model of a biological depollution process where du is the pollutant inflow while $d(x_2+x_3)$ is the residual pollution outflow. It is readily seen to be a special case of the general mass-balance model (3.11) with the following definitions :

$$r(x) = \begin{pmatrix} +k_1x_1x_2 \\ -k_1x_1x_2 + k_2x_1x_3 \\ -k_2x_1x_3 \end{pmatrix} \quad q(x) = \begin{pmatrix} dx_1 \\ dx_2 \\ dx_3 \end{pmatrix} \quad Bu = \begin{pmatrix} 0 \\ 0 \\ du \end{pmatrix}$$

The Hamiltonian representation is :

$$F(x) = \begin{pmatrix} 0 & k_1x_1x_2 & 0 \\ -k_1x_1x_2 & 0 & k_2x_1x_3 \\ 0 & -k_2x_1x_3 & 0 \end{pmatrix} \quad D(x) = \begin{pmatrix} dx_1 & 0 & 0 \\ & dx_2 & 0 \\ 0 & 0 & dx_3 \end{pmatrix}$$

The compartmental matrix is :

$$G(x) = \begin{pmatrix} -d & k_1x_1 & 0 \\ 0 & -d-k_1x_1 & k_2x_1 \\ 0 & 0 & -d-k_2x_1 \end{pmatrix}$$

The compartmental network of the biochemical process is shown in Fig.3.4 where it can be seen that the system is fully outflow connected.

3.9 Example 2 : a grinding process

An industrial grinding circuit, as represented in Fig.3.5 is made up of the interconnection of a mill and a separator. The mill is fed with raw material.

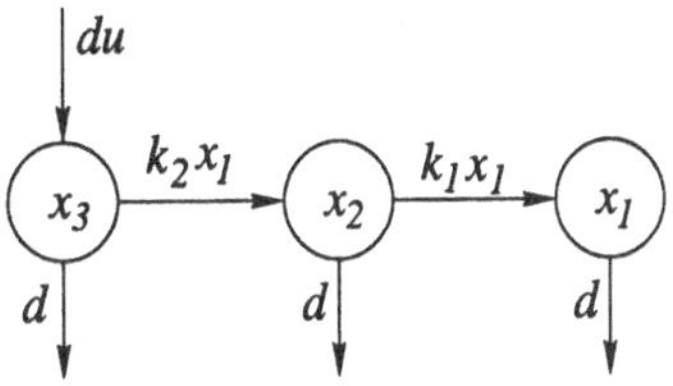

Fig. 3.4. Compartmental network of the biochemical process model

After grinding, the material is introduced in a separator where it is separated in two classes : fine particles which are given off and oversize particles which are recycled to the mill. A simple dynamical model has been proposed for this system in [5]:

$$\begin{aligned}
\dot{x}_1 &= -\gamma_1 x_1 + (1-\alpha)\phi(x_3) \\
\dot{x}_2 &= -\gamma_2 x_2 + \alpha\phi(x_3) \\
\dot{x}_3 &= \gamma_2 x_2 - \phi(x_3) + u \\
\phi(x_3) &= k_1 x_3 e^{-k_2 x_3}
\end{aligned}$$

with the following notations and definitions :

x_1 = hold-up of fine particles in the separator
x_2 = hold-up of oversize particles in the separator
x_3 = hold-up of material in the mill
u = inflow rate
$\gamma_1 x_1$ = outflow rate of fine particles
$\gamma_2 x_2$ = flowrate of recycled particles
$\phi(x_3)$ = outflowrate from the mill = grinding function
α = separation constant $(0 < \alpha < 1)$
$\gamma_1, \gamma_2, k_1, k_2$ = characteristic positive constant parameters

This model is readily seen to be a special case of the general mass-balance model (3.11) with the following definitions :

$$r(x) = \begin{pmatrix} (1-\alpha)\phi(x_3) \\ -\gamma_2 x_2 + \alpha\phi(x_3) \\ \gamma_2 x_2 - \phi(x_3) \end{pmatrix} \quad q(x) = \begin{pmatrix} -\gamma_1 x_1 \\ 0 \\ 0 \end{pmatrix} \quad Bu = \begin{pmatrix} 0 \\ 0 \\ u \end{pmatrix}$$

The Hamiltonian representation is :

$$F(x) = \begin{pmatrix} 0 & 0 & (1-\alpha)\phi(x_3) \\ 0 & 0 & -\gamma_2 x_2 + \alpha\phi(x_3) \\ -(1-\alpha)\phi(x_3) & \gamma_2 x_2 - \alpha\phi(x_3) & 0 \end{pmatrix}$$

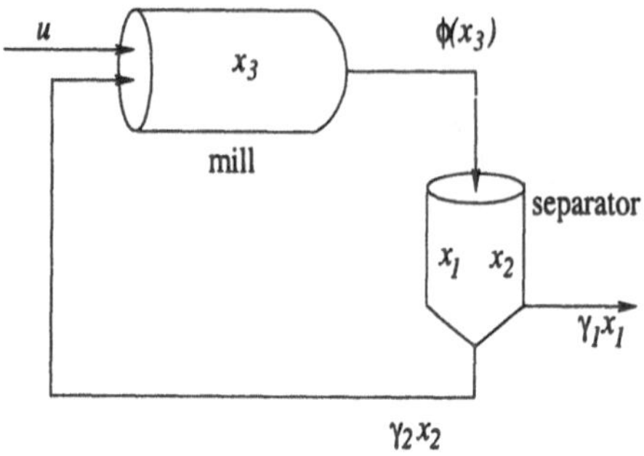

Fig. 3.5. Grinding circuit

$$D(x) = \begin{pmatrix} \gamma_1 x_1 & 0 & 0 \\ 0 & 0 & 0 \\ 0 & 0 & 0 \end{pmatrix}$$

The compartmental matrix is :

$$G(x) = \begin{pmatrix} -\gamma_1 & 0 & (1-\alpha)k_1 e^{-k_2 x_3} \\ 0 & -\gamma_2 & \alpha k_1 e^{-k_2 x_3} \\ 0 & +\gamma_2 & -k_1 e^{-k_2 x_3} \end{pmatrix}$$

The compartmental network of the grinding process is shown in Fig.3.6 where it can be seen that the system is fully outflow connected.

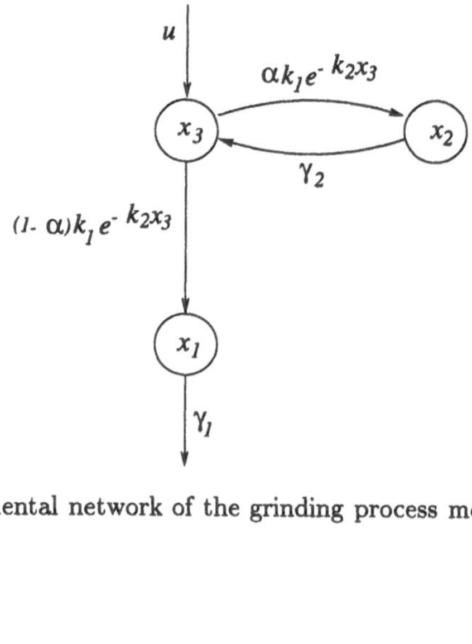

Fig. 3.6. Compartmental network of the grinding process model

3.10 Bounded input - bounded state

Obviously, the state x of a mass-balance system is bounded if and only if the total mass $M(x) = \sum_i x_i$ is itself bounded. The dynamics of the total mass is written as :

$$\dot{M} = -\sum_i q_i(x) + \sum_{i,k} b_{ik} u_k \tag{3.14}$$

From this expression, a natural condition for state boundedness is clearly that the total outflow $\sum_i q_i(x)$ should exceed the total inflow $\sum_{i,k} b_{ik} u_k$ when the total mass $M(x)$ is big enough (in order to make the right hand side of (3.14) negative). This intuitive condition is made technically precise as follows.

Property 2. Assume that :

(A1) the input $u(t)$ is bounded :

$$0 \leq u_k(t) \leq u_k^{\max} \quad \forall t \quad \forall k = 1, \ldots, m$$

(A2) There exists a constant M_0 such that

$$\sum_i q_i(x) \geq \sum_{i,k} b_{ik} u_k^{\max}$$

when $M(x) \geq M_0$

Then, the state of the system (3.11) is bounded and the simplex

$$\Delta = \{x \in R_+^n : M(x) \leq M_0\}$$

is invariant.

The system is BIBS if condition (A2) holds for any $u^{\max}$, for example if each $q_i(x) \to \infty$ as $x_i \to \infty$. ■

As a matter of illustration, it is readily checked that the biochemical process of Example 1 is BIBS. Indeed in this example we have

$$\sum_i q_i(x) = d \sum_i x_i = dM(x)$$

and therefore $M_0 = \frac{u^{\max}}{d}$

In contrast, the grinding process of Example 2 is *not* BIBS. Even worse, the state variable x_3 may be unbounded for any value of $u^{\max} > 0$ as we shall see in Section 12. This means that the process is *globally unstable* for any bounded input.

3.11 Systems without inflows

Consider the case of systems *without inflows* ($u = 0$) which are written in compartmental form

$$\dot{x} = G(x)x \tag{3.15}$$

Obviously, the origin $x = 0$ is an equilibrium of the system.

Property 3. If the compartmental matrix $G(x)$ is full rank for all $x \in R^n_+$ (equivalently if the system is fully outflow connected), then the origin $x = 0$ is a globally asymptotically stable (GAS) equilibrium of the unforced system $\dot{x} = G(x)x$ in the non negative orthant, with the total mass $M(x) = \sum_i x_i$ as Lyapunov function. ■

Indeed, for such systems, the total mass can only decrease along the system trajectories since there are outflows but no inflows :

$$\dot{M} = -\sum_i q_i(x)$$

Property 3 says that the total mass $M(x)$ and the state x will decrease until the system is empty if there are no inflows and the compartmental matrix is nonsingular for all x. A proof of this property and other related results can be found in [2].

3.12 A fundamental control problem

Obviously, the normal productive mode of operation of inflow-controlled mass balance systems is to have non zero inflows of raw material : $u(t) > 0$. Let us consider the case of constant inputs denoted $\bar{u}$:

$$\dot{x} = r(x) - q(x) + B\bar{u} \tag{3.16}$$

An equilibrium of this system is a state vector $\bar{x}$ which satisfies the equilibrium equation :

$$r(\bar{x}) - q(\bar{x}) + B\bar{u} = 0$$

In general, mass balance systems (3.16) have multiple equilibria. One of these equilibria is the operating point of interest. It is generally locally asymptotically stable. This means that an open loop operation may be acceptable in practice. But if big enough disturbances occur, it may arise that the system is driven too far from the operating point towards a region of the state space which is outside of its basin of attraction. From time to time, the process may therefore be lead by accident to a behaviour which may be undesirable or even catastrophic. We illustrate the point with our two examples.

Example 1 : The biochemical process

For a constant inflow rate $\bar{u} > \frac{d}{k_1}$, the biochemical process has three equilibria (see Fig.3.7). Two of these equilibria (E_1, E_2) are solutions of the following equations :

$$\bar{x}_2 = \frac{d}{k_1} \qquad \bar{x}_1 + \bar{x}_3 = \bar{u} - \frac{d}{k_1} \qquad \bar{x}_3(d + k_2\bar{x}_1) = d\bar{u}$$

The third equilibrium ($E3$) is

$$\bar{x}_1 = 0 \quad \bar{x}_2 = 0 \quad \bar{x}_3 = \bar{u}$$

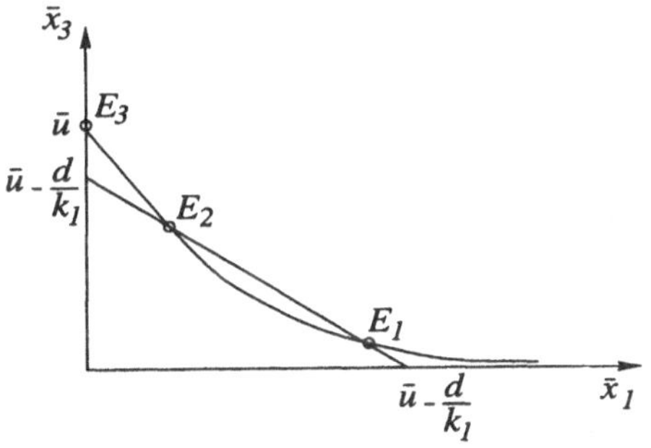

Fig. 3.7. Equilibria of the biochemical process

We know that the system is BIBS : the total mass $M(x)$ and hence all the trajectories are bounded. By computing the Jacobian matrix, it can be easily checked that $E1$ and $E3$ are asymptotically stable while $E2$ is unstable.

$E1$ is the normal operating point corresponding to a high conversion of substrate x_3 into product x_1. It is stable and the process can be normally operated at this point. But there is another stable equilibrium $E3$ called "wash-out steady state" which is highly undesirable because it corresponds to a complete loss of productivity : $\bar{x}_1 = 0$. The pollutant just goes through the tank without any degradation.

The problem is that an intermittent disturbance (like for instance a pulse of toxic matter) may irreversibly drive the process to this wash-out steady-state, making the process totally unproductive.

Example 2 : The grinding process

The equilibria of the grinding process $(\bar{x}_1, \bar{x}_2, \bar{x}_3)$ are parametrized by a constant input flowrate $\bar{u}$ as follows :

$$\bar{x}_1 = \frac{\gamma_1}{\bar{u}} \qquad \bar{x}_2 = \frac{\alpha\bar{u}}{\gamma_2(1-\alpha)} \qquad \phi(\bar{x}_3) = \frac{\bar{u}}{(1-\alpha)}$$

In view of the shape of $\phi(x_3)$ as illustrated in Fig.3.8, there are two distinct equilibria if :

$$\bar{u} < (1-\alpha)\phi_{max}$$

The equilibrium $E1$ on the left of the maximum is stable and the other

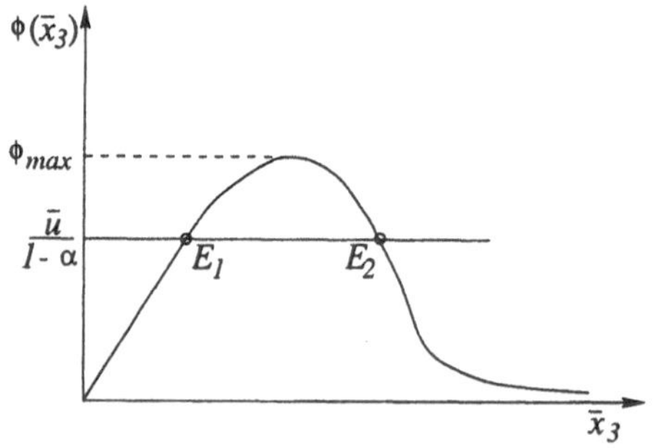

Fig. 3.8. Equilibria of the grinding process

one $E2$ is unstable. Furthermore, for any value of $\bar{u}$, the trajectories become unstable as soon as the state enters the set D defined by :

$$D \begin{cases} (1-\alpha)\phi(x_3) < \gamma_1 x_1 < \bar{u} \\ \alpha\phi(x_3) < \gamma_2 x_2 \\ \partial\phi/\partial x_3 < 0 \end{cases}$$

Indeed, it can be shown that this set D is positively invariant and if $x(0) \in D$ then $x_1 \to 0$ $x_2 \to 0$ $x_3 \to \infty$. In some sense, the system is *Bounded Input - Unbounded State (BIUS)*. This means that there can be an irreversible accumulation of material in the mill with a decrease of the production to zero. In the industrial jargon, this is called *mill plugging*. In practice, the state may be lead to the set D by intermittent disturbances like variations of hardness of the raw material. ■

In both examples we thus have a stable open loop operating point with a potential process destabilisation which can take two forms :

- drift of the state x towards another (unproductive) equilibrium

- unbounded increase of the total mass $M(x)$

The control challenge is then to design a feedback controller which is able to prevent the process from such undesirable behaviours.

Ideally a good control law should meet the following specifications :

S1. The feedback control action is positive and bounded;

S2. The closed loop system has a single equilibrium in the positive orthant which is globally asymptotically stable;

S3. The single closed-loop equilibrium may be assigned by an appropriate set point.

Moreover, it could be desirable that the feedback stabilisation be robust against modelling uncertainties regarding $r(x)$ which is the most uncertain term of the model in many applications.

This is indeed a vast problem which is far to be completely explored. Hereafter, we limit ourselves to the presentation of two very limited solutions of this problem namely (i) the output feedback control of a class of single input mass balance systems that are BIBS, relative degree one and minimum phase; (ii) the state feedback stabilisation of the total mass in BIUS systems.

3.13 Robust output feedback control of BIBS minimum phase systems

We consider single-input BIBS mass-balance systems of the form :

$$\begin{aligned} \dot{x}_i &= r_i(x) - a_i x_i \quad i = 1, \dots, n-1 \\ \dot{x}_n &= r_n(x) - a_n x_n + u \end{aligned}$$

with $a_i > 0 \ \forall i$. With the notations :

$$\xi = (x_1, \dots, x_{n-1})^T \qquad y = x_n$$

and appropriate definitions of φ and ψ, this system is rewritten as :

$$\dot{\xi} = \varphi(\xi, y) \tag{3.17}$$

$$\dot{y} = -\psi(\xi, y) - a_n y + u \tag{3.18}$$

The goal is to regulate the measured output y at a given set point $y^* > 0$ by *output feedback*.

In order to achieve this objective, the following dynamic controller is proposed in [6] :

$$u = \sigma(y) + \mu(y)\lambda(\theta) \tag{3.19}$$

$$\dot{\theta} = k_i(y^* - y) + \lambda(\theta) - \theta \tag{3.20}$$

with the following definitions :

$$\sigma(y) = \underset{[0,\sigma_m]}{\text{sat}} [k_0 + k_p(y^* - y)] \qquad \sigma_m = k_0 + k_p y^*$$

$$\mu(y) = \underset{[0,1]}{\text{sat}} \left[1 + \frac{k_p}{k_0}(y^* - y)\right]$$

$$\lambda(\theta) = \underset{[0,\theta_m]}{\text{sat}} [\theta]$$

In this control law :

- the function $\sigma(y)$ is a proportional action, with offset k_0, saturated between 0 and σ_m;
- the function $\lambda(\theta)$ is an integral action saturated between 0 and θ_m; the computation of the integral θ is provided with an anti-windup term $(\lambda(\theta) - \theta)$ which limits the excursions of θ outside the interval $[0, \theta_m]$.

The stabilisation properties of this control law will be analysed under the following assumptions :

A1The function $\psi(\xi, y)$ is non negative :

$$\psi(\xi, y) \geq 0 \quad \forall(\xi, y) \in R^n_+$$

A2The zero dynamics $\dot{\xi} = \varphi(\xi, y^*)$ have a single equilibrium $\bar{\xi} \in R^{n-1}_+$ which is GAS in the non negative orthant with a Lyapunov function denoted $W(\xi)$.

Assumption A1 expresses that species x_n can only be consumed inside the system but not produced. Assumption A2 is a global minimum phase condition. We have the following property.

Property 4 Under Assumptions A1-A2, the closed loop system (3.17)-(3.18)-(3.19)-(3.20) has the following properties :

1. The control input is positive and bounded :

$$0 \leq u(t) \leq \sigma_m + \theta_m \quad \forall t$$

2. There exist $k_0 > 0, k_p > 0, k_i > 0$ and $\theta_m > 0$ such that the closed loop system has a single equilibrium $(\bar{\xi}, y^*, \bar{\theta} = \psi(\bar{\xi}, y^*) + dy^*)$ which is GAS in the non negative orthant with Lyapunov function

$$V(\xi, y, \theta) = W(\xi) + \frac{1}{2}(y - y^*)^2 + \frac{1}{k_i} \int_{\bar{\theta}}^{\theta} (\lambda(\tau) - \bar{\theta}) d\tau$$

■

The proof of this result can be found in [6]. (See also [7] for related results). It is worth noting that the control law (3.19)-(3.20) is independent of $r(x)$. This means that the stabilisability is robust against modelling uncertainties regarding $r(x)$ provided the conditions of positivity and mass conservation are preserved. Some qualitative knowledge of $r(x)$ may nevertheless be useful for the tuning of the design parameters k_0, k_p, k_i.

3.14 Robust state feedback stabilisation of the total mass

We now consider single-input mass balance systems of the form :

$$\dot{x}_i = r_i(x) - q_i(x) + b_i u \quad i = 1, \ldots, n \tag{3.21}$$

with $b_i \geq 0 \;\; \forall i, \sum_i b_i > 0$

This system may be globally unstable (bounded input/unbounded state). The symptom of this instability is an unbounded accumulation of mass inside the system like for instance in the case of the grinding process of Example 2.

One way of approaching the problem is to consider that the control objective is to globally stabilise the total mass $M(x)$ at a given set point $M^* > 0$ in order to prevent the unbounded mass accumulation.

In order to achieve this control objective, the following *positive* control law is proposed in [1] :

$$u(x) = \max(0, \tilde{u}(x)) \tag{3.22}$$

$$\tilde{u}(x) = \left(\sum_i b_i\right)^{-1} \left[\sum_i q_i(x) + \lambda(M^* - M(x))\right] \tag{3.23}$$

where $\lambda > 0$ is an arbitrary design parameter. The stabilising properties of this control law are as follows.

Property 5 If the system (3.21) is fully outflow connected, then the closed loop system (3.21)-(3.22)-(3.23) has the following properties for any initial condition $x(0) \in R^n_+$:

1. the set $\Omega = \{x \in R^n_+ : M(x) = M^*\}$ is positively invariant
2. the state $x(t)$ is bounded for all $t \geq 0$ and $\lim_{t\to\infty} M(x) = M^*$.

■

The proof of this property can be found in [1]. It is worth noting that the control law (3.22)-(3.23) is independent from the internal transformation term $r(x)$. This means that the feedback stabilisation is robust against a full modelling uncertainty regarding $r(x)$ provided it satisfies the conditions of positivity and mass conservativity.

3.15 Concluding remarks

In this paper we have focused our attention on inflow controlled systems. But there are many engineering applications where outflow controlled systems or systems controlled by the internal transformation rates (like distillation columns for instance, see [11]) are relevant as well.

We have presented two very specific solutions for single input mass balance systems. But it is obvious that the fundamental control problem we have formulated is far from being solved and deserves deeper investigations. In particular a special interest should be devoted to control design methodologies which explicitely account for the structural specificities (Hamiltonian and Compartmental) of mass balance systems and rely on the construction of physically based Lyapunov functions.

Acknowledgements

This paper presents research results of the Belgian Program on Interuniversity Attraction Poles, Prime Minister's Office, Science Policy Programming.

Appendix : stability conditions

In this appendix some interesting stability results for mass balance systems with constant inputs are collected. These results can be useful for Lyapunov control design or for the stability analysis of zero-dynamics.

Compartmental Jacobian matrix

We consider the general case of inflow controlled mass balance systems with constant inflows :

$$\dot{x} = r(x) - q(x) + p(\bar{u})$$

The Jacobian matrix of the system is defined as :

$$J(x) = \frac{\partial}{\partial x}[r(x) - q(x)]$$

When this matrix has a compartmental structure, we have the following stability result.

Property A1

a) If $J(x)$ is a compartmental matrix $\forall\ x \in R^n_+$, then all bounded orbits tend to an equilibrium in R^n_+.

b) If there is a *bounded* closed convex set $D \subseteq R^n_+$ which is *positively invariant* and if $J(x)$ is a *non singular* compartmental matrix $\forall x \in D$, then there is a *unique* equilibrium $\bar{x} \in D$ which is GAS in D with Lyapunov function $V(x) = \sum_i |r_i(x) - q_i(x) + p_i(\bar{u})|$.

■

A proof of part a) can be found in [3] Appendix 4 while part b) is a concise reformulation of a theorem by Rosenbrock [10].

The assumption that $J(x)$ is compartmental $\forall x \in R^n_+$ is fairly restrictive. For instance, this assumption is *not* satisfied neither for the grinding process nor for the biochemical processes that we have used as examples in this paper. A simple sufficient condition to have $J(x)$ compartmental for all x is as follows.

Property A2 The Jacobian matrix $J(x) = \frac{\partial}{\partial x}[r(x) - q(x)]$ is compartmental $\forall x \in R^n_+$ if the functions $r(x)$ and $q(x)$ satisfy the following *monotonicity* conditions :

$$1) \frac{\partial q_i}{\partial x_i} \geq 0 \quad \frac{\partial q_i}{\partial x_k} = 0 \quad k \neq i$$

$$2) \frac{\partial r_{ij}}{\partial x_i} \geq 0 \quad \frac{\partial r_{ij}}{\partial x_j} \leq 0 \quad \frac{\partial r_{ij}}{\partial x_k} = 0 \quad k \neq i \neq j$$

■

In the next two sections, we describe two examples of systems that have a single GAS equilibrium in the nonnegative orthant although their Jacobian matrix is not compartmental.

The Gouz's condition

We consider a class of mass-balance systems of the form :

$$\dot{x}_i = \sum_{j \neq i} [r_{ji}(x_j) - r_{ij}(x_i)] - dx_i + \bar{u}_i \tag{3.24}$$

where d is a positive constant, the outflow rates $q_i(x_i) = dx_i$ depend linearly on x_i only and the transformation rates $r_{ij}(x_i)$ depend on x_i only.

For example this can be the model of a stirred tank chemical reactor with monomolecular reactions as explained in [4] (see also [11]).

The set $\Omega = \{x \in R^n_+ : M(x) = d^{-1} \sum_i \bar{u}_i\}$ is bounded, convex, compact and invariant. By the Brouwer fixed point theorem, it contains at least an equilibrium point $\bar{x} = (\bar{x}_1, \bar{x}_2, \ldots, \bar{x}_n)$ which satisfies the set of algebraic equations :

$$\sum_{j \neq i} [r_{ji}(\bar{x}_j) - r_{ij}(\bar{x}_i)] - d\bar{x}_i + \bar{u}_i = 0$$

The following property then gives a condition for this equilibrium to be unique and GAS in the non negative orthant.

Property A3 If $(r_{ij}(x_i) - r_{ij}(\bar{x}_i))(x_i - \bar{x}_i) \geq \ \forall x_i \geq 0$, then the equilibrium $(\bar{x}_1, \ldots, \bar{x}_n)$ of the system (3.24) is GAS in the non negative orthant with Lyapunov function.

$$V(x) = \sum_i |x_i - \bar{x}_i|$$

■

The proof of this property is given in [4]. The interesting feature is that the rate functions $r_{ij}(x_i)$ can be *non-monotonic* (which makes the Jacobian matrix non-compartmental) in contrast with the assumptions of Property A2.

Conservative Lotka-Volterra systems

We consider now a class of Lotka-Volterra ecologies of the form :

$$\dot{x}_i = x_i \left(\sum_{j \neq i} a_{ij} x_j - a_{i0} \right) + \bar{u}_i \quad i = 1, \ldots, n \tag{3.25}$$

with $a_{i0} > 0$ the natural mortality rates;
$a_{ij} = -a_{ij} \ \forall i \neq j$ the predation coefficients (i.e. $A = [a_{ij}]$ is skew symmetric);

$\bar{u}_i \geq 0$ the feeding rate of species x_i with $\sum_i \bar{u}_i > 0$.

This is a mass balance system with a bilinear Hamiltonian representation :

$$F(x) = [a_{ij}x_ix_j] \quad D(x) = (\text{diag } a_{i0}x_i)$$

Assume that the system has an equilibrium in the positive orthant int$\{R^n_+\}$ i.e. there is a strictly positive solution $(\bar{x}_1, \bar{x}_2, \ldots, \bar{x}_n)$ to the set of algebraic equations :

$$a_{i0} = \sum_{j \neq i} a_{ij}\bar{x}_j + \frac{\bar{u}_i}{\bar{x}_i} \quad i = 1, \ldots, n$$

Assume that this equilibrium $(\bar{x}_1, \bar{x}_2, \ldots, \bar{x}_n)$ is the only trajectory in the set :

$$D = \{x \in \text{ int}\{R^n_+\} : \bar{u}_i(x_i - \bar{x}_i) = 0 \forall i\}$$

Then we have the following stability property.

Property A4 The equilibrium $(\bar{x}_1, \bar{x}_2, \ldots, \bar{x}_n)$ of the Lotka-Volterra system (3.25) is unique and GAS in the positive orthant with Lyapunov function

$$V(x) = \sum_i (x_i - \bar{x}_i ln x_i)$$

■

The proof is established, as usual, by using the time derivative of V:

$$\dot{V}(x) = -\sum_i \left[\frac{\bar{u}_i\bar{x}_i}{x_i} \left(1 - \frac{x_i}{\bar{x}_i}\right)^2 \right]$$

and the La Salle's invariance principle.

References

1. G. Bastin and L. Praly, Feedback stabilisation with positive control of a class of mass-balance systems, to be presented at the IFAC World Congress, Beijing, July 1999.

2. J. Eisenfeld, On washout in nonlinear compartmental systems, Mathematical Biosciences, Vol. 58, pp. 259 - 275, 1982.

3. J.A. Jacquez and C.P. Simon, Qualitative theory of compartmental systems, SIAM Review, Vol. 35(1), pp. 43 - 79, 1993.

4. J-L Gouzé, Stability of a class of stirred tank reactors, CD-Rom Proceedings ECC97, Paper FR-A-C7, Brussels, July 1995.

5. F. Grognard, F. Jadot, L. Magni, G. Bastin, R. Sepulchre and V. Wertz, Robust global state feedback stabilisation of cement mills, Cesame Report 99-02, University of Louvain la Neuve, January 1999.

6. F. Jadot, Dynamics and robust nonlinear PI control of stirred tank reactors, PhD Thesis, CESAME, University of Louvain la Neuve, Belgium, June 1996.

7. F. Jadot, G. Bastin and F. Viel,Robust global stabilisation of stirred tank reactors by saturated output feedback, CESAME Report 98-34, University of Louvain la Neuve, July 1998.

8. B.M. Maschke and A. J. van der Schaft, Port controlled Hamiltonian systems : modeling origins and system theoretic properties, Proc. IFAC NOLCOS 92, pp. 282 - 288, Bordeaux, June 1992.

9. B.M. Maschke, R. Ortega and A. J. van der Schaft, Energy based Lyapunov functions for Hamiltonian systems with dissipation, Proc. IEEE Conf. Dec. Control, Tampa, December 1998.

10. H. Rosenbrock, A Lyapounov function with applications to some nonlinear physical problems, Automatica, Vol. 1, pp. 31 - 53, 1962.

11. P. Rouchon, Remarks on some applications of nonlinear control techniques to chemical processes, Proc. IFAC NOLCOS 92, pp. 37-42, Bordeaux, June 1992.

4. Control of dynamic bifurcations

Nils Berglund and Klaus R. Schneider

Weierstraß-Institut für Angewandte Analysis und Stochastik
Mohrenstraße 39
D-10117 Berlin, Germany
berglund@hilbert.wias-berlin.de
schneide@wias-berlin.de

Summary.

We consider differential equations $\dot{x} = f(x, \lambda)$ where the parameter $\lambda = \varepsilon t$ moves slowly through a bifurcation point of f. Such a dynamic bifurcation is often accompanied by a potentially dangerous jump transition. We construct smooth scalar feedback controls which avoid these jumps. For transcritical and pitchfork bifurcations, a small constant additive control is usually sufficient. For Hopf bifurcations, we have to construct a more elaborate control creating a suitable bifurcation with double zero eigenvalue.

4.1 Introduction

Consider the nonlinear control system

$$\frac{\mathrm{d}x}{\mathrm{d}t} = f(x, u, \lambda), \tag{4.1}$$

with state $x \in \mathbb{R}^n$ and control $u \in \mathbb{R}^k$, which depends on some parameter $\lambda \in \mathbb{R}^p$. Assume that the uncontrolled system

$$\frac{\mathrm{d}x}{\mathrm{d}t} = f(x, 0, \lambda) \equiv f_0(x, \lambda) \tag{4.2}$$

changes its qualitative behavior when λ passes λ_0, i.e., $\lambda = \lambda_0$ is a bifurcation point for (4.2). We are interested in bifurcations involving an exchange of stabilities between a family $x^\star(\lambda)$ of "nominal" equilibria of (4.2) and another family of attractors. These attractors are either other equilibria or periodic orbits (Poincaré–Andronov–Hopf bifurcation).

The motivation to study control systems whose state is close to a bifurcation point comes from the well-known fact that the performance of a control system can be improved if it is maintained to operate at high loading levels, that is, near a stability boundary (see for example [1, 35]).

The existence of a bifurcation in the uncontrolled system (4.2) raises the following questions:

1. How does this bifurcation influence the controllability of (4.1)?
2. How can we control an exchange of stabilities?

The first problem has been investigated by the means of control sets, see e.g. [14] for one-dimensional systems, [13] for Hopf bifurcations and [19] for a Takens–Bogdanov singularity (i.e., when a Hopf and a saddle–node bifurcation curve intersect).

The second question is related to the problem of controlling the direction of the bifurcation. In order to avoid escaping trajectories, one usually tries to render the bifurcation supercritical, that is, a stable equilibrium or limit cycle should exist for $\lambda > \lambda_0$, which attracts the orbits departing from the nominal equilibrium $x^\star(\lambda)$. To do this, one has to find a control stabilizing the critical steady state of (4.2) for $\lambda = \lambda_0$. This problem has been solved by using a smooth state feedback, see [4, 3, 2] for the continuous-time case and [25] for the discrete-time case.

In what follows, we are concerned with the problem of *dynamic* exchange of stabilities. In contrast to static bifurcation theory, the theory of dynamic bifurcations considers a process in which the parameter λ depends on time, where one usually assumes that this dependence is slow [6]. Such a situation occurs for instance if the device modelled by the equation is ageing, so that its characteristics are slowly modified.

Instead of (4.2), we thus consider an uncontrolled system of the form

$$\frac{\mathrm{d}x}{\mathrm{d}t} = f_0(x, \varepsilon t), \qquad 0 < \varepsilon \ll 1. \tag{4.3}$$

Basically, an exchange of stability in the static system (4.2) may result in two types of behaviour for (4.3): immediate exchange or delayed exchange. In the first case, the solution of (4.3) tracks the stable branch emerging from the bifurcation point immediately after the bifurcation [22, 23, 27, 7]. In the second case, the solution tracks the *unstable* branch for some time before jumping on the stable equilibrium (see [32, 28, 29, 5, 6, 20] for the Hopf bifurcation and [18, 15, 10, 11, 26] for pitchfork and transcritical bifurcations). Since a jump of a state variable may have catastrophic consequences for the device, our goal is to construct a control ensuring an immediate exchange of stabilities. Note that this feature may be used to detect the bifurcation point. We restrict our analysis to affine scalar feedback controls of the form

$$\frac{\mathrm{d}x}{\mathrm{d}t} = f_0(x, \varepsilon t) + b\, u(x, \varepsilon t), \tag{4.4}$$

where b is a fixed vector in $\mathbb{R}^n$, and u is a scalar function.

This paper is organized as follows. In Section 4.2, we present a few elements of the theory of dynamic bifurcations, and show how the center manifold theorem can be used to reduce the dimension of the system. In Section 4.3, we consider one-dimensional cases such as the transcritical and pitchfork bifurcation, which are relatively easy to control. Section 4.4 is devoted to two-dimensional bifurcations. We first discuss the Hopf bifurcation, which displays a delay which is more robust than for one-dimensional bifurcations. To suppress this delay, we have to shift the eigenvalues' imaginary parts in order to produce a double zero eigenvalue, for which we present a result on immediate exchange of stability.

Acknowledgment: NB was supported by the Nonlinear Control Network of the European Community, Grant ERB FMRXCT–970137.

4.2 Dynamic bifurcations

Consider a one-parameter family of dynamical systems

$$\frac{dx}{dt} = f(x, \lambda), \qquad x \in \mathbb{R}^n, \lambda \in \mathbb{R}. \tag{4.5}$$

In the theory of dynamic bifurcations, one is concerned with the slowly time-dependent system

$$\frac{dx}{dt} = f(x, \varepsilon t), \qquad 0 < \varepsilon \ll 1, \tag{4.6}$$

that one wants to study on the time scale ε^{-1}. It is convenient to introduce the slow time $\tau = \varepsilon t$, in order to transform (4.6) into the singularly perturbed system

$$\varepsilon \frac{dx}{d\tau} = f(x, \tau). \tag{4.7}$$

The basic idea is to use information on the bifurcation diagram of (4.5) in order to analyse solutions of (4.7).

Assume first that for $\lambda \in [a, b]$, (4.5) admits a family of asymptotically stable equilibria $x^\star(\lambda)$. That is, we require that $f(x^\star(\lambda), \lambda) = 0$ and that all eigenvalues of the Jacobian matrix $A(\lambda) = \partial_x f(x^\star(\lambda), \lambda)$ have real parts smaller than some $K < 0$, uniformly for $\lambda \in [a, b]$. It is known [31, 16, 34, 8] that all solutions of (4.7) starting at $\tau = a$ in a sufficiently small neighbourhood of $x^\star(a)$ will reach an $\mathcal{O}(\varepsilon)$–neighbourhood of $x^\star(\tau)$ after a slow time of order $\varepsilon|\ln \varepsilon|$ and remain there until $\tau = b$. Thus, a sufficiently slow drift of the parameter λ will cause the system to track the nominal equilibrium $x^\star(\lambda)$ as closely as desired.

A new situation arises when $x^\star(\lambda)$ undergoes a bifurcation. Assume that at $\lambda = 0$, the Jacobian matrix $A(0)$ has m eigenvalues with zero real parts and $n - m$ eigenvalues with negative real parts. We can introduce coordinates $(y, z) \in \mathbb{R}^{n-m} \times \mathbb{R}^m$ such that (4.6) can be written in the form

$$\begin{aligned} dy/dt &= A_- y + g_-(y, z, \tau) \\ dz/dt &= A_0 z + g_0(y, z, \tau) \\ d\tau/dt &= \varepsilon \\ d\varepsilon/dt &= 0, \end{aligned} \tag{4.8}$$

where all eigenvalues of A_- have negative real parts, and all eigenvalues of A_0 have zero real parts. The functions g_- and g_0 vanish at $\tau = 0$ together with their derivatives with respect to y and z. Thus, at the bifurcation point z can be considered as a slow variable as well as τ. By the center manifold theorem [12] there exists a locally invariant manifold $y = h(z, \tau, \varepsilon)$, on which the dynamics is governed by the m-dimensional equation

$$\varepsilon \frac{dz}{d\tau} = A_0 z + g_0(h(z, \tau, \varepsilon), z, \tau). \tag{4.9}$$

Moreover, trajectories starting close to this manifold are locally attracted by it with an exponential rate (see Lemma 1, p. 20 in [12]).

This observation allows us to restrict the analysis of (4.7) near the bifurcation point to the analysis of the lower-dimensional equation (4.9). Note, however, that we have to pay attention to the following points:

1. If we add a control to (4.7), we will modify the shape of the center manifold.
2. The center manifold is not analytic in general.

To simplify the discussion, we will only consider the low-dimensional systems on the center manifold. The above remarks imply that some additional verifications are necessary before conclusions about the reduced equations can be carried over to the general ones.

4.3 One-dimensional center manifold

We consider the scalar equation

$$\varepsilon \frac{dx}{d\tau} = f_0(x, \tau) + u(x, \tau), \qquad x \in \mathbb{R}. \tag{4.10}$$

4.3.1 Transcritical bifurcation

Assume that the uncontrolled vector field $f_0(x, \tau)$ has two families of equilibria $x = \varphi_1(\tau)$ and $x = \varphi_2(\tau)$ intersecting at $\tau = 0$. The family $\varphi_1(\tau)$ is stable for $\tau < 0$, while the family $\varphi_2(\tau)$ is stable for $\tau > 0$. This kind of bifurcation is referred to as transcritical bifurcation. We introduce the so-called singular stable solution

$$\varphi(\tau) = \begin{cases} \varphi_1(\tau) & \text{if } \tau < 0, \\ \varphi_2(\tau) & \text{if } \tau > 0. \end{cases} \tag{4.11}$$

Our goal is to find a control u such that the solution of (4.10) starting at $\tau_0 < 0$ in the basin of attraction of $\varphi_1(\tau)$ always stays in a small neighborhood of the singular stable solution $\varphi(\tau)$ for $\tau > \tau_0$. It turns out that the dynamics depends essentially on the values of $\varphi_1'(0)$ and $\varphi_2'(0)$. We discuss three representative cases.

Example 4.3.1 (Immediate exchange of stability).

Assume that the uncontrolled system has the form

$$\varepsilon \frac{\mathrm{d}x}{\mathrm{d}\tau} = (x + \tau)(\tau - x). \tag{4.12}$$

Then we have $\varphi_1(\tau) = -\tau$, $\varphi_2(\tau) = \tau$ and $\varphi(\tau) = |\tau|$. It is shown in [22] that the solutions starting above $\varphi_2(\tau_0)$ at $\tau_0 < 0$ will track the singular stable solution $\varphi(\tau)$, so that no control is necessary. More precisely, it follows from [7] that (4.12) admits a particular solution $x(\tau)$ satisfying

$$|x(\tau) - \varphi(\tau)| \leqslant \begin{cases} M\varepsilon|\tau|^{-1} & \text{if } \varepsilon^{1/2} \leqslant |\tau| \leqslant T, \\ M\varepsilon^{1/2} & \text{if } |\tau| \leqslant \varepsilon^{1/2}, \end{cases} \tag{4.13}$$

for some positive M and T, which attracts nearby solutions exponentially fast (Fig. 4.1a). Moreover this exchange of stability is robust in the following sense: it is shown in [11] that there exists a constant $c > 0$ such that solutions still track the stable equilibrium curve if we add a constant term $u_0 > -c\varepsilon$ to (4.12) (Fig. 4.1b). The same is true for more general bifurcations, for which $\varphi_1'(0) < 0$ and $\varphi_1'(0) < \varphi_2'(0)$.

Example 4.3.2 (Delayed exchange of stability).

The uncontrolled system

$$\varepsilon \frac{\mathrm{d}x}{\mathrm{d}\tau} = x(\tau - x) \tag{4.14}$$

has the equilibria $\varphi_1(\tau) = 0$ and $\varphi_2(\tau) = \tau$. This happens to be an explicitly solvable Bernoulli equation. The important fact is that the solution starting at $\tau_0 < 0$ at some $x_0 > 0$ remains close to the origin for $\tau_0 < \tau < -\tau_0$ as $\varepsilon \to 0$, and jumps to the branch $\varphi_2(\tau)$ near $\tau = -\tau_0$ (Fig. 4.1c).

Is is relatively easy to find a control which guarantees that the solution remains close to $\varphi(\tau)$. This is due to the fact that the vector field $-x^2$ is a codimension two singularity with unfolding

$$\frac{\mathrm{d}x}{\mathrm{d}t} = x(\lambda - x) + \mu. \tag{4.15}$$

If μ is a positive constant, this equation has two families of equilibria which do not intersect. The family located in the half plane $x > 0$ is asymptotically stable and lies at a distance of order $\mu^{1/2}$ from $x = \varphi(\tau)$. This implies that the solution of the initial value problem

$$\varepsilon\frac{\mathrm{d}x}{\mathrm{d}\tau} = x(\tau - x) + u_0, \qquad x(\tau_0) > \tau_0,\ \tau_0 < 0,\ u_0 > 0 \tag{4.16}$$

stays near $\varphi(\tau)$ provided ε is sufficiently small (Fig. 4.1d). More precisely, there exists a continuous function $\delta(\varepsilon)$ with $\lim_{\varepsilon\to 0}\delta(\varepsilon) = 0$ such that the solution will track the upper equilibrium if $u_0 > \delta(\varepsilon)$. Thus, the smaller the drift velocity ε, the weaker the control has to be. It is known [15] that $\delta(\varepsilon)$ goes to zero faster than any power law.

Example 4.3.3 (Diverging solutions).

For the equation

$$\varepsilon\frac{\mathrm{d}x}{\mathrm{d}\tau} = (x - 2\tau)(\tau - x), \tag{4.17}$$

we have $\varphi_1(\tau) = \tau$ and $\varphi_2(\tau) = 2\tau$. Solutions of this equation diverge for some $\tau \leqslant 0$ (Fig. 4.1e). This can be avoided by adding a control

$$\varepsilon\frac{\mathrm{d}x}{\mathrm{d}\tau} = (x - 2\tau)(\tau - x) + u_0, \tag{4.18}$$

which splits the equilibrium branches if $u_0 > 0$ (Fig. 4.1f). In this case, we must have $u_0 > \delta(\varepsilon) = \varepsilon$. Note that if $u_0 = \varepsilon$, the change of variables $y = x - \tau$ transforms (4.18) into (4.14). The same qualitative features hold if $\varphi_2'(0) > \varphi_1'(0) > 0$.

4.3.2 Pitchfork bifurcation

Similar results hold for the pitchfork bifurcation. Assume that the uncontrolled vector field $f_0(x,\tau)$ has a family of equilibria $x = \varphi_0(\tau)$ which is stable for $\tau < 0$ and unstable for $\tau > 0$. For positive τ, there exist two additional stable equilibria $\varphi_\pm(\tau) = \pm c\sqrt{\tau} + \mathcal{O}(\tau)$.

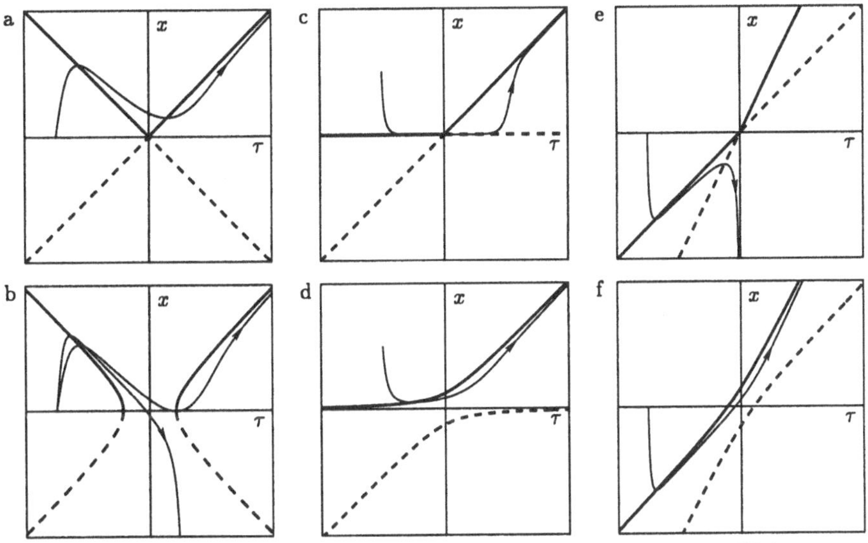

Fig. 4.1. Exchange of stability for dynamic transcritical bifurcations. Light curves represent solutions of the time-dependent equation, heavy curves represent stable (full) and unstable (broken) equilibria of the static system. (a) Solutions of (4.12) track the singular stable solution $\varphi = |\tau|$. (b) This behaviour subsists if we add a negative constant to (4.12), provided ε is large enough. If ε is too small, the solution slips through the gap. (c) The solution of (4.14) with initial condition $x_0 > 0$ at $\tau_0 < 0$ exhibits a jump at $\tau = -\tau_0$. (d) This jump is suppressed if we add a small positive control. (e) The uncontrolled system (4.17) has diverging solutions. (f) A sufficiently large additive control suppresses this divergence.

Example 4.3.4 (Immediate exchange of stability).

Assume that the uncontrolled system has the form

$$\varepsilon \frac{dx}{d\tau} = (x - \tau)(\tau - x^2). \tag{4.19}$$

It is shown in [23] that the solutions will track the lower branch $\varphi_-(\tau)$ after the bifurcation (Fig. 4.2a). More precisely, let

$$\varphi(\tau) = \begin{cases} \varphi_0(\tau) & \text{if } \tau < 0, \\ \varphi_-(\tau) & \text{if } \tau > 0. \end{cases} \tag{4.20}$$

In [7] we obtained the existence of an attracting particular solution $x(\tau)$ satisfying

$$|x(\tau) - \varphi(\tau)| \leqslant \begin{cases} M\varepsilon|\tau|^{-1} & \text{if } -T \leqslant \tau \leqslant -\varepsilon^{1/2}, \\ M\varepsilon^{1/2} & \text{if } -\varepsilon^{1/2} \leqslant \tau \leqslant \varepsilon, \\ M\tau^{1/2} & \text{if } \varepsilon \leqslant \tau \leqslant \varepsilon^{1/2}, \\ M\varepsilon|\tau|^{-3/2} & \text{if } \varepsilon^{1/2} \leqslant \tau \leqslant T, \end{cases} \tag{4.21}$$

for some positive M and T.

One may wish to make the solution track the *upper* equilibrium $\varphi_+(\tau)$ after the bifurcation. This can be achieved by adding a constant control of the form

$$\varepsilon\frac{\mathrm{d}x}{\mathrm{d}\tau} = (x-\tau)(\tau-x^2)+u_0, \tag{4.22}$$

with $u_0 > \delta(\varepsilon) = \varepsilon$ (Fig. 4.2b).

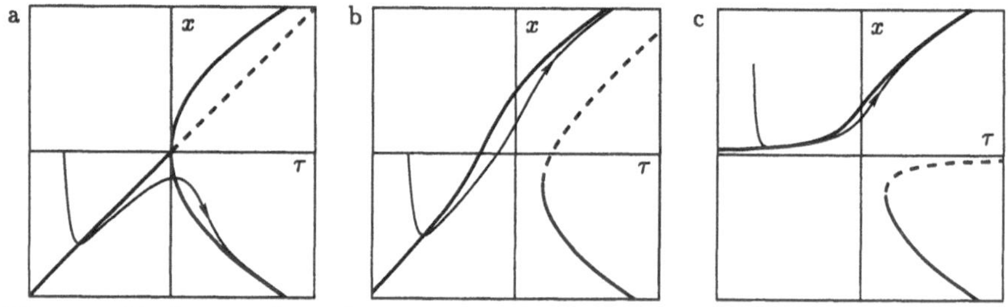

Fig. 4.2. Exchange of stability for pitchfork bifurcations. (a) Solutions of (4.19) track the lower stable equilibrium. (b) A sufficiently large positive control makes the system follow the upper equilibrium. (c) The same occurs for system (4.23).

Example 4.3.5 (Delayed exchange of stability).

The uncontrolled system

$$\varepsilon\frac{\mathrm{d}x}{\mathrm{d}\tau} = \tau x - x^3 \tag{4.23}$$

displays a bifurcation delay similar to Example 4.3.2. One can provoke an immediate exchange of stability by adding a constant control u_0; there exists a function $\delta(\varepsilon)$ such that solutions track the upper equilibrium $\tau^{1/2}$ if $u_0 > \delta(\varepsilon)$ (Fig. 4.2c) and the lower equilibrium $-\tau^{1/2}$ if $u_0 < -\delta(\varepsilon)$. The function $\delta(\varepsilon)$ goes to zero faster than any power law [15].

4.3.3 Bifurcations with identically zero equilibrium

One can encounter systems of the form (4.10) for which $f(0,\tau) = 0$ for all τ. This happens, for instance, when f is symmetric under the transformation $x \to -x$. In such a case, we can write the uncontrolled system in the form

$$\varepsilon\frac{\mathrm{d}x}{\mathrm{d}\tau} = a(\tau)x + g(x,\tau), \tag{4.24}$$

where $|g(x,\tau)| \leqslant Mx^2$ for $|x| \leqslant d$. Assume that we start in the basin of attraction of the origin at a time τ_0 at which $a(\tau_0) < 0$. Then one can show [7, 26] that

$$x(\tau) = \mathcal{O}(\varepsilon) \quad \text{for } \tau_0 + \mathcal{O}(\varepsilon|\ln\varepsilon|) \leqslant \tau \leqslant \Pi(\tau_0) + \mathcal{O}(\varepsilon|\ln\varepsilon|), \tag{4.25}$$

where $\Pi(\tau_0) > \tau_0$ is the first time such that

$$\int_{\tau_0}^{\Pi(\tau_0)} a(\tau)\,\mathrm{d}\tau = 0. \tag{4.26}$$

If, for instance, $a(\tau)$ is negative for $\tau < 0$ and positive for $\tau > 0$, the delay time $\Pi(\tau_0)$ is obtained by making equal the areas delimited by the τ-axis, the curve $a(\tau)$ and the times τ_0, 0 and $\Pi(\tau_0)$, see Fig. 4.3a.

The delay can be suppressed by adding a constant control as in Examples 4.3.2 and 4.3.5. If, however, one does not wish to destroy the equilibrium $x = 0$, it is possible to influence the delay by a linear control $u(x) = cx$. The behaviour will strongly depend on the initial condition.

Example 4.3.6 (Pitchfork bifurcation).

Consider again the equation of Example 4.3.5, but with a linear control

$$\varepsilon\frac{\mathrm{d}x}{\mathrm{d}\tau} = \tau x - x^3 + cx. \tag{4.27}$$

If we start at some distance of $x = 0$ at $\tau_0 < 0$, the bifurcation is translated to the time $-c$, while the delay time is given by $\Pi(\tau_0) = -2c - \tau_0$. The effect of the control is thus twice as large as expected from the static theory.

Example 4.3.7 (Relaxation oscillations).

If $f_0(x,\tau)$ depends periodically on time, the system may exhibit *relaxation oscillations*, which are periodic solutions with alternating slow and fast motions [24]. This happens for instance for the equation

$$\varepsilon\frac{\mathrm{d}x}{\mathrm{d}\tau} = (A + \sin\tau)x - x^3, \tag{4.28}$$

if $0 < A < 1$. The oscillations may be suppressed by adding a linear control $u(x) = -Ax$, although from the static theory one would expect that a control $u(x) = -(A+1)x$ were necessary.

4.4 Two-dimensional center manifold

We consider now the two-dimensional version of (4.4). By an appropriate choice of variables, it can be written as

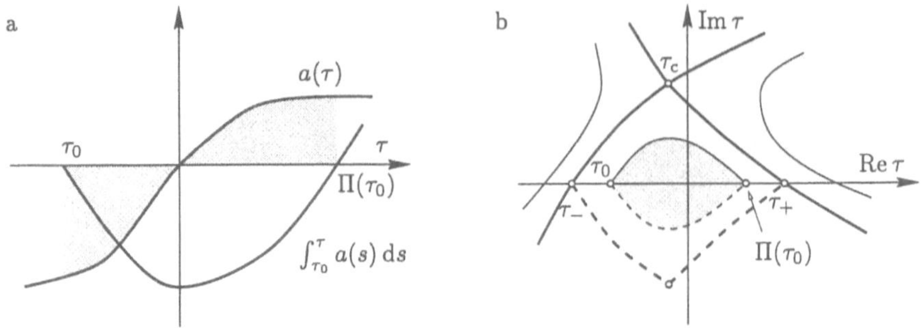

Fig. 4.3. Determination of the bifurcation delay time. (a) In the case of the one-dimensional equation (4.24), the delay is simply obtained by making two areas the same. (b) In the case of a Hopf bifurcation, there is a maximal delay τ_+ given by the largest real time which can be connected to the negative real axis by a path with constant Re Ψ. This path must have certain properties described in [30], in particular the equation should be analytic in the shaded region.

$$\begin{aligned} \varepsilon \frac{\mathrm{d}x}{\mathrm{d}\tau} &= f_1(x, y, \tau) \\ \varepsilon \frac{\mathrm{d}y}{\mathrm{d}\tau} &= f_2(x, y, \tau) + u(x, y, \tau). \end{aligned} \tag{4.29}$$

4.4.1 Hopf bifurcation

We assume that the static uncontrolled system

$$\begin{aligned} \frac{\mathrm{d}x}{\mathrm{d}t} &= f_1(x, y, \lambda) \\ \frac{\mathrm{d}y}{\mathrm{d}t} &= f_2(x, y, \lambda) \end{aligned} \tag{4.30}$$

admits a family of equilibria such that the eigenvalues of the linearization are of the form $a(\lambda) \pm \mathrm{i}\omega(\lambda)$, with $a(0) = 0$, $a'(0) > 0$ and $\omega(0) \neq 0$. It is known that the system can in general be controlled by a smooth feedback in such a way that the bifurcation is supercritical, that is, a stable periodic orbit exists for positive λ [4].

When $\lambda = \varepsilon t$ is made slowly time-dependent and the right-hand side of (4.30) is analytic, the bifurcation is delayed, as has been proved by Neishtadt [28, 29] (Fig. 4.5a). Unlike in the case of pitchfork bifurcations, this delay also exists when the equilibrium depends on λ, and is stable with respect to analytic deterministic perturbations.

Example 4.4.1 (Hopf bifurcation).

Consider the system

$$\begin{aligned}\varepsilon\frac{\mathrm{d}x}{\mathrm{d}\tau} &= \tau(x-\tau)+\omega_0 y-\big[(x-\tau)^2+y^2\big](x-\tau)\\ \varepsilon\frac{\mathrm{d}y}{\mathrm{d}\tau} &= -\omega_0(x-\tau)+\tau y-\big[(x-\tau)^2+y^2\big]y,\end{aligned} \tag{4.31}$$

which admits the family of equilibria $(\tau, 0)$. The linearization around them has eigenvalues $\tau \pm \mathrm{i}\omega_0$. The complex variable $\zeta = x - \tau + \mathrm{i}y$ satisfies the equation

$$\varepsilon\frac{\mathrm{d}\zeta}{\mathrm{d}\tau} = (\tau - \mathrm{i}\omega_0)\zeta - |\zeta|^2\zeta - \varepsilon. \tag{4.32}$$

The delay phenomenon can be understood by considering the linearization of (4.32), which admits the solution

$$\begin{aligned}\zeta(\tau) &= \mathrm{e}^{[\Psi(\tau)-\Psi(\tau_0)]/\varepsilon}\,\zeta(\tau_0) - \int_{\tau_0}^{\tau}\mathrm{e}^{[\Psi(\tau)-\Psi(s)]/\varepsilon}\,\mathrm{d}s,\\ \Psi(\tau) &= \int_0^{\tau}(s-\mathrm{i}\omega_0)\,\mathrm{d}s = \frac{1}{2}\tau^2 - \mathrm{i}\omega_0\tau.\end{aligned} \tag{4.33}$$

The first term is small for $\tau_0 < \tau < -\tau_0$, as in the case of the pitchfork bifurcation. A crucial role is played by the second term, which is due to the τ-dependence of the equilibria. It can be evaluated using a deformation of the integration path into the complex plane. The function $\Psi(\tau)$ can be extended to complex τ and we have

$$\operatorname{Re}\Psi(\tau) = \frac{1}{2}\big[(\operatorname{Re}\tau)^2 - (\operatorname{Im}\tau - \omega_0)^2 + \omega_0^2\big]. \tag{4.34}$$

The level lines of this function are hyperbolas centered at $\tau = \mathrm{i}\omega_0$. The integral in (4.33) is small if we manage to connect τ_0 and τ by a path on which $\operatorname{Re}\Psi(s) \geqslant \operatorname{Re}\Psi(\tau)$, i.e., if we never go uphill in the landscape of $\operatorname{Re}\Psi(s)$. This is possible if

$$\tau \leqslant \hat{\tau} = \min\{-\tau_0, \omega_0\}. \tag{4.35}$$

The existence of the maximal delay $\tau = \omega_0$ is a nonperturbative effect, entirely determined by the linearization around the equilibria.

The computation of the delay in the general case is discussed in [30]. It is given by the formula

$$\hat{\tau} = \min\{\Pi(\tau_0), \tau_+\}, \tag{4.36}$$

where $\Pi(\tau_0)$ is defined by (4.26), and the maximal delay τ_+ can be determined by the level lines of $\operatorname{Re}\Psi(\tau)$ (Fig. 4.3b).

This shows in particular that the delay is robust, and the jump transition occurring at the delay time $\hat{\tau}$ cannot be avoided by adding a small constant control, as in the case of the pitchfork bifurcation. We may, of course, use a linear control which shifts the real part of the eigenvalues of the linearization, in order to increase the delay as in Section 4.3.3. This, however, will only postpone the problem to some later time, if the real part of the linearization is monotonically increasing.

Here we propose a different strategy to avoid a jump. We would like to provoke an immediate exchange of stability in order to detect the bifurcation point before it is too late. Expression (4.36) for the delay time shows that this can only be done by decreasing the buffer time, which might be achieved by shifting the *imaginary* part of the eigenvalues. This will create a bifurcation with double zero eigenvalue, which we study below.

4.4.2 Double zero eigenvalue

We start by analysing the autonomous control system

$$\frac{\mathrm{d}z}{\mathrm{d}t} = f(z,\lambda) + b\,u(z,\lambda), \qquad z \in \mathbb{R}^2, \tag{4.37}$$

where $f(z,\lambda)$ admits an equilibrium branch $z^\star(\lambda)$ such that the linearization $\partial_z f(z^\star(\lambda),\lambda)$ has eigenvalues $a(\lambda) \pm \mathrm{i}\omega(\lambda)$, with $a(0) = 0$, $a'(0) > 0$ and $\omega(0) = 1$ (this value of $\omega(0)$ may be achieved by a rescaling of time). Let $F(z,\lambda) = f(z,\lambda) + b\,u(z,\lambda)$. The scalar feedback $u(z,\lambda)$ is determined by two requirements:

1. The matrix $\partial_z F(0,0)$ should have a double zero eigenvalue.
2. In analogy with works on stabilization of bifurcations [4], the origin should be a stable equilibrium of (4.37) when $\lambda = 0$.

After a suitable affine transformation, we can write (4.37) as

$$\begin{aligned} \frac{\mathrm{d}x}{\mathrm{d}t} &= a(\lambda)x + \omega(\lambda)y + g_1(x,y,\lambda) \\ \frac{\mathrm{d}y}{\mathrm{d}t} &= -\omega(\lambda)x + a(\lambda)y + g_2(x,y,\lambda) + \tilde{u}(x,y,\lambda), \end{aligned} \tag{4.38}$$

where g_1 and g_2 are of order x^2+y^2. We have used the fact that the linear part is rotation invariant, so that we may take $b = \left(\begin{smallmatrix}0\\1\end{smallmatrix}\right)$.

For $\lambda = 0$, we propose the control

$$\tilde{u}(x,y,0) = x + v_1 x^2 + v_2 xy + v_3 y^2 + v_4 x^3, \tag{4.39}$$

where the coefficients $v_1,\dots,v_4$ have yet to be determined. The system (4.38) takes the form

$$\begin{aligned}\frac{dx}{dt} &= y + c_1x^2 + c_2xy + c_3y^2 + c_4x^3 + \cdots \\ \frac{dy}{dt} &= (d_1 + v_1)x^2 + (d_2 + v_2)xy + (d_3 + v_3)y^2 + (d_4 + v_4)x^3 + \cdots ,\end{aligned} \tag{4.40}$$

where c_i and d_i are the Taylor coefficients of g_1 and g_2 at the origin, respectively. A normal form of (4.40) is

$$\begin{aligned}\frac{dx}{dt} &= y \\ \frac{dy}{dt} &= \gamma x^2 + \delta xy + \alpha x^2 y + \beta x^3 + \mathcal{O}(\|z\|^4),\end{aligned} \tag{4.41}$$

where the coefficients α, β, γ and δ are algebraic functions of c_i, d_i and v_i. This system has already been studied by Takens [33, 17]. If $\gamma \neq 0$ or $\delta \neq 0$, the origin is an unstable Bogdanov–Takens singularity. We thus require that $\gamma = \delta = 0$ (which amounts to imposing that $v_1 = -d_1$, $v_2 = -d_2 - 2c_1$). For the origin to be stable, we require moreover that $\alpha, \beta < 0$, which imposes some inequalities on v_3 and v_4 (in fact, this requires that $c_1(d_3 + v_3) + c_4 < 0$ and $2c_1^2 + d_4 + v_4 < 0$, these conditions can be satisfied if $c_1 \neq 0$ or if $c_4 < 0$).

For general values of λ, we choose a control of the form

$$\tilde{u}(x, y, \lambda) = (1 + C\lambda)\tilde{u}(x, y, 0). \tag{4.42}$$

After inserting this into (4.38), carrying out a linear transformation and computing the normal form, we get the system

$$\begin{aligned}\frac{dx}{dt} &= y \\ \frac{dy}{dt} &= \mu(\lambda)x + 2a(\lambda)y + \gamma(\lambda)x^2 + \delta(\lambda)xy - x^2y - x^3 + \mathcal{O}(\|z\|^4),\end{aligned} \tag{4.43}$$

where $\gamma(0) = \delta(0) = 0$, and

$$\mu(\lambda) = [C - \omega'(0)]\lambda + \mathcal{O}(\lambda^2) \tag{4.44}$$

can be influenced by the choice of C. This equation happens to be a codimension-four unfolding of the singular vector field $(y, -x^2y - x^3)$ which has been studied in detail, see [21, 36] and references therein. The bifurcation diagram in the section $\gamma = \delta = 0$ has already been studied in [33], it is shown in Fig. 4.4.

We now consider the time-dependent version of (4.37),

$$\varepsilon \frac{dz}{d\tau} = f(z, \tau) + b\, u(z, \tau). \tag{4.45}$$

It can be shown that similar transformations as above yield the equation

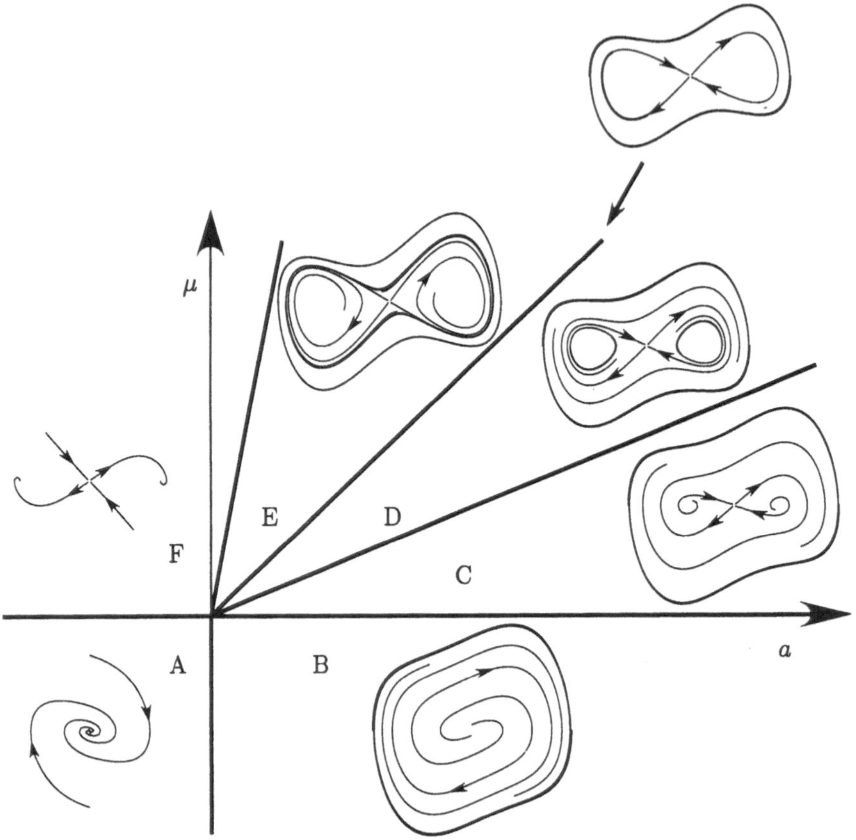

Fig. 4.4. Schematic bifurcation diagram of equation (4.43) in the plane $\gamma = \delta = 0$. The transition A-B is the original Hopf bifurcation. By moving the eigenvalues' imaginary parts to 0, we change the function $\mu(\lambda)$ in such a way that $\mu(0) = 0$. This produces new bifurcation lines. The transition A-F is a supercritical saddle-node bifurcation, the transition C-B a subcritical one. D-C is a subcritical Hopf bifurcation, D-E a homoclinic bifurcation and E-F a saddle-node bifurcation of periodic orbits.

$$\begin{aligned}
\varepsilon \frac{\mathrm{d}x}{\mathrm{d}\tau} &= y \\
\varepsilon \frac{\mathrm{d}y}{\mathrm{d}\tau} &= \mu(\tau)x + 2a(\tau)y + \gamma(\tau)x^2 + \delta(\tau)xy - x^2y - x^3 \\
&\quad + \mathcal{O}(\|z\|^4) + \varepsilon R(x, y, \tau, \varepsilon),
\end{aligned} \tag{4.46}$$

where $R(0, 0, \tau, 0)$ is directly related to the drift $\frac{\mathrm{d}}{\mathrm{d}\tau} z^\star(\tau)$ of the nominal equilibrium.

The dynamics of (4.46) depends essentially on the path $(a(\tau), \mu(\tau))$ through the bifurcation diagram of Fig. 4.4, the effect of $\gamma(\tau)$ and $\delta(\tau)$ is small in a neighbourhood of the bifurcation point. Various typical solutions are shown in Fig. 4.5. If we go from region A to region B, the Hopf bifurcation induces the usual delayed appearance of oscillations (Fig. 4.5a). If we go into region C, the delay is suppressed, but we still have oscillations (Fig. 4.5b,c). If, however, $\mathrm{d}\mu/\mathrm{d}a(0)$ is large enough to reach one of the regions D, E or F, there is an immediate exchange of stabilities with a stable focus (Fig. 4.5d).

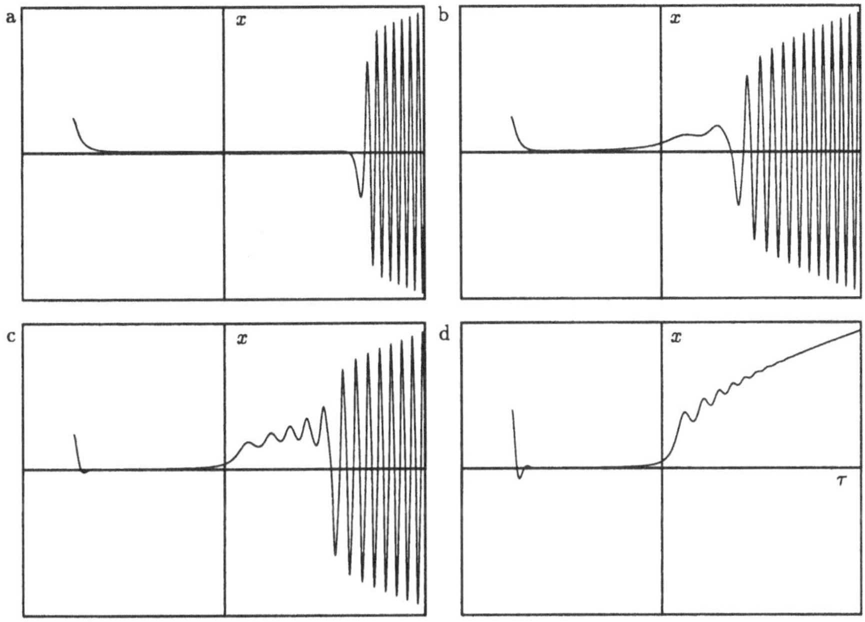

Fig. 4.5. Solutions $x(\tau)$ of equation (4.46) in the case $\gamma = \delta \equiv 0$, $R \equiv 1$, $a(\tau) = 2\tau$ and different functions $\mu(\tau)$. (a) $\mu(\tau) = -0.2$: We traverse the bifurcation diagram of Fig. 4.4 from region A to region B. The system undergoes a Hopf bifurcation, which results in the delayed appearance of large amplitude oscillations. (b) $\mu(\tau) = 0$: The delay is suppressed, but we still have oscillations. (c) $\mu(\tau) = a(\tau)$: We cross the bifurcation diagram from region A to region C. The trajectory starts by following the unstable focus, before being attracted by the limit cycle. (d) $\mu(\tau) = 2.5a(\tau)$: Theorem 4.4.1 applies, there is immediate exchange of stabilities between the nominal equilibrium and a stable focus.

In [9] we prove the following result on exchange of stabilities:

Theorem 4.4.1. *Assume that* $\mu'(0) > 0$. *There exist positive constants* d, T, M, κ *and a neighbourhood* $\mathcal{M}$ *of the origin in* $\mathbb{R}^2$ *with the following property. For every* $\tau_0 \in [-T, 0)$, *there is a constant* $c_1 > 0$ *such that for sufficiently small* ε, *any solution of* (4.46) *with initial condition* $(x, y)(\tau_0) \in \mathcal{M}$ *satisfies*

$$|x(\tau)| \leqslant M\frac{\varepsilon}{|\tau|}, \qquad |y(\tau)| \leqslant M\frac{\varepsilon}{|\tau|^{1/2}}, \qquad \tau_1(\varepsilon) \leqslant \tau \leqslant -\Big(\frac{\varepsilon}{d}\Big)^{2/3}, \tag{4.47}$$

$$|x(\tau)| \leqslant M\varepsilon^{1/3}, \qquad |y(\tau)| \leqslant M\varepsilon^{2/3}, \qquad -\Big(\frac{\varepsilon}{d}\Big)^{2/3} \leqslant \tau \leqslant \Big(\frac{\varepsilon}{d}\Big)^{2/3}, \tag{4.48}$$

where $\tau_1(\varepsilon) = \tau_0 + c_1\varepsilon|\ln\varepsilon|$. *If, moreover, the relations*

$$\mu'(0) > 2a'(0), \qquad R(0,0,0,0) \neq 0 \tag{4.49}$$

hold, then for $(\varepsilon/d)^{2/3} \leqslant \tau \leqslant T$ *we have*

$$\begin{aligned} |x(\tau) - x_+(\tau)| &\leqslant M\Big[\frac{\varepsilon}{\tau} + \frac{\varepsilon^{1/2}}{\tau^{1/4}}\,\mathrm{e}^{-\kappa\tau^2/\varepsilon}\Big], \\ |y(\tau)| &\leqslant M\Big[\frac{\varepsilon}{\tau^{1/2}} + \varepsilon^{1/2}\tau^{1/4}\,\mathrm{e}^{-\kappa\tau^2/\varepsilon}\Big], \end{aligned} \tag{4.50}$$

where

$$x_+(\tau) = \begin{cases} \sqrt{\mu} + \mathcal{O}(\tau), & \text{if } R(0,0,0,0) > 0, \\ -\sqrt{\mu} + \mathcal{O}(\tau), & \text{if } R(0,0,0,0) < 0 \end{cases} \tag{4.51}$$

are equilibria of (4.43), *i.e., the right-hand side of* (4.43) *vanishes when* $x = x_+$ *and* $y = 0$.

In [9] we also prove that similar properties hold in the n-dimensional case. The control that we have constructed is robust in the following sense. If the coefficients in the feedback (4.39) are not perfectly adjusted, the functions $\mu(\tau)$, $\gamma(\tau)$ and $\delta(\tau)$ will not vanish exactly at the same time. This means that the bifurcation diagram of Fig. 4.4 will be traversed on a line which misses the origin. Depending on whether the path passes above or below the origin, solutions will either track a stable branch emerging from a pitchfork bifurcation, or start oscillating, but with a relatively small amplitude. This can be considered as an almost immediate transfer of stability from the nominal equilibrium to the limit cycle.

References

1. E.H. Abed, *Bifurcation-theoretic issues in the control of voltage collapse, Dynamical systems approaches to nonlinear problems in systems and circuits*, Proc. Conf. Qual. Methods Anal. Nonlinear Dyn., Henniker/UK 1986 (1988)

2. E.H. Abed, *Local bifurcation control*, in J.H. Chow (Ed.), *Systems and control theory for power systems* (Springer, New York, 1995).

3. E.H. Abed, J.-H. Fu, *Local feedback stabilization and bifurcation control. I. Hopf bifurcation*, Systems Control Lett. **7**:11–17 (1986).

4. D. Aeyels, *Stabilization of a class of nonlinear systems by a smooth feedback control*, Systems Control Lett. **5**:289–294 (1985).

5. S.M. Baer, T. Erneux, J. Rinzel, *The slow passage through a Hopf bifurcation: delay, memory effects, and resonance*, SIAM J. Appl. Math. **49**:55–71 (1989).

6. E. Benoît (Ed.), *Dynamic Bifurcations, Proceedings, Luminy 1990* (Springer–Verlag, Lecture Notes in Mathematics 1493, Berlin, 1991).

7. N. Berglund, *Adiabatic Dynamical Systems and Hysteresis*, Thesis EPFL no 1800 (1998). Available at
`http://dpwww.epfl.ch/instituts/ipt/berglund/these.html`

8. N. Berglund, *On the Reduction of Adiabatic Dynamical Systems near Equilibrium Curves*, Proceedings of the International Workshop *"Celestial Mechanics, Separatrix Splitting, Diffusion"*, Aussois, France, June 21–27, 1998.

9. N. Berglund, *Control of Dynamic Hopf Bifurcations*, preprint `WIAS-479`, `mp-arc/99-89` (1999). To be published.

10. N. Berglund, H. Kunz, *Chaotic Hysteresis in an Adiabatically Oscillating Double Well*, Phys. Rev. Letters **78**:1692–1694 (1997).

11. N. Berglund, H. Kunz, *Memory Effects and Scaling Laws in Slowly Driven Systems*, J. Phys. A **32**:15–39 (1999).

12. J. Carr, *Applications of Centre Manifold Theory* (Springer–Verlag, New York, 1981).

13. F. Colonius, G. Häckl, W. Kliemann, *Controllability near a Hopf bifurcation*, Proc. 31st IEEE Conf. Dec. Contr., Tucson, Arizona (1992).

14. F. Colonius, W. Kliemann, *Controllability and stabilization of one-dimensional systems near bifurcation points*, Syst. Control Lett. **24**:87–95 (1995).

15. T. Erneux, P. Mandel, *Imperfect bifurcation with a slowly-varying control parameter*, SIAM J. Appl. Math. **46**:1–15 (1986).

16. N. Fenichel, *Geometric singular perturbation theory for ordinary differential equations*, J. Diff. Eq. **31**:53–98 (1979).

17. J. Guckenheimer, P. Holmes, *Nonlinear Oscillations, Dynamical Systems, and Bifurcations of Vector Fields* (Springer–Verlag, New York, 1983).

18. R. Haberman, *Slowly varying jump and transition phenomena associated with algebraic bifurcation problems*, SIAM J. Appl. Math. **37**:69–106 (1979).

19. G. Häckl, K.R. Schneider, *Controllability near Takens–Bogdanov points*, J. Dy-

nam. Control Systems **2**:583–598 (1996).

20. L. Holden, T. Erneux, *Slow passage through a Hopf bifurcation: From oscillatory to steady state solutions*, SIAM J. Appl. Math. **53**:1045–1058 (1993).

21. A.I. Khibnik, B. Krauskopf, C. Rousseau, *Global study of a family of cubic Liénard equations*, Nonlinearity **11**:1505–1519 (1998).

22. N.R. Lebovitz, R.J. Schaar, *Exchange of Stabilities in Autonomous Systems*, Stud. in Appl. Math. **54**:229–260 (1975).

23. N.R. Lebovitz, R.J. Schaar, *Exchange of Stabilities in Autonomous Systems II*, Stud. in Appl. Math. **56**:1–50 (1977).

24. E.F. Mishchenko, N.Kh. Rozov, *Differential Equations with Small Parameters and Relaxations Oscillations* (Plenum, New York, 1980).

25. W. Müller, K.R. Schneider, *Feedback Stabilization of Nonlinear Discrete-Time Systems*, J. Differ. Equations Appl. **4**:579–596 (1998).

26. N.N. Nefedov, K.R. Schneider, *Delayed exchange of stabilities in singularly perturbed systems*, Zeitschr. Angewandte Mathematik und Mechanik (ZAMM) **78**:199–202 (1998).

27. N.N. Nefedov, K.R. Schneider, *Immediate exchange of stabilities in singularly perturbed systems*, Diff. Integr. Equations, 1999 (to appear).

28. A.I. Neishtadt, *Persistence of stability loss for dynamical bifurcations I*, Diff. Equ. **23**:1385–1391 (1987).

29. A.I. Neishtadt, *Persistence of stability loss for dynamical bifurcations II*, Diff. Equ. **24**:171–176 (1988).

30. A.I. Neishtadt, *On Calculation of Stability Loss Delay Time for Dynamical Bifurcations* in D. Jacobnitzer Ed., *XI*th *International Congress of Mathematical Physics* (International Press, Boston, 1995).

31. L.S. Pontryagin, L.V. Rodygin, *Approximate solution of a system of ordinary differential equations involving a small parameter in the derivatives*, Dokl. Akad. Nauk SSSR **131**:237–240 (1960).

32. M.A. Shishkova, *Examination of system of differential equations with a small parameter in highest derivatives*, Dokl. Akad. Nauk SSSR **209**:576–579 (1973). [English transl.: Soviet Math. Dokl. **14**:384–387 (1973)].

33. F. Takens, *Forced oscillations and bifurcations*, Comm. Math. Inst., Rijksuniversiteit Utrecht **3**:1–59 (1974).

34. A.B. Vasil'eva, V.F. Butusov, L.V. Kalachev, *The Boundary Function Method for Singular Perturbation Problems* (SIAM, Philadelphia, 1995).

35. V. Venkatasubramanian, H. Schaettler, J. Zaborszky, *On the dynamics of differential-algebraic systems such as the balanced large electric power system*, in J.H. Chow (Ed.), *Systems and control theory for power systems* (Springer, New York, 1995).

36. E.V. Volokitin, S.A. Treskov, *Parametric portrait of the Fitz-Hugh differential system*, Math. Modelling **6**:65–78 (1994). (in Russian)

5. Extension of Popov criterion to time-varying nonlinearities: LMI, frequential and graphical conditions

Pierre-Alexandre Bliman

I.N.R.I.A. - Rocquencourt
Domaine de Voluceau, B.P. 105
78153 Le Chesnay Cedex, France
Phone: (33) 1 39 63 55 68, Fax: (33) 1 39 63 57 86
pierre-alexandre.bliman@inria.fr

Summary.

We present here some simple and tractable extensions of the classical absolute stability Popov criterion to multivariable systems with time-varying memoryless nonlinearities subject to sector conditions. The results apply to rational systems and delay systems. The proposed sufficient conditions are expressed in the frequency domain, a form well-suited for robustness issues, and lead to simple graphical interpretations for scalar systems. Apart from the usual conditions, the results assume basically a generalized sector condition on the derivative of the nonlinearities with respect to time. Results for local and global stability are given. For rational transfers, the frequency domain conditions are equivalent to some easy-to-check Linear Matrix Inequalities; this leads, for delay systems, to a tractable method of numerical resolution by approximation.

5.1 Introduction

We consider in this paper multivariable nonlinear control systems given by one of the following differential and functional differential equations

$$\dot{x} = Ax + Bu, \quad u = -\psi(t, y), \quad y = Cx \tag{5.1}$$

$$\dot{x} = \sum_{l=0}^{L} A_l x(t - h_l) + Bu, \quad u = -\psi(t, y) \quad y = \sum_{l=0}^{L} C_l x(t - h_l) \tag{5.2}$$

where $n, p \in \mathbb{N} \setminus \{0\}$, $L \in \mathbb{N}$, $x \in \mathbb{R}^n$, $y \in \mathbb{R}^p$, $A, A_l \in \mathbb{R}^{n \times n}$, $B \in \mathbb{R}^{n \times p}$, $C, C_l \in \mathbb{R}^{p \times n}$, $0 \leq h_0 < \cdots < h_L = h$. One denotes by H the matrix transfer function corresponding to the system under study, namely

$$H(s) = C(sI - A)^{-1}B\,, \tag{5.3}$$

$$H(s) = (\sum_{l=0}^{L} C_l e^{-h_l s})(sI - \sum_{l=0}^{L} A_l e^{-h_l s})^{-1}B\,. \tag{5.4}$$

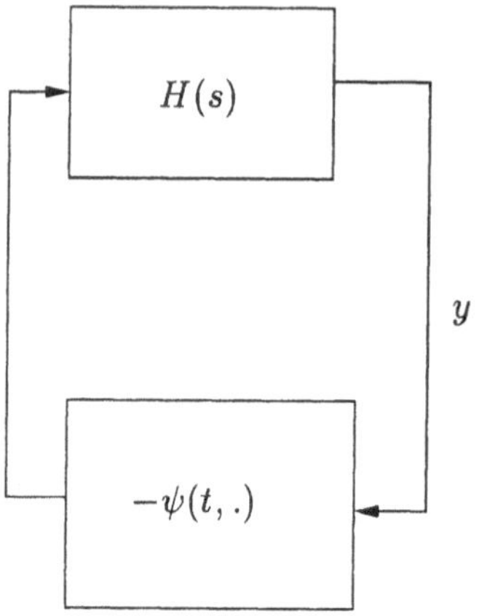

Fig. 5.1. The systems under study

One supposes throughout that the chosen representations are minimal [14]. Let the nonlinearity $\psi : \mathbb{R}^+ \times \mathbb{R}^p \to \mathbb{R}^p$ be time-dependent, *decentralized* [15] (that is: $\forall i \in \{1,\ldots,p\}$, $\psi_i(t,y) = \psi_i(t,y_i)$), and let it fulfill the following *sector condition*

$$\forall (t,y) \in \mathbb{R}^+ \times \mathbb{R}^p,\ \psi(t,y)^T(\psi(t,y) - Ky) \le 0 \tag{5.5}$$

for a certain diagonal matrix $K = \operatorname{diag}\{K_i\} \ge 0$. Our goal is to find sufficient conditions for absolute stability (resp. for absolute stability with finite domain [15]), that is conditions under which the uniform global (resp. local) asymptotic stability of the null equilibrium of equations (5.1) and (5.2) is guaranteed, for any nonlinearity ψ fulfilling (5.5).

As is well known, asymptotic stability of all the *linear time-invariant* systems obtained when choosing

$$\psi(t,y) = \operatorname{diag}\{k_i\}y, \quad 0 \le k_i \le K_i$$

is not even sufficient to deduce the stability of the *stationnary* systems fulfilling (5.5), see counterexamples with rational systems in [23] or [2, p. 86-88]. Circle criterion [8] furnishes sufficient conditions of absolute stability expressed in the frequency domain, namely

$$I + KH(s) \quad \text{is Strictly Positive Real (SPR)} \; . \tag{5.6}$$

Popov criterion [24], [25, 18] for delay systems, constitutes a refinement of the latter, valid only for *stationnary* nonlinearities fulfilling (5.5). Instead of assumption (5.6), it supposes that there exists a diagonal matrix η such that

$$I + (I + \eta s)KH(s) \quad \text{is SPR} \; . \tag{5.7}$$

Concerning absolute stability of systems with *time-varying nonlinearities*, one may consult [19] for a review of the period 1968-1977, surveying an important number of contributions, especially from Eastern Europe. Pyatnitskii has shown [26] that, at least for rational systems, this property is *equivalent* to the asymptotic stability of all the *time-varying linear* systems of the considered class, that is for the maps

$$\psi(t, y) = \operatorname{diag}\{k_i(t)\}y, \quad 0 \leq k_i(t) \leq K_i \; .$$

This characterization gives rise to a class of sufficient conditions for absolute stability of rational systems, (see e.g. [20], and also [16, 3] for some frequential conditions without restrictions on the rate of variation of the nonlinearity), but such an approach seems to have remained unexploited for delay systems.

On the other hand, it is possible to obtain absolute stability for smaller classes of time-varying nonlinearities, by making restrictions on $\frac{\partial \psi}{\partial t}$, an a priori knowledge which may be available, according to the system under study. This is for instance the case when one checks the stability of limit cycles, as some estimates on the periodic orbit may be available. An example of application of such results comes from the control of chaos [9]: in order to stabilize an unstable periodic orbit of a strange attractor, Pyragas [27, 28] proposed to use a feedback control built on the difference between the actual value and the delayed value of the output, with delay equal to the period of the cycle. The analysis of the corresponding closed loop system requires stability results for nonstationnary nonlinear delay systems.

The idea of restricting the variation of ψ has been applied in some papers, see e.g. [21, 32, 33], the second one in the context of delay systems. Criteria have been proposed in the context of rational systems by Rekasius *et al.* [30] and Hul'chuk *et al.* [13], and Bertoni *et al.* [4] proposed a slightly weaker result. In these papers, the restriction on the derivative of the nonlinearity wrt time takes the form of a *generalized sector condition* (see formula (5.11) below), and the generalization of Popov criterion is made by addition in (5.7) of some terms related to this sector.

In the present paper, we propose some simple and tractable absolute stability criteria, valid for systems with time-varying nonlinearities whose variation wrt time is constrained in the same manner than in [30, 13]. The presentation is unified for both rational and delay systems, of the type (5.1) and (5.2) respectively. More precisely, our main contribution is the following

- We provide for rational systems (5.1) the criterion given in [30, 13], under a form suitable for local stability results too. It is expressed as a Linear Matrix Inequality (LMI), a standard class of problems for which sound numerical methods have been developed [7].
- We extend the frequency domain criteria given in [30, 13, 5] to delay systems (5.2). We also show that for rational systems, the assumed frequency condition is equivalent to a LMI condition, slightly weaker than the previous one. This furnishes a tractable method of numerical resolution for the delay system (5.2), by rational approximation of the transfer (5.4).

Secondarily,

- The graphical criterion for scalar systems given in [4] is extended, and we provide a new one, which is weaker but situated in the Popov plane.

The results provided permit to link circle and Popov criteria when *no* variation wrt time of ψ is permitted, the nonlinearity is time-invariant, and Popov criterion applies; when *any* variation is permitted, circle criterion applies. The results herein fulfill the gap: they give sufficient conditions of stability adapted to the magnitude of $\frac{\partial \psi}{\partial t}$.

The paper is organized as follows. The LMI criterion for rational systems is stated in Section 5.2. The frequency domain criterion, valid for delay systems as well, is stated in Section 5.3, together with its LMI expression for rational systems. The weaker frequency domain criterion, which may be interpreted in the Popov plane, is stated in Section 5.4. All three criteria contain Popov and circle criteria as subcases. Then, some extensions and remarks are given in Section 5.5. The study of the absolute stability of a third order rational system is provided as an illustration in Section 5.6 (an illustrative example for delay systems may be found in [6]). Extended sketch of the proofs is presented in Section 5.7.

We do not consider here problems of existence and uniqueness of the solutions, as these questions have been extensively studied. Hence, in all the sequel, we only assume that there exists *global solutions of* (5.1) (resp. (5.2)), that is, *by definition*: for all $\phi \in \mathbb{R}^n$ (resp. for all $\phi \in \mathcal{C}([-h,0];\mathbb{R}^n)$), there exists a continuous function x defined on $[0,+\infty)$ (resp. on $[-h,+\infty)$), absolutely continuous [31] on $[0,+\infty)$, such that $x(0)=\phi$ (resp. $x|_{[-h,0]}=\phi$) and (5.1) (resp. (5.2)) is fulfilled almost everywhere on $[0,+\infty)$. The stability results given below concern the asymptotic behavior of these global solutions.

Notations. In all the paper, $||.||$ denotes the euclidian norm or the associated matrix norm, the asterisk * denotes complex conjugation, I_r stands for the

$r \times r$ identity matrix (simply I when no misunderstanding is possible) . For $z \in \mathbb{R}$, one denotes by $\operatorname{sgn} z$ the sign of z ($\operatorname{sgn} 0 = -1$ or $+1$ indifferently), and

$$|z|_+ \stackrel{\text{def}}{=} \sup\{z, 0\}, \quad |z|_- \stackrel{\text{def}}{=} \sup\{-z, 0\} .$$

At last, by convention, one extends the action of any map acting on scalar or scalar-valued functions to an operator acting on matrices or matrix-valued functions, obtained by componentwise application of the map. As an exemple, for any diagonal matrix η, one has

$$|\eta|_\pm = \sup\{\pm\eta, 0\} = \operatorname{diag}\{\sup\{\pm\eta_i, 0\}\} = \operatorname{diag}\{|\eta_i|_\pm\} .$$

5.2 A LMI criterion for rational systems

Theorem 5.2.1. [5] *Assume that there exists a convex open neighborhood $\mathcal{O}$ of 0 in $\mathbb{R}^p$ for which the following assumptions hold.*

(H0) *The function ψ is measurable and, for any $y \in \mathcal{O}$, $t \mapsto \psi(t, y)$ is locally Lipschitz (and hence t-a.e. differentiable), with a Lipschitz constant locally integrable wrt $y \in \mathcal{O}$.*

(H1) *The nonlinearity ψ is decentralized and there exists a diagonal matrix $K \geq 0$ such that*

$$\forall (t, y) \in \mathbb{R}^+ \times \mathcal{O}, \quad \psi(t, y)^T(\psi(t, y) - Ky) \leq 0 . \tag{5.8}$$

Assume that there exists diagonal matrices $D_j = \operatorname{diag}\{D_{j,i}\}$, $j \in \{1, 2, 3\}$, such that the following LMI is feasible

$$P > 0, \ \eta \stackrel{\text{def}}{=} \operatorname{diag}\{\eta_i\} \geq 0, \ R \stackrel{\text{def}}{=} R_P + \begin{pmatrix} 2C^T\eta D_1 KC & C^T K D_2 \eta \\ \eta D_2 KC & 2\eta D_3 K \end{pmatrix} < 0 \tag{5.9}$$

$$R_P \stackrel{\text{def}}{=} \begin{pmatrix} A^T P + PA & -PB + C^T K + A^T C^T K\eta \\ -B^T P + KC + \eta KCA & -2I - \eta KCB - B^T C^T K\eta \end{pmatrix} \tag{5.10}$$

and such that the following Hypothesis is fulfilled (with the same η and D_j)

(H2) *There exists $\gamma : \mathbb{R}^+ \to \mathbb{R}$ with $\lim_{z \to 0} \gamma(z) = 0$ such that for almost any $t \in \mathbb{R}^+$, $\forall y \in \mathcal{O}$, $\forall i \in \{1, \ldots, p\}$,*

$$\eta_i \left(\int_0^{y_i} \frac{\partial \psi_i}{\partial t}(t, z)\, dz - D_{1,i} y_i^2 - D_{2,i} y_i \psi_i(t, y_i) - D_{3,i} \psi_i(t, y_i)^2 \right) \tag{5.11}$$

$$\leq \|y\|^2\ \gamma(\|y\|)\ .$$

Then, the origin of system (5.1) *is uniformly locally asymptotically stable. Moreover, if* $\gamma \equiv 0$ *and* $\mathcal{O} = \mathbb{R}^p$*, then the origin of system* (5.1) *is uniformly globally asymptotically stable.*

Circle criterion is found as a particular case of Theorem 5.2.1 for $\eta = 0$, and Popov criterion when $\frac{\partial\psi}{\partial t} = 0$, taking $D_j = 0$ (in both cases, *via* Kalman-Yakubovich-Popov Lemma [17]).

Multiplication by η_i in (5.11) indicates that this constraint is inactive when $\eta_i = 0$. An important case where (5.11) is fulfilled is the case where there exists a measurable map Δ with diagonal matrix values, such that

$$\text{for almost any } t \in \mathbb{R}^+,\ \forall y \in \mathcal{O} \tag{5.12}$$

$$y^T\eta\left(\frac{\partial\psi}{\partial t}(t,y) - \Delta(t)y\right) \leq \|y\|^2\ \gamma(\|y\|)$$

and the matrices D_j are then given by

$$D_1 = \frac{1}{2}\sup_{t\geq 0}\text{ ess}\,\{\Delta(t)\},\ D_2 = D_3 = 0\ .$$

Matrix D_1 as given by the previous formula is nonnegative, otherwise sector condition (H1) would be violated. Condition (5.12) is a "local" sector condition (fulfilled e.g. if $\frac{\partial^2\psi_i}{\partial y_i\partial t}(t,0)$ exists a.e. and is equal to $\Delta_i(t)$), "global" when $\gamma \equiv 0$ and $\mathcal{O} = \mathbb{R}^p$.

5.3 A frequency domain criterion and its graphical interpretation

Solvability of (5.9), (5.10) is clearly a consequence of the solvability of the LMI

$$P > 0,\ \eta = \text{ diag}\{\eta_i\} \geq 0,\ R_P + \begin{pmatrix} 2C^T\eta|D_1|_+KC & C^TKD_2\eta \\ \eta D_2KC & 2\eta D_3K \end{pmatrix} < 0 \tag{5.13}$$

as $\eta, K \geq 0$. The previous LMI may indeed be *equivalently* transformed into a frequency condition, and this is the contents of the following Theorem. Kalman-Yakubovich-Popov (KYP) Lemma (as stated e.g. in [17, Theorems 1.10.1 and 1.11.1]) is used as a central argument in the demonstration. An important feature of Theorem 5.3.1 lies in the fact that this result is true for delay systems too.

Theorem 5.3.1. *Assume that there exists a convex open neighborhood $\mathcal{O}$ of 0 in $\mathbb{R}^p$ for which Hypotheses* (H0), (H1) *hold, together with*

(H3) *There exists $\alpha > 0$ such that the poles of the transfer H have real part smaller than $-\alpha$.*

Assume that there exists diagonal matrices $\eta \geq 0$ and D_j, $j \in \{1,2,3\}$, such that the transfer function matrix

$$I - \eta D_3 K + (I + \eta(sI + D_2))KH(s) - H^*(s)\eta|D_1|_+ KH(s) \qquad (5.14)$$

is SPR and such that Hypothesis (H2) *holds (with the same η and D_j). Then,*

- *In the case of the rational system* (5.1), *the hypotheses of Theorem* 5.2.1 *are fulfilled, and hence its conclusions hold.*
- *In the case of the delay system* (5.2), *the origin is uniformly locally asymptotically stable. Moreover, if $\gamma \equiv 0$ and $\mathcal{O} = \mathbb{R}^p$, then the origin of system* (5.2) *is uniformly globally asymptotically stable.*

The global stability result for rational systems expressed in Theorem 5.3.1 is the same than in [30, 13]. It is slightly weaker than Theorem 5.2.1: the results are indeed equivalent when $D_1 \geq 0$.

For a scalar system, $p = 1$, and condition (5.14) is equivalent to

$$\exists \eta \geq 0,\ \forall \omega \in \mathbb{R},\ \frac{1}{K} + \operatorname{Re} H(j\omega) - \eta \Big(D_3 - D_2 \operatorname{Re} H(j\omega) + \omega \operatorname{Im} H(j\omega) + |D_1|_+ |H(j\omega)|^2 \Big) \geq 0\,. \qquad (5.15)$$

The graphical interpretation generalizes the interpretation given in [4]:

> *If, apart from the regularity, sector and stability conditions* (H0), (H1), (H2), (H3), *there exists a line of slope $1/\eta \in \mathbb{R}^+ \cup \{+\infty\}$ passing through the point $(-\frac{1}{K}, 0)$ and lying to the left of the locus $(\operatorname{Re} H(j\omega), D_3 - D_2 \operatorname{Re} H(j\omega) + \omega \operatorname{Im} H(j\omega) + |D_1|_+ |H(j\omega)|^2)$ without intersecting it, then the uniform local stability property holds. If $\gamma \equiv 0$ and $\mathcal{O} = \mathbb{R}^p$, then the uniform global stability property holds.*

One may effectively check the applicability of Theorem 5.3.1 for rational systems, as this relies to solve LMI (5.10), (5.13). So, in order to apply the result to delay systems, it suffices to approximate the actual transfer by rational transfers, using e.g. techniques developed in [22, 12]. A proper statement of the transfer approximation property is provided in [6].

5.4 A weaker frequency domain criterion and its interpretation in Popov plane

An interesting problem is, given K, to determinate the largest incertitude on $\frac{\partial\psi}{\partial t}$ under which absolute stability may be proved. In this case, the graphical criterion deduced from Theorem 5.3.1 can hardly be used, as the ordinate changes with the D_j. To overcome this drawback, condition (5.14) should rather be seen as a geometrical condition in the 3-dimensional space $(\operatorname{Re} H(j\omega), \operatorname{Im} H(j\omega), \omega \operatorname{Im} H(j\omega))$ obtained as the product of Nyquist and Popov planes, a condition not easy to interpret. We present in the sequel a weaker but simpler condition, located in the Popov plane. As usual, $\mathcal{H}_\infty$-norm of the transfer function H is denoted by $\|H\|_\infty \stackrel{\text{def}}{=} \sup\{\|H(s)\| \ : \ \operatorname{Re} s > 0\}$ (when H is stable and proper, this is equal to $\sup\{\|H(j\omega)\| \ : \ \omega \in \mathbb{R}\}$).

Theorem 5.4.1. *Assume that there exists a convex open neighborhood $\mathcal{O}$ of 0 in $\mathbb{R}^p$ for which Hypotheses* (H0), (H1), (H3) *hold. Assume that there exists diagonal matrices $\eta \geq 0$ and D_j, $j \in \{1,2,3\}$, such that the transfer function matrix*

$$I - \eta K D_3 + (I + \eta(sI + D_2))KH(s) - \eta|D_1|_+ K\|H\|_\infty^2 \quad \textit{is SPR} \quad (5.16)$$

and such that Hypothesis (H2) *holds. Then, the conclusions of Theorem* 5.3.1 *hold.*

The proof is straightforward, and left to the reader. One shows as in Section 5.3, that condition (5.16) is equivalent to the feasibility of the LMI

$$P > 0, \ \eta \geq 0, \ R_P + \begin{pmatrix} 0 & C^T K D_2 \eta \\ \eta D_2 K C & 2\eta(D_3 + |D_1|_+ \|H\|_\infty^2)K \end{pmatrix} < 0 \qquad (5.17)$$

for the rational system (5.1).

As an example let us examine the case of a scalar system fulfilling (5.12), the general case (5.11) is similar. Formula (5.16) is equivalent to

$$\exists \eta \geq 0, \ \forall \omega \in \mathbb{R}, \ \frac{1}{K} + \operatorname{Re} H(j\omega) \\ - \eta \left(\omega \operatorname{Im} H(j\omega) + \frac{1}{2} \operatorname*{sup\ ess}_{t \geq 0} \{\Delta(t)\} \ \|H\|_\infty^2 \right) \geq 0 \quad (5.18)$$

and this has a clear interpretation

> *If, apart from the regularity, sector and stability conditions* (H0), (H1), (H2), (H3), *a line of slope $1/\eta \in \mathbb{R}^+ \cup \{+\infty\}$ passing through*

the point $(-\frac{1}{K}, 0)$ *lies above the Popov locus and may be translated vertically towards the locus by a distance*

$$\frac{1}{2} \sup_{t \geq 0} \text{ess} \, \{\Delta(t)\} \, \|H\|_\infty^2$$

without intersecting it, then the uniform local stability property holds. If $\gamma \equiv 0$ *and* $\mathcal{O} = \mathbb{R}^p$*, then the uniform global stability property holds.*

This is illustrated in Figure 5.2: in the Popov diagram, (5.18) holds if

$$\sup_{t \geq 0} \text{ess} \, \Delta(t) < \frac{2d}{\|H\|_\infty^2} .$$

In the configuration shown in Figure 5.2, the quantity d involved is indeed

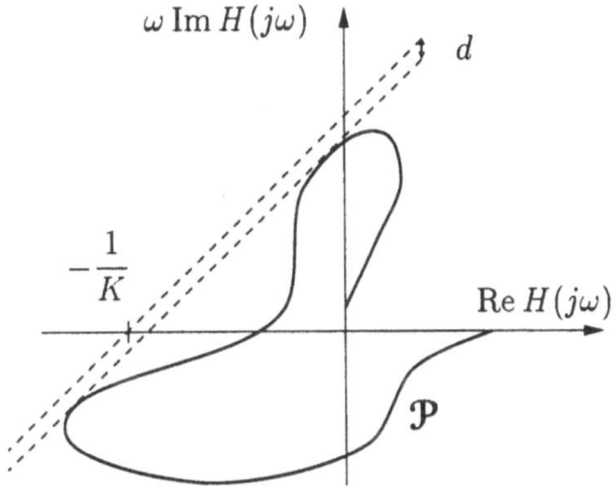

Fig. 5.2. Graphical stability criterion in the Popov plane

the least $z > 0$ *such that the point* $(-1/K, -z)$ *belongs to the convex hull of the Popov locus* $\mathcal{P}$, see [6].

5.5 Extensions and remarks

When Hypothesis (H2) is fulfilled with a diagonal matrix η which is not nonnegative, one may consider instead of $\psi(t, y)$ the new input

$$\hat{\psi}(t, y) \stackrel{\text{def}}{=} \text{sgn}\, \eta \; \psi(t, y) + \frac{1}{2}(I - \text{sgn}\, \eta) K y$$

and apply all the previous results to the transformed system [29]. Direct computations show that this amounts to apply the criteria stated above to the system obtained when replacing

$$
\begin{array}{lll}
\eta & & \hat{\eta} \stackrel{\text{def}}{=} |\eta| \,. \\
A & & \hat{A} \stackrel{\text{def}}{=} A - B\frac{I-\operatorname{sgn}\eta}{2}KC \\
A_l & & \hat{A}_l \stackrel{\text{def}}{=} A_l - B\frac{I-\operatorname{sgn}\eta}{2}KC_l \\
B & \text{by} & \hat{B} \stackrel{\text{def}}{=} B \operatorname{sgn}\eta \\
D_1 & & \hat{D}_1 \stackrel{\text{def}}{=} \operatorname{sgn}\eta\ D_1 - \frac{I-\operatorname{sgn}\eta}{2}K(D_2+KD_3) \\
D_2 & & \hat{D}_2 \stackrel{\text{def}}{=} D_2 + (I - \operatorname{sgn}\eta)KD_3 \\
D_3 & & \hat{D}_3 \stackrel{\text{def}}{=} \operatorname{sgn}\eta\ D_3 \,.
\end{array}
$$

To verify this [6], check that

$$
Ax - B\psi(t,y) = \hat{A}x - \hat{B}\hat{\psi}(t,y)
$$

$$
(\text{resp. } \sum_{l=0}^{L} A_l x(t-h_l) - B\psi(t,y) = \sum_{l=0}^{L} \hat{A}_l x(t-h_l) - \hat{B}\hat{\psi}(t,y)\ )
$$

when $y = Cx$ (resp. $y = \sum_{l=0}^{L} C_l x(t-h_l)$), and that the following identities are valid

$$
\psi(t,y)^T(\psi(t,y) - Ky) = \hat{\psi}(t,y)^T(\hat{\psi}(t,y) - Ky)
$$

$$
\eta_i \frac{\partial \psi_i}{\partial t}(t,y) = |\eta_i| \frac{\partial \hat{\psi}_i}{\partial t}(t,y)
$$

$$
\begin{aligned}
&\eta_i(D_{1,i}y_i^2 + D_{2,i}y_i\psi_i(t,y_i) + D_{3,i}\psi_i(t,y_i)^2) \\
&\qquad = |\eta_i|(\hat{D}_{1,i}y_i^2 + \hat{D}_{2,i}y_i\hat{\psi}_i(t,y_i) + \hat{D}_{3,i}\hat{\psi}_i(t,y_i)^2) \,.
\end{aligned}
$$

An interesting feature is the possibility to express frequency condition (5.14) in terms of the data rather than the hatted quantities [6], under the form

$$
\begin{aligned}
&I - \eta KD_3 + (I + \eta(sI + D_2))KH(s) \\
&\qquad - H^*(s) \sup\left\{\eta D_1; \frac{|\eta| - \eta}{2}K(D_2 + KD_3)\right\} KH(s) \quad \text{is SPR} \,.
\end{aligned}
$$

Remark 5.5.1. • The hypotheses of the criteria do not imply continuity wrt y neither of ψ (except in 0), nor of $\frac{\partial \psi}{\partial t}$. • When using Hypothesis (H2), one may wish to consider matrices D_j, $j \in \{1,2,3\}$ fulfilling (5.11) only on $[t_0, +\infty)$ for a certain $t_0 > 0$, or even "at infinity", that is taking $t_0 \to +\infty$. Due to the strict inequalities involved, this is licit, provided, for the local stability results, that the flow is continuous in the neighborhood of 0.

5.6 Computation of stability margin for a 3rd order rational system

We present here an example for rational systems. Let us consider the system

$$\dddot{y} + 6\ddot{y} + 11\dot{y} + 6y = -\psi(t, y) \tag{5.19}$$

which may be realized as in (5.1) with

$$A \stackrel{\text{def}}{=} \begin{pmatrix} 0 & 1 & 0 \\ 0 & 0 & 1 \\ -6 & -11 & -6 \end{pmatrix}, B \stackrel{\text{def}}{=} \begin{pmatrix} 0 \\ 0 \\ 1 \end{pmatrix}, C \stackrel{\text{def}}{=} (1\ 0\ 0) \ .$$

One supposes that ψ fulfills Hypotheses (H0), (H1). Application of circle criterion provides stability if (see Figure 5.3)

$$K < \frac{1}{0.03578} \simeq 27.95 \ .$$

On the other hand, Popov criterion guarantees stability if $\psi(t, y) = \psi(y)$

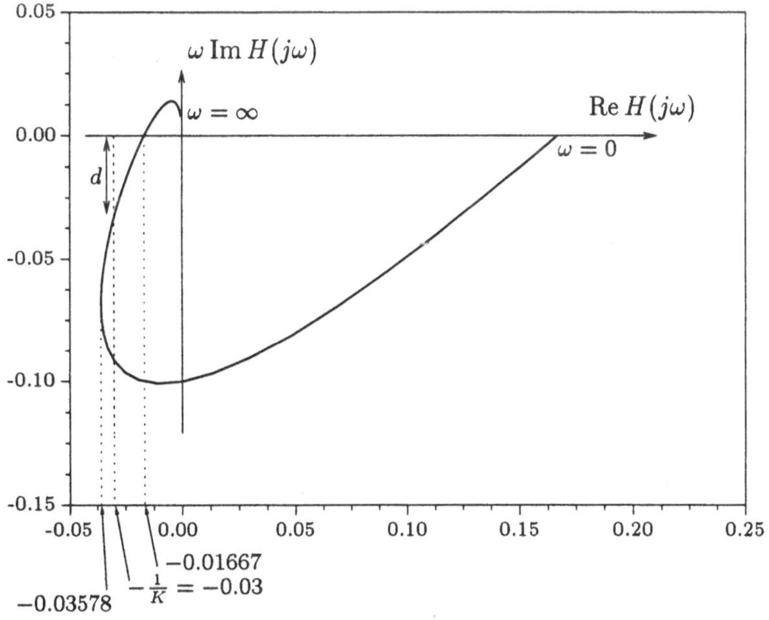

Fig. 5.3. Popov locus of system (5.19)

and

$$K < 60 \simeq \frac{1}{0.01667} .$$

In the remaining of the section, one studies the stability of (5.19) for time-varying nonlinearities with sector

$$K = 1/0.0300 \simeq 33.33 \in [27.95, 60] .$$

In view of the previously exposed results, one is looking for bounds on $\Delta(t)$ ensuring stability. All computations to be presented have been achieved using the Scilab package LMITOOL[1].

Using Theorems 5.2.1 or 5.3.1 and proceeding by dichotomy, one may solve the LMI (5.9), (5.10) if

$$\sup_{t\geq 0} \text{ess } \Delta(t) \leq \delta_{1,2} \stackrel{\text{def}}{=} 35.66 .$$

The graphical interpretation of the criterion may be seen in Figure 5.4.

Now, one computes a bound using Theorem 5.4.1. First, one evaluates either graphically (see Figure 5.3), or using a solver of algebraic equations, the quantity d. One gets

$$d \simeq 0.03325 .$$

One then computes $||H||_\infty$, and obtains

$$||H||_\infty \simeq 0.1667 .$$

The bound is now given by

$$\sup_{t\geq 0} \text{ess } \Delta(t) \leq \delta_3 \stackrel{\text{def}}{=} \frac{2d}{||H||_\infty^2} \simeq 2.394 .$$

This may be computed independently by use of the LMI (5.10), (5.17).

One verifies that the ordering

$$\delta_{1,2} > \delta_3$$

is consistent with the increasing conservativeness of the criteria.

Let us give a sample of the results that may be obtained. Let $\mathcal{O}$ be a convex open neighborhood of 0 in $\mathbb{R}$, such that

[1] Scilab is a free software developed by INRIA, which is distributed with all its source code. For the distribution and details, see Scilab's homepage on the web at the address `http://www-rocq.inria.fr/scilab/`

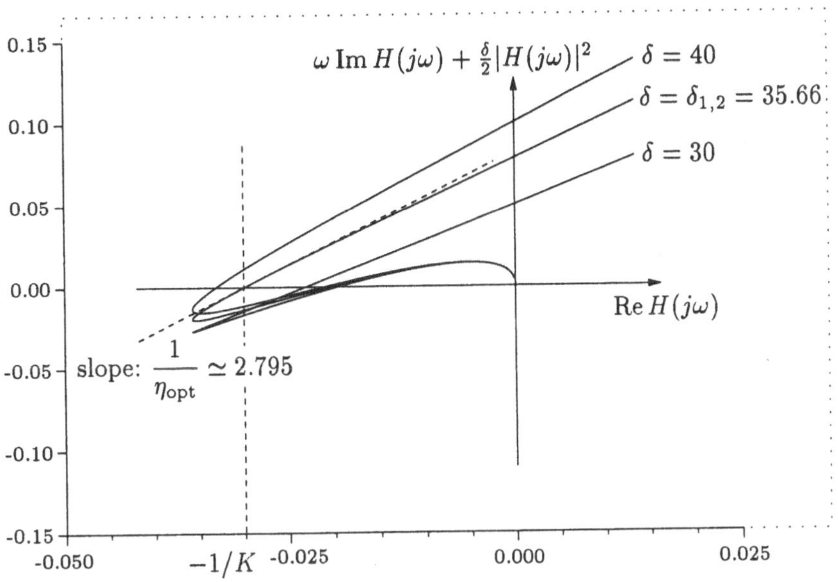

Fig. 5.4. Graphical interpretation of Theorem 5.3.1 for system (5.19)

- There exists $L \in L^1_{loc}(\mathcal{O})$ such that, for all $t, t' \in \mathbb{R}^+$, for all $y \in \mathcal{O}$, $|\psi(t,y) - \psi(t',y)| \leq L(y)|t - t'|$ (condition (H0)).
- For all $t \in \mathbb{R}^+$, for all $y \in \mathcal{O} \setminus \{0\}$, $0 \leq \dfrac{\psi(t,y)}{y} \leq \dfrac{1}{0.03}$, and $\psi(t,0) \equiv 0$ (condition (H1)).

Assume that there exists global solutions of system (5.19). The origin of system (5.19) is uniformly locally stable if

$$\limsup_{y \to 0} \frac{1}{y} \frac{\partial \psi}{\partial t}(t,y) \leq \delta_{1,2} \quad t - \text{a.e.}$$

for example if $\frac{\partial^2 \psi}{\partial y \partial t}(t,0)$ exists t-a.e. and verifies

$$\frac{\partial^2 \psi}{\partial y \partial t}(t,0) \leq \delta_{1,2} \;\; t - \text{a.e.}$$

The origin of system (5.19) is uniformly globally stable if $\mathcal{O} = \mathbb{R}$ and

$$\forall y \in \mathbb{R} \setminus \{0\} \quad \frac{1}{y} \frac{\partial \psi}{\partial t}(t,y) \leq \delta_{1,2} \;\; t - \text{a.e.}$$

5.7 Extended sketch of the proofs

We give in the sequel extended sketch of the proofs of the local stability results. The global stability results are proved along the same lines.

Proof of Theorem 5.2.1

Denote $K = \operatorname{diag}\{K_i\}$, and consider the candidate Lyapunov function

$$V(t,x) \stackrel{\text{def}}{=} x^T P x + 2\sum_{i=1}^{p} \eta_i K_i \int_0^{(Cx)_i} \psi_i(t,z)\, dz\ . \tag{5.20}$$

Using (H1) and denoting abusively $\dot{V} = \frac{d}{dt}[V(t,x(t))]$, one deduces that, as long as $y(t) \in \mathcal{O}$,

$$\dot{V} \leq \begin{pmatrix} x \\ \psi(t,y) \end{pmatrix}^T R_P \begin{pmatrix} x \\ \psi(t,y) \end{pmatrix} + 2\sum_{i=1}^{p} \eta_i K_i \int_0^{y_i(t)} \frac{\partial \psi_i}{\partial t}(t,z)\, dz \quad t\text{-a.e.},$$

where R_P is defined by (5.10). When proving Popov criterion, only the quadratic term is present in the derivative of V, and application of Kalman-Yakubovich-Popov (KYP) Lemma then leads to (5.7).

Using now Hypothesis (H2) shows that, as long as $y(t) \in \mathcal{O}$,

$$\dot{V} \leq \begin{pmatrix} x \\ \psi(t,y) \end{pmatrix}^T R \begin{pmatrix} x \\ \psi(t,y) \end{pmatrix} + o(\|y\|^2) \quad t\text{-a.e.},$$

where R is defined in (5.9). On the other hand, for any $(t,x) \in \mathbb{R}^+ \times \mathbb{R}^n$,

$$x^T P x \leq V(t,x) \leq x^T (P + C^T K \eta K C) x\ . \tag{5.21}$$

One then deduces that there exists $\varepsilon > 0$ such that,

$$\dot{V} + \varepsilon V \leq 0$$

as long as $Cx(t) \in \mathcal{O}$. The hypotheses imply that $t \mapsto V(t,x(t))$ is absolutely continuous, because $\forall t, t' \in \mathbb{R}^+$ s.t. $y(t), y(t') \in \mathcal{O}$, $\forall i \in \{1,\dots,p\}$,

$$\left| \int_0^{y_i(t)} \psi_i(t,z)\, dz - \int_0^{y_i(t')} \psi_i(t',z)\, dz \right| \leq |t-t'| \int_0^{y_i(t)} \lambda_i(z)\, dz$$
$$+ K_i \max\{|y_i(t)|, |y_i(t')|\}\ |y_i(t) - y_i(t')|$$

where λ_i is the Lispchitz constant of ψ, defined by Hypothesis (H1). This implies

$$V(t,x(t))\, e^{\varepsilon t} - V(0,x(0)) = \int_0^t (\dot{V} + \varepsilon V)\, e^{\varepsilon\tau}\, d\tau \leq 0,$$

as long as $Cx(t) \in \mathcal{O}$. The hypotheses on the set $\mathcal{O}$, together with (5.21), permit to conclude that system (5.1) is uniformly locally exponentially stable.

Proof of Theorem 5.3.1 for rational systems

The principle of the proof consists in showing that (5.14) is realized for a certain $\eta \geq 0$ *iff* LMI (5.10), (5.13) is feasible.

Application of KYP Lemma (see [17, Theorems 1.10.1 and 1.11.1]) to this LMI shows that its solvability is equivalent to frequency domain condition (5.14). It hence suffices to show the applicability of KYP Lemma to the present situation. This is done in the sequel.

The pair $\{A, B\}$ is controllable, due to the minimality of the representation (A, B, C). Let us now check the observability of the pair $\{KC + \eta KCA + \eta D_2 KC, A\}$. It is obvious that the map

$$\eta \mapsto \det \begin{bmatrix} KC + \eta KCA + \eta D_2 KC \\ (KC + \eta KCA + \eta D_2 KC)A \\ \cdots \\ (KC + \eta KCA + \eta D_2 KC)A^{n-1} \end{bmatrix}$$

defined for diagonal matrices η, is indeed a polynomial function of the diagonal elements of η. This polynomial is not zero, as it is worth

$$\det \begin{bmatrix} KC \\ KCA \\ \cdots \\ KCA^{n-1} \end{bmatrix}$$

when $\eta = 0$, a quantity which is nonzero due to observability of the pair $\{C, A\}$ and invertibility of K. One deduces that the preceding map takes on nonzero values on any neighborhood of the set of the nonnegative diagonal matrices: it then suffices to take an η close enough to the one verifying (5.14) to conclude in favour of the asymptotic stability.

Proof of Theorem 5.3.1 for delay systems

The proof of Theorem 5.3.1 is inspired from [25, 11], with adequate improvements. A complete version may be found in [6]. The principle is the following. Define ψ_T, x_T, y_T as follows

$$\psi_T(t) = \psi(t, y(t)) \text{ if } 0 \leq t \leq T,\ \psi_T(t) = 0 \text{ if } -h \leq t < 0 \text{ or } t > T$$

$$\dot{x}_T = \sum_{l=0}^{L} A_l x_T(t - h_l) - B\psi_T, \quad y_T = \sum_{l=0}^{L} C_l x_T(t - h_l), \quad x_T|_{[-h,0]} = 0 \ .$$

By linearity, we have

$$\dot{x} - \dot{x}_T = \sum_{l=0}^{L} A_l(x(t-h_l) - x_T(t-h_l)) \quad \text{for } t \in [0,T]$$

$$(x - x_T)|_{[-h,0]} = \phi$$

¿From Hypothesis (H3), one deduces that there exists $c_1 > 0$, $\alpha' \in (0,\alpha)$, independent of ϕ and T, such that

$$\forall t \in [-h,T], \quad \|x(t) - x_T(t)\| \leq c_1 e^{-\alpha' t} \|\phi\|_{C([-h,0])} \tag{5.22a}$$

$$\forall t \geq T, \quad \|x_T(t)\| \leq c_1 e^{-\alpha'(t-T)} \|x_T(T+\cdot)\|_{C([-h,0])} \tag{5.22b}$$

$$\forall t \in [0,T], \quad \|y(t)\| \leq c_1 e^{-\alpha' t} \|\phi\|_{C([-h,0])} + \|y_T(t)\| \,. \tag{5.22c}$$

As in the proof of Theorem 5.2.1, the map $T \mapsto \int_0^{y_i(T)} \psi_i(T,z)\, dz$ is absolutely continuous, whence the identity

$$\begin{aligned}
&\int_0^{y_i(T)} \psi_i(T,z)\, dz \\
&= \int_0^T \frac{d}{dt}\left[\int_0^{y_i(t)} \psi_i(t,z)\, dz\right] dt + \int_0^{y_i(0)} \psi_i(0,z)\, dz \\
&= \int_0^T \left(\dot{y}_i(t)\psi_{T,i}(t) + \int_0^{y_i(t)} \frac{\partial \psi_i}{\partial t}(t,z)\, dz\right) dt + \int_0^{y_i(0)} \psi_i(0,z)\, dz \,.
\end{aligned}$$

Let $\rho_y > 0$ be such that the open ball of $\mathbb{R}^p$ centered in 0 and of radius ρ_y is included in $\mathcal{O}$. ¿From the inequality

$$\int_0^T (K_i y_i(t) - \psi_i(t, y_i(t)))\psi_i(t, y_i(t))\, dt + \eta_i K_i \int_0^{y_i(T)} \psi_i(T,z)\, dz \geq 0$$

valid for any $T > 0$, any $i \in \{1,\ldots,p\}$, as soon as

$$\forall t \in [0,T], \quad \|y(t)\| < \rho_y \tag{5.23}$$

and from estimates (5.22a) to (5.22c), one deduces by summation on the index i, that

$$\begin{aligned}
&\int_0^{+\infty} \Big(\psi_T^T(t)(Ky_T(t) - \psi_T(t)) + \psi_T^T(t)\eta K \dot{y}_T(t) \\
&\qquad + y_T^T(t)\eta |D_1|_+ K y_T(t) + \psi_T^T(t)\eta K(D_2 y_T(t) + D_3\psi_T(t))\Big)\, dt \\
&\geq c_2 \Bigg(\|\phi\|_{C([-h,0])}\left(\|\phi\|_{C([-h,0])} + \sup_{t\in[0,T]} \|y_T(t)\|\right) \\
&\qquad + \sup_{t\in[0,T]} \gamma(\|y(t)\|) \int_0^T \|y_T(t)\|^2\Bigg) \,.
\end{aligned}$$

In the previous indefinite integral, the terms with $\psi_T(t)$ vanishe on $[T, +\infty)$, and one has bounded D_1 by $|D_1|_+$. Performing the Fourier transform of the left hand side of the previous inequality, using (H2) and the identity $\tilde{y}_T(\omega) = -H(j\omega)\tilde{\psi}_T(\omega)$ linking Fourier transforms, gets $\exists \varepsilon > 0$ such that

$$\frac{\varepsilon}{\|H\|_\infty^2} \int_0^{+\infty} \|y_T(t)\|^2 \, dt \leq \varepsilon \int_0^{+\infty} \|\psi_T(t)\|^2 \, dt$$
$$\leq c_2 \left(\|\phi\|_{C([-h,0])} \left(\|\phi\|_{C([-h,0])} + \sup_{t\in[0,T]} \|y_T(t)\| \right) \right.$$
$$\left. + \sup_{t\in[0,T]} \gamma(\|y(t)\|) \int_0^T \|y_T(t)\|^2 \right)$$

as long as (5.23) holds. Suppose additionally that

$$\forall t \in [0,T], \; c_2\gamma(\|y(t)\|) \leq \frac{\varepsilon}{2\|H\|_\infty^2} \,. \tag{5.24}$$

Then

$$\int_0^T \|y_T(t)\|^2 \, dt, \; \int_0^T \|\dot{y}_T(t)\|^2 \, dt$$
$$\leq c_3 \|\phi\|_{C([-h,0])} \left(\|\phi\|_{C([-h,0])} + \sup_{t\in[0,T]} \|y_T(t)\| \right) \,. \tag{5.25}$$

The estimate on y_T in (5.25) is obtained directly, and the estimate on $\dot{y}_T$ is then deduced, with the help of sector estimate (5.8) and the fact that H is strictly proper. One infers, using Cauchy-Schwarz inequality and $y_T(0) = 0$, that (5.23), (5.24) imply, for any $t \in [0,T]$:

$$\|y_T(t)\|^2 \leq c_3 \|\phi\|_{C([-h,0])} \left(\|\phi\|_{C([-h,0])} + \sup_{t\in[0,T]} \|y_T(t)\| \right) \,.$$

Solving the polynomial inequality leads to

$$T \text{ fulfills (5.23), (5.24)} \;\Rightarrow\; \sup_{t\in[0,T]} \|y_T(t)\|, \; \sup_{t\in[0,T]} \|y(t)\| \leq c_4 \|\phi\|_{C([-h,0])} \,.$$

Now, let $\rho_x > 0$ be such that (recall that $\gamma(z) \to 0$ when $z \to 0$)

$$\rho_x \leq \frac{\rho_y}{2c_4} \quad \text{and} \quad \forall z \in \mathbb{R}^+, \; z \leq c_4\, \rho_x \;\Rightarrow\; c_2\gamma(z) \leq \frac{\varepsilon}{4\|H\|_\infty^2} \,.$$

For $\phi \in C([-h,0];\mathbb{R}^n)$ with $\|\phi\|_{C([-h,0])} < \rho_x$, the previous computations show that, as long as (5.23), (5.24) are verified, one has

$$\sup_{t\in[0,T]} \|y(t)\| \leq c_4 \|\phi\|_{C([-h,0])} \leq c_4\ \rho_x$$

so

$$c_2 \sup_{t\in[0,T]} \gamma(\|y(t)\|) \leq \frac{\varepsilon}{4\|H\|_\infty^2} < \frac{\varepsilon}{2\|H\|_\infty^2}$$

and

$$\sup_{t\in[0,T]} \|y(t)\| \leq c_4\ \rho_x \leq \frac{\rho_y}{2} < \rho_y\ .$$

Hence, (5.23), (5.24) are verified for any $T > 0$, so

$$\|\phi\|_{C([-h,0])} < \rho_x \Rightarrow \forall T \geq 0 \quad \sup_{t\in[0,T]} \|y_T(t)\| \leq c_4 \rho_x\ .$$

This in turn implies by (5.25), that, for any $T > 0$,

$$\int_0^T \|y_T(t)\|^2\ dt,\ \int_0^T \|\dot{y}_T(t)\|^2\ dt \leq c_5 \|\phi\|_{C([-h,0])}^2\ .$$

From (5.22c), one deduces that similar inequalities hold for y and $\dot{y}$. One concludes that $y(t) \to 0$ when $t \to +\infty$, which expresses the uniform local asymptotic stability of the origin.

References

1. M.A. Aizerman, On the effect of nonlinear functions of several variables on the stability of automatic control problems, *Avtomat. i telemeh.*, **8**, 1, 20-29, 1947 [Russian]

2. M.A. Aizerman, F.R. Gantmacher, *Absolute stability of regulator systems*, Holden-Day Inc., 1964

3. N.E. Barabanov, New frequency criteria for absolute stability and instability of automatic-control systems with nonstationary nonlinearities, *Differential Equations* **25**, no 4, 367-373, 1989

4. G. Bertoni, C. Bonivento, E. Sarti, A graphical method for investigating the absolute stability of time-varying systems, *Atti Accad. Sci. Ist. Bologna Cl. Sci. Fis. Rend.* (12) 7, fasc. 1, 54-71, 1969/1970

5. P.-A. Bliman, A.M. Krasnosel'skii, Popov absolute stability criterion for time-varying multivariable nonlinear systems, to appear in the *Proc. of 5th Eur. Cont. Conf.*, Karlsruhe (Germany), 1999

6. P.-A. Bliman, *Extension of Popov absolute stability criterion to nonautonomous systems with delays*, INRIA Report no 3625 (available in `http://www.inria.fr/RRRT/RR-3625.html`), February 1999

7. S. Boyd, L. El Ghaoui, E. Feron, V. Balakrishnan, *Linear matrix inequalities in system and control theory*, SIAM Studies in Applied Mathematics vol. 15, 1994

8. C.A. Desoer, M. Vidyasagar, *Feedback systems: input-output properties*, Academic Press, 1975

9. A.L. Fradkov, A.Yu. Pogromsky, *Introduction to control of oscillation and chaos*, World Scientific Publishing Co, 1998

10. Lj.T. Grujić, P. Borne, J.C. Gentina, Matrix approaches to the absolute stability of time-varying Lurie-Postnikov systems, *Internat. J. Control* **30**, no 6, 967-980, 1979

11. A. Halanay, *Differential equations: stability, oscillations, time lags*, Academic Press, New York-London, 1966

12. A. Halanay, Vl. Răsvan, Approximations of delays by ordinary differential equations, *Recent advances in differential equations*, R. Conti ed., Academic Press, New York-London, 155-197, 1981

13. H.H. Hul'chuk, M.M. Lychak, Absolute stability of nonlinear control systems with nonstationary nonlinearities and tachometer feedback, *Soviet Automat. Control* **5**, no 4, 6-9, 1972

14. T. Kailath, *Linear systems*, Prentice-Hall, 1980

15. H.K. Khalil, *Nonlinear systems*, Macmillan Publishing Company, 1992

16. G.A. Leonov, Extension of Popov's frequency criterion for nonstationary nonlinearities, *Automat. Remote Control* **41** (1980), no 11, part 1, 1494-1499, 1981

17. G.A. Leonov, D.V. Ponomarenko, V.B. Smirnova, *Frequency-domain methods for nonlinear analysis: theory and applications*, World Scientific Publishing Co, 1996

18. X.-J. Li, On the absolute stability of systems with time lags, *Chinese Math.* **4**, 609-626, 1963

19. M.R. Liberzon, New results on absolute stability of nonstationary controlled systems (survey), *Automat. Remote Control* **40** (1979), no 8, part 1, 1124-1140, 1980

20. M.R. Liberzon, The inner feature of absolute stability of nonstationary systems. *Automat. Remote Control* **50**, no 5, part 1, 608-613, 1989

21. K.S. Narendra, J.H. Taylor, *Frequency domain criteria for absolute stability*, Academic Press, New-York London, 1973

22. J.R. Partington, K. Glover, H.J. Zwart, R.F. Curtain, L_∞ approximation and nuclearity of delay systems, *Systems Control Lett.* **10**, no 1, 59-65, 1988

23. V.A. Pliss, *Certain problems in the theory of stability in the whole*, Publisher:

Leningrad State University (LUG), 1958 [Russian]

24. V.M. Popov, Absolute stability of nonlinear systems of automatic control, *Automat. Remote Control*, **22**, 857-875, 1961

25. V.M. Popov, A. Halanay, On the stability of nonlinear automatic control systems with lagging argument, *Automat. Remote Control* **23**, 783-786, 1962

26. E.S. Pyatnitskii, Absolute stability of nonstationary nonlinear systems. *Automat. Remote Control*, no 1, 1-9, 1970

27. K. Pyragas, Continuous control of chaos by self-controlling feedback, *Phys. Lett. A* **170**, 421-428, 1992

28. K. Pyragas, Control of chaos via extended delay feedback, *Phys. Lett. A* **206**, no 5-6, 323-330, 1995

29. Z.V. Rekasius, J.E. Gibson, Stability analysis of nonlinear control systems by the second method of Lyapunov, *IRE Trans. Automatic Control* **AC-7**, no 1, 3-15, 1962

30. Z.V. Rekasius, J.R. Rowland, A stability criterion for feedback systems containing a single time-varying nonlinear element, *IEEE Trans. Automatic Control*, 352-354, 1965

31. W. Rudin, *Real and complex analysis*, 3rd edition, McGraw-Hill, 1987

32. J.A. Walker, Stability of feedback systems involving time delays and a time-varying nonlinearity, *Int. J. of Control* **6**, no 4, 365-372, 1967

33. J.A. Walker, Liapunov functions for feed-back systems containing a single time-varying non-linearity, *Int. J. of Control* **7**, no 2, 171-174. 1968

6. Uniqueness of control sets for perturbations of linear systems

Fritz Colonius and Marco Spadini***

*Institut für Mathematik
Universität Augsburg
86135 Augsburg, Germany
colonius@math.uni-augsburg.de

**Dipartimento di Matematica Applicata 'G. Sansone'
Via S. Marta 3
I-50139 Firenze, Italy
spadini@alibaba.math.unifi.it

Summary.

Linear systems with controllable (A, B) and bounded control range have a unique control set. This control set is bounded if and only if A is hyperbolic. Then uniqueness also remains valid under small nonlinear perturbations. Examples show that for nonhyperbolic A small nonlinear perturbations may lead to infinitely many (invariant) control sets.

6.1 Introduction

In this paper, we study the control sets for small nonlinear perturbations of linear control processes. More precisely we consider the maximal subsets of the state space $\mathbf{R}^d$ where complete controllability of the following perturbation of a linear control process (with restricted controls) holds

$$\dot{x}(t) = Ax(t) + Bu(t) + \varepsilon F\big(u(t), x(t), \varepsilon\big), \qquad u(t) \in U, \tag{6.1}$$

where U is a compact and convex subset of $\mathbf{R}^m$ with nonvoid interior, and A and B are constant matrices of respective dimensions $d \times d$ and $m \times d$. We assume that the pair (A, B) is controllable, i.e., rank $\left[B, AB, \ldots, A^{d-1}B\right] = d$, and that F is a C^1-function. We also assume that

$$\|D_1 F\| \leq M_1 \text{ and } \|D_2 F\| \leq M_2 \text{ uniformly.} \tag{6.2}$$

Throughout we assume that for all $x_0 \in \mathbf{R}^d$ and all controls u there exists a unique solution $\varphi(t, x_0, u)$, $t \in \mathbf{R}$, of (6.1) with initial value $\varphi(0, x_0, u) = x_0$.

The term $u(t)$ may be interpreted as a control function or as a time varying perturbation acting on the system. Control sets are of interest, in particular, since they contain all limit sets of the trajectories as time tends to infinity. Furthermore, they are related to the support of invariant measures for associated stochastic systems, compare [4]. This paper is focused on control sets with nonempty interior, as it is known that control sets which do not enjoy this property may have a very complicated structure (see e.g. [3] for examples). It is known, that the unperturbed equation (with $\varepsilon = 0$) has a unique control set (with nonempty interior), if the pair (A, B) is controllable and $0 \in \operatorname{int} U$. As shown by simple examples (see Section 2, below), the number of the control sets of (6.1) may vary dramatically when ε changes from zero to non zero values.

The main aim of this paper is to give conditions ensuring the existence of exactly one control set with nonvoid interior when ε is small enough. It will turn out that hyperbolicity of the matrix A is the crucial assumption.

As an application, we consider the following control process:

$$\dot{x}(t) = Ax(t) + Bu(t) + G(u(t), x(t)), \qquad u(t) \in U, \tag{6.3}$$

where U is compact and convex in $\mathbf{R}^m$ and $G : \mathbf{R}^m \times \mathbf{R}^d \to \mathbf{R}^d$ is C^1, and we prove that if there exist $M_1 > 0$ and $M_2 > 0$ (depending only on A and B) such that, for any G which satisfies $\|D_1 G\| \leq M_1$ and $\|D_2 G\| \leq M_2$ uniformly, (6.3) admits exactly one control set with nonvoid interior. Further applications will be shown in a forthcoming paper.

In Section 2, we recall the definition of control sets and give conditions which imply that in the interior of a control set there exists a periodic trajectory corresponding to a continuous control. Then we give a number of examples which show that (6.1) may admit multiple control sets for any $\varepsilon > 0$, while it has a unique control set for $\varepsilon = 0$. In Section 3 we discuss properties of the unique control set for the linear system. In particular, we give conditions ensuring its boundedness. In Section 4, the nonlinear problem is discussed.

Notation. We denote by $C_T(\mathbf{R}^d)$, $T > 0$, the space of continuous T-periodic function $y : \mathbf{R} \to \mathbf{R}^d$ endowed with the sup-norm $\|y\|_0 := \max\{|y(t)|, \ t \in [0, T]\}$. Similarly, $C_T^1(\mathbf{R}^d)$ is the space of T-periodic continuously differentiable functions $y : \mathbf{R} \to \mathbf{R}^d$ endowed with the norm $\|y\|_1 := \max\{\|y\|_0, \|\dot{y}\|_0\}$.

6.2 Problem formulation and examples

In this section, we give some definitions and prove preliminary results on control sets. Then some examples and counterexamples are discussed.

Consider the system

$$\dot{x}(t) = f(x(t), u(t)), \qquad u(t) \in U, \tag{6.4}$$

where $U \subset \mathbf{R}^m$ is bounded and f is C^1. We assume that unique solutions $\varphi(t, x_0, u)$, $t \in \mathbf{R}$, exist for all $x_0 \in \mathbf{R}^d$ and all measurable control functions u. A useful notion that we use in the sequel is local accessibility, i.e. the system (6.4) is locally accessible if, for all $T > 0$ and x

$$\operatorname{int}\{\varphi(t, x, u),\ T \geq t > 0 \text{ and } u : \mathbf{R} \to U, \text{ piecewise continuous}\} \neq \emptyset.$$

In the sequel, we show that for $\varepsilon > 0$ small enough one always has the local accessibility of (6.1) (see Remark 6.4.1). We start with the following definition.

Definition 6.2.1. *A subset D of $\mathbf{R}^d$ with nonvoid interior is a control set of* (6.4) *if for all $x \in D$ one has*

$$D \subset \operatorname{cl}\Big\{\varphi(t, x, u),\ t > 0 \text{ and } u : \mathbf{R} \to U, \text{ piecewise continuous}\Big\},$$

and D is a maximal subset of $\mathbf{R}^d$ with this property.

This definition does not change if piecewise continuous controls are replaced by locally integrable ones (cp. [3], Section 3.2). If local accessibility is assumed, exact controllability in the interior of control sets holds. Thus for all x, $y \in \operatorname{int} D$ there are $T > 0$ and a piecewise continuous control u such that $\varphi(T, x, u) = y$. However, in the next section we will need this property for a continuous control function. We can guarantee this under a controllability condition for the linearized system.

Proposition 6.2.1. *Let D be a control for* (6.4) *set with nonvoid interior, and assume that local accessibility holds in D. Suppose that there is a point $x_0 \in \operatorname{int} D$, for which there are a constant control $u_0 \in \operatorname{int} U$ and a time $T_0 > 0$ such that the linearized control system*

$$\dot{y} = D_1 f(\varphi(t, x_0, u_0), u_0) y + D_2 f(\varphi(t, x_0, u_0), u_0) u(t), \quad u(t) \in \mathbf{R}^m,$$

is controllable on every interval $[0, T]$, $T_0 \geq T > 0$.

Then there are $T_1 > 0$ and a continuous control function $u_1 \in \mathcal{U}$ such that $(\varphi(\cdot, x_0, u_1), u_1)$ is T_1-periodic.

Proof. As in [7], Section 3.7, Th. 7, the map

$$\alpha : L_\infty([0,T], \mathbf{R}^m) \to \mathbf{R}^d, \quad u \mapsto \varphi(T, x_0, u)$$

is continuously differentiable. By the controllability assumption, it follows that the restriction

$$\alpha : \{u \in C([0,T], \mathbf{R}^m),\ u(0) = u(T) = u_0\} \to \mathbf{R}^d, \quad u \mapsto \varphi(T, x_0, u),$$

has a surjective derivative at $u(t) \equiv u_0 \in \operatorname{int} U$ (this can be derived from [7] sec. 2.8, Th. 1). Hence, by the Surjective Mapping Theorem (see e.g. [5]), the set

$$Q := \left\{ y \in \mathbf{R}^d, \begin{array}{c} \text{there is a continuous control } u \in \mathcal{U} \text{ with} \\ u(0) = u(T) = u_0 \text{ and } y = \varphi(T, x_0, u) \end{array} \right\}$$

has nonvoid interior. Without loss of generality, we may assume that $T > 0$ is small enough such that $Q \subset \operatorname{int} D$.

Pick $y \in \operatorname{int} Q$. By the local accessibility assumption, controllability in the interior of D holds. Hence one finds a (piecewise constant) control v and $S > 0$ with $\varphi(S, y, v) = x_0$. Since the final value problem depends continuously on the right hand side, one also finds a continuous control $w \in \mathcal{U}$ with $w(0) = w(S) = u_0$ and $z \in \operatorname{int} Q$ with $\varphi(S, z, w) = x_0$. By the definition of Q there is a continuous control $u \in \mathcal{U}$ with $u(0) = u(T) = u_0$ and $\varphi(T, x_0, u) = z$.

Concatenation of v and w and periodic continuation yields a continuous $(T+S)$-periodic control u_1 with $\varphi(T + S, x_0, u_1) = x_0$. With $T_1 := T + S$, the corresponding trajectory is T_1-periodic .

We note the following consequence for a control system of the form (6.1).

Proposition 6.2.2. *Consider the control system* (6.1), *assume that* (A, B) *is controllable and that* U *is bounded. Then there exists a constant* c_* *depending only on* A, B, M_1, M_2 *(an explicit expression for* c_* *will be given in Remark 6.4.1) such that for all* $\varepsilon \in (0, c_*]$, *and every control set* D *with nonvoid interior the following holds. For every* $x_0 \in \operatorname{int} D$ *there are* $T > 0$ *and a continuous control function* $u_0 \in \mathcal{U}$ *such that* $(\varphi(\cdot, x_0, u_0), u_0)$ *is* T*-periodic.*

Proof. The controllability assumption together with the boundedness of the derivative of F implies that there exists a constant $c_* > 0$ depending only on A, B, M_1 and M_2 such that, for $\varepsilon \in (0, c_*]$, and $u_0 \in \operatorname{int} U$ the linearized system is controllable on arbitrarily short time intervals. Hence the assertion follows from the preceding proposition.

Next we turn to a number of examples which illustrate the behavior of control sets for $\varepsilon = 0$ and $\varepsilon > 0$. They show that the unique control set of the unperturbed system may split into different control sets for positive ε.

Example 6.2.1. Let

$$F(x,\varepsilon) = \begin{cases} \dfrac{1-\cos(\varepsilon\pi x)}{\varepsilon} & \text{for } x \in [-2/\varepsilon, 2/\varepsilon] \text{ and } \varepsilon \neq 0, \\ 0 & \text{otherwise}\ , \end{cases}$$

and consider the scalar control process $\dot{x}(t) = u(t) + \varepsilon F(x(t), \varepsilon)$ with $u(t) \in [-1, 1]$. The linear system $\dot{x}(t) = u$ is obviously controllable and, while for $\varepsilon = 0$ the only control set is $\mathbf{R}$, for $\varepsilon > 0$ there are 3 control sets, namely two unbounded intervals and one bounded interval.

In Example 6.2.1, F depends explicitly on ε. The next example shows that things may go wrong even for ε-independent F's.

Example 6.2.2. For $n \in \mathbf{N} \cup \{0\}$, let

$$F_n(x) = \begin{cases} n - n\cos\left(\dfrac{\pi(x-2^n)}{2^{n-1}}\right) & \text{for } x \in [2^n, 2^{n+1}], \\ 0 & \text{otherwise}, \end{cases}$$

and

$$F(x) = \sum_{n=1}^{\infty} F_n(x).$$

Consider the scalar control process $\dot{x}(t) = u(t) + \varepsilon F\big(x(t)\big)$ with $u(t) \in [-1, 1]$. Then, for $\varepsilon = 0$ the only control set is $\mathbf{R}$; whereas, for $\varepsilon > 0$ there are infinitely many control sets with nonempty interior.

In both Examples 6.2.1 and 6.2.2, the matrix A is singular. Below, we exhibit an example with nonsingular A, where multiple birth of control sets does arise. To do that, we have to increase the dimension by one.

Example 6.2.3. Consider the control process (6.1) with

$$A = \begin{pmatrix} 0 & 1 \\ -1 & 0 \end{pmatrix}, \qquad B = \begin{pmatrix} 1 & 0 \\ 0 & 1 \end{pmatrix},$$

where U is the closure of the unit ball in $\mathbf{R}^2$ with center at $(0, 0)$, and

$$F(x) = \rho\left(\sqrt{x^2+y^2}\right)(x, y)\,, \text{ with } \quad \rho(r) = \sum_{n=1}^{\infty} F_n(r),$$

where F_n is defined as in Example 6.2.2. With this choice of A, B, F and U, for $\varepsilon = 0$ there exists only one control set, and, although (A, B) is controllable and A is nonsingular, for $\varepsilon > 0$ the control process (6.1) admits infinitely many control sets with nonempty interior.

Remark 6.2.1. Example 5.5 in [1] shows that for $\varepsilon \to 0^+$ the number of control sets near an equilibrium of the system may tend to infinity.

Remark 6.2.2. The countably many control sets occurring for positive ε in Example 6.2.2 are in fact invariant, i.e. they satisfy

$$D = \mathrm{cl}\left\{\varphi(t,x,u) : t > 0,\ u : \mathbf{R} \to U \text{piecewise continuous}\right\}.$$

Hence this system has countably many different generic limit behaviours, see [2], [3] for precise statements.

For the analysis of associated stochastic systems the invariant control sets are in 1-1 correspondence to the invariant measures (see [4] and [2]). Thus Example 6.2.2 gives a 'bifurcation' result for the associated stochastic systems.

6.3 Periodic solutions of linear systems and control sets

In this section we focus on the linear control process (with restricted controls) in $\mathbf{R}^d$:

$$\begin{cases} \dot{x}(t) = Ax(t) + Bu(t), \\ u \in \mathcal{U}, \end{cases} \tag{6.5}$$

where $A : \mathbf{R}^d \to \mathbf{R}^d$ and $B : \mathbf{R}^m \to \mathbf{R}^d$ are (constant) linear operators,

$$\mathcal{U} = \{u : \mathbf{R} \to U \text{ is continuous }\},$$

and U is a compact convex subset of $\mathbf{R}^m$. We will prove some results on the boundedness and uniqueness of control sets of (6.5). For related topics on linear control processes with restricted controls see e.g. [6], Sec. 5.3, or [7], Sec. 3.6. The results that will be presented go beyond their intrinsic interest as the underlying idea is at the root of the corresponding proofs for nonlinear perturbations provided in the next section.

Let us consider the periodic solutions of the following linear differential equation in $\mathbf{R}^d$

$$\dot{x} = Ax + y, \tag{6.6}$$

where A is a hyperbolic matrix and y is a given periodic function. In particular, we will prove that there exists $K > 0$ such that, if y is T-periodic, then, for every $T > 0$ given, the T-periodic solution x of (6.6) (which is unique by the hyperbolicity of A) is bounded by a constant depending on A and $\|y\|_0$.

Theorem 6.3.1. *Let A be hyperbolic (i.e. such that $\sigma(A) \cap i\mathbf{R} = \emptyset$). Then there exists $K > 0$, depending only on A, such that for any $T > 0$ and $y \in C_T(\mathbf{R}^d)$, the T-periodic solution x of (6.6) satisfies $\|x\|_1 < K \|y\|_0$.*

The proof of this theorem relies on several lemmas discussing the behavior with respect to Jordan blocks.

Lemma 6.3.1. *Let $a \neq 0$ and consider the scalar differential equation*

$$\dot{x} = ax + y. \tag{6.7}$$

Then for every $T > 0$ and $y \in C_T(\mathbf{R})$, there exists $K > 0$ such that, if x is the unique T-periodic solution of (6.7), then $\|x\|_1 < K \|y\|_0$.

Proof. Assume first $a < 0$. The unique T-periodic solution of (6.7) is given by

$$x(t) = \frac{e^{ta}}{1 - e^{Ta}} \int_0^T e^{(T-s)a} y(s)\, ds + \int_0^t e^{(t-s)a} y(s)\, ds,$$

hence

$$|x(t)| \leq -\frac{2}{a} \|y\|_0 .$$

¿From (6.7) we find $|\dot{x}(t)| \leq |a| \|x\|_0 + \|y\|_0 \leq 2 \|y\|_0$. And finally

$$\|x\|_1 \leq (2 - 2/a) \|y\|_0 .$$

Assume now $a > 0$ and consider the equation

$$\dot{x} = -ax - \tilde{y}, \quad \tilde{y}(t) := -y(-t).$$

Any T-periodic solution of this last equation is a time reversed T-periodic solution of (6.7). Hence the assertion follows from the first part of the proof.

Lemma 6.3.2. *Assume that $A = \mathrm{diag}\{\alpha_1, \dots, \alpha_d\}$ with $\alpha_i \in \mathbf{R} \setminus \{0\}$. Then for every $T > 0$ and $y \in C_T(\mathbf{R}^d)$, there exists $K > 0$ such that, if x is the unique T-periodic solution of (6.6), then $\|x\|_1 < K \|y\|_0$.*

Proof. Let $y(t) = (y_1(t), \dots, y_d(t))$, equation (6.6) splits in the following system of d uncoupled linear differential equations

$$\begin{cases} \dot{x}_1 = \alpha_1 x_1 + y_1, \\ \quad \vdots \\ \dot{x}_d = \alpha_d x_d + y_d. \end{cases}$$

The assertion follows applying Lemma 6.3.1 to each one of the equations above.

Lemma 6.3.3. *Assume that the square $d \times d$ matrix A has the following form:*

$$A = \begin{pmatrix} \alpha & 1 & 0 & \cdots & 0 \\ 0 & \alpha & 1 & & \vdots \\ \vdots & & \ddots & \ddots & 0 \\ 0 & \cdots & 0 & \alpha & 1 \\ 0 & \cdots & \cdots & 0 & \alpha \end{pmatrix},$$

with $\alpha \neq 0$. Then for every $T > 0$ and $y \in C_T(\mathbf{R}^d)$, there exists $K > 0$ such that, if x is the unique T-periodic solution of (6.6), then $\|x\|_1 < K\|y\|_0$.

Proof. With $y(t) = (y_1(t), \ldots, y_d(t))$, the d-th component of equation (6.6) takes the form

$$\dot{x}_d(t) = \alpha x_d(t) + y_d(t).$$

Hence by Lemma 6.3.1 we get that there exists $K_d > 0$ such that

$$\|x_d\|_1 \leq K_d \|y_d\|_0 \leq K_d \|y\|_0.$$

The $(d-1)$-st component of (6.6) has the form

$$\dot{x}_{d-1}(t) = \alpha x_{d-1}(t) + x_d(t) + y_{d-1}(t).$$

Applying Lemma 6.3.1 again we get the existence of $K_{d-1} > 0$ such that

$$\|x_{d-1}\|_1 \leq K_{d-1} \|x_d + y_{d-1}\|_0 \leq (1 + K_d)K_{d-1} \|y\|_0.$$

Analogously we can then estimate $\|x_{d-2}\|_1$ and so on. Hence, in a finite number of steps, we get an estimate for every component of x.

Lemma 6.3.4. *Let $a \neq 0$ and*

$$A = \begin{pmatrix} a & b \\ -b & a \end{pmatrix}.$$

Then for every $T > 0$ and $y \in C_T(\mathbf{R}^2)$, there exists $K > 0$ such that, if x is the unique T-periodic solution of (6.6), then $\|x\|_1 < K\|y\|_0$.

Proof. Consider the complex-valued differential equation

$$\dot{z} = (a + ib)z + \eta, \tag{6.8}$$

with $\eta(t) = y_1(t) + iy_2(t)$. Clearly it is enough to show that there exists a positive number K such that for any T-periodic y_1 and y_2 the T-periodic solution z of (6.8) satisfies

$$|z| \leq K \sup_{t\in[0,T]} |\eta(t)|.$$

Assume first $a < 0$. As in the case of equation (6.7), the unique T-periodic solution of (6.8) is given by:

$$z(t) = \frac{e^{t(a+ib)}}{1-e^{T(a+ib)}} \int_0^T e^{(T-s)(a+ib)}\eta(s)\, ds + \int_0^t e^{(t-s)(a+ib)}\eta(s)\, ds,$$

Analogously to the proof of Lemma 6.3.1 and taking into account that $\left|e^{t(a+ib)}\right| = e^{ta}$ we get

$$|z| \leq 2\,|a|^{-1} \sup_{t\in[0,T]} |\eta(t)|.$$

In the case when $a > 0$, the proof is performed, analogously to Lemma 6.3.1, by time reversal.

Lemma 6.3.5. *For $a \neq 0$ define*

$$\mathcal{A} = \begin{pmatrix} a & b \\ -b & a \end{pmatrix}, \qquad I = \begin{pmatrix} 1 & 0 \\ 0 & 1 \end{pmatrix},$$

and let A be the $(2d \times 2d)$-matrix given by

$$A = \begin{pmatrix} \mathcal{A} & I & & \\ & \mathcal{A} & \ddots & \\ & & \ddots & I \\ & & & \mathcal{A} \end{pmatrix}.$$

Then for every $T > 0$ and $y \in C_T(\mathbf{R}^{2d})$, there exists $K > 0$ such that, if x is the unique T-periodic solution of (6.6), then $\|x\|_1 < K\,\|y\|_0$.

Proof. Let $y(t) = (y_1(t), \ldots, y_{2d}(t))$. Equation (6.6) splits into the following system of 2-dimensional differential equations

$$\begin{cases} \dot{\xi}_1 = \mathcal{A}\xi_1 + \xi_2 + \eta_1, \\ \dot{\xi}_2 = \mathcal{A}\xi_2 + \xi_3 + \eta_2, \\ \quad\vdots \\ \dot{\xi}_d = \mathcal{A}\xi_d + \eta_d, \end{cases}$$

where, for $j = 1, \ldots, d$, we put $\eta_j = (y_{2j-1}, y_{2j})$ and $\xi_j = (x_{2j-1}, x_{2j})$.

Applying Lemma 6.3.4 to the d-th equation of the system above, we get the existence of a positive constant K_d such that $\|\xi_d\|_1 < K_d\,\|\eta_d\|_0$. Following the same argument of the proof of Lemma 6.3.3, we get an estimate of any component of x.

Proof (Proof of Theorem 6.3.1). Up to a coordinate change we can assume that the matrix A is in real Jordan canonical form. Hence, equation (6.6) splits up into independent linear subsystems of the forms considered in Lemmas 6.3.2 – 6.3.5.

Theorem 6.3.1 enables us to prove some facts about the control sets of (6.5). Defining $\mathcal{U}_T = \mathcal{U} \cap C_T(\mathbf{R}^m)$, we have that

$$\operatorname{int} \mathcal{U}_T = \{u \in \mathcal{U}_T : u(t) \in \operatorname{int} U \text{ for all } t \in \mathbf{R}\}.$$

We will need the following consequence of Proposition 6.2.2. Observe that local accessibility holds for (6.5) by the controllability assumption.

Lemma 6.3.6. *Let D be a control set of* (6.5) *with non empty interior and let p belong to* $\operatorname{int} D$. *Then there exists $u_0 \in \operatorname{int} \mathcal{U}_T$ such that $\dot{x}(t) = Ax(t) + Bu_0(t)$ has a periodic orbit whose image contains p.*

The remaining part of this section will be devoted to proving, as applications of Theorem 6.3.1, some facts about the boundedness of control sets with non empty interiors (Theorems 6.3.2 and 6.3.3, below), and a uniqueness result for such control sets (Theorem 6.3.4, below). These results are to be compared with those contained in [3], Chapter 3, where a uniqueness result for the control sets of a linear systems is proved assuming that U contains the origin in its interior.

Notice that in our theorems below, we always assume the hyperbolicity of the matrix A. This assumption cannot be dropped since we are considering sets U which do not necessarily contain the origin. In fact, for such U's we do not necessarily have the existence of control sets. The following simple example from [3] illustrates this fact:

Example 6.3.1. Consider the scalar control process

$$\dot{x}(t) = u(t), \qquad u(t) \in U \subset \mathbf{R}.$$

If $U \subset (0, \infty)$ then there are no control sets at all. If $U = \{0\}$ then every point is a control set.

Theorem 6.3.2. *Let A be hyperbolic. Then the control sets of (6.5) which have non empty interior are bounded.*

Proof. Assume by contradiction that it exists an unbounded control set with non empty interior. Then by Lemma 6.3.6, there exists an unbounded sequence of periodic solutions of (6.5). This contradicts Theorem 6.3.1 since $\sup_{u \in \mathcal{U}} \|u\|_0 \leq \max\{|v| : v \in U\}$ is finite.

We will also need the following

Lemma 6.3.7. *Assume that the pair* (A, B) *is controllable. Given* $T > 0$ *and* $\bar{u} \in \operatorname{int}\mathcal{U}_T$*, every* T*-periodic solution of*

$$\dot{x}(t) = Ax(t) + B\bar{u}(t), \tag{6.9}$$

is contained in the interior of a control set of (6.5).

Proof. Denote by $\varphi(\cdot, u, p)$ the solution of the Cauchy problem

$$\begin{cases} \dot{x}(t) = Ax(t) + Bu(t), \\ x(0) = p, \end{cases}$$

and let p_0 be the starting point of a T-periodic solution of (6.9). Denote by E_T the space of continuous functions v such that $v(0) = v(T) = 0$, and take

$$\mathcal{V} = \{v \in E_T : \bar{u}(t) + v(t) \in \operatorname{int}(U)\}.$$

Obviously $\mathcal{V}$ is an open subset of the Banach space E_T. Define $\Theta : \mathcal{V} \to \mathbf{R}^m$ as

$$\Theta(v) = \varphi(T, \bar{u} + v, p_0) = e^{TA}p_0 + \int_0^T e^{(T-s)A} B\left(\bar{u}(s) + v(s)\right)\, ds.$$

Notice that $\Theta(0) = p_0$. For any $\omega \in E_T$, we have

$$\Theta'(0)\omega = \int_0^T e^{(T-s)A} B\omega(s)\, ds.$$

The controllability assumption implies that $\Theta'(0)$ is surjective. The Surjective Mapping Theorem (see e.g. [5]) implies that there exists a neighborhood V_0 of p_0 which is made up of images of Θ. In particular, p_0 can be driven to any point of V_0.

Applying the same argument to the time reversed control process, we have that there exists a neighborhood V_1 of p_0, any point of which can be driven to p_0. Hence $V_0 \cap V_1$ is contained in control set.

Take now any point $q \in \varphi([0, T], \bar{u}, p_0)$ and let $t_0 \in [0, T]$ be such that $q = \varphi(t_0, \bar{u}, p_0)$. By the continuity of $\varphi(t_0, \bar{u}, \cdot)$ there exists a neighborhood W of q such that

$$\varphi(t_0, \bar{u}, \cdot)^{-1}(W) \subset V_0 \cap V_1.$$

Analogously, by the continuity of the time reversed system, shrinking W if necessary, we can assume that

$$\varphi(t_0, \bar{u}, W) \subset V_0 \cap V_1.$$

Hence, any point of W can be driven to any other point of W. That is W is contained in a control set. The assertion now follows from the compactness of $\varphi([0, T], \bar{u}, p_0)$.

This lemma can be used to prove two remarkable facts:

Theorem 6.3.3. *Assume the pair (A, B) is controllable and that U is convex with $0 \in \operatorname{int}(U)$. Then the control sets of (6.5) which have nonempty interior are bounded if and only if A is hyperbolic.*

Proof. We already know that, if A is hyperbolic, the control sets with nonempty interior are bounded.

Let A be non hyperbolic. If $\det A = 0$ then $\ker A \neq \{0\}$. Any point of $\ker A$ is a periodic solution of (6.5), which, by Lemma 6.3.7 is contained in the interior of a control set.

If $\det A \neq 0$ there exists a pair of conjugate imaginary eigenvalues, say $\pm i\beta$, $\beta \neq 0$. By the Jordan real canonical form of the matrix A, there exists a 2-dimensional subspace V of $\mathbf{R}^m$ such that

$$A|_V = \begin{pmatrix} 0 & \beta \\ -\beta & 0 \end{pmatrix}.$$

Hence each point of V is the starting point of a periodic solution of $\dot{x} = Ax$ (that is of equation (6.5) with control function $u(t) \equiv 0$) and period $2\pi/\beta$. Thus, by Lemma 6.3.7 each point of V is contained in the interior of a control set.

Theorem 6.3.4. *Assume that the pair (A, B) is controllable, A is hyperbolic and U is convex. Then there exists a unique control set with non empty interior of (6.5).*

Proof. Let $T > 0$ and $\bar{u} \in \operatorname{int}\mathcal{U}_T$. The hyperbolicity of A guarantees the existence of a T-periodic solution of $\dot{x}(t) = Ax(t) + B\bar{u}(t)$, whose image is, by Lemma 6.3.7, contained in the interior of a control set. This proves the existence part of the assertion.

Let us now prove the uniqueness. Assume by contradiction that there exist two control sets D_0 and D_1 with nonempty interior. By Lemma 6.3.6 we know that there exist periodic continuous controls u_0 and u_1 (say T_0- and T_1-periodic, respectively) which take values in the interior of U and such that the T_i-periodic solution of

$$\dot{x}(t) = Ax(t) + Bu_i(t),$$

is contained in D_i, $i = 0, 1$. Define

$$u_\lambda(t) = \lambda u_1\left(\frac{T_1}{\lambda T_1 + (1-\lambda)T_0}\, t\right) + (1-\lambda)u_0\left(\frac{T_0}{\lambda T_1 + (1-\lambda)T_0}\, t\right).$$

Since U is convex by assumption, one has $u_\lambda \in \operatorname{int} \mathcal{U}_{T_\lambda}$, with $T_\lambda = \lambda T_1 + (1-\lambda)T_0$. The equation

$$\dot{x}(t) = Ax(t) + Bu_\lambda(t),$$

has a unique T_λ-periodic solution $x_\lambda(\cdot)$ whose image is contained in the interior of a control set by Lemma 6.3.7. We want to show that these images constitute a continuum joining D_0 and D_1. This will yield the desired contradiction.

Consider the time transformed system

$$\dot{\xi}(\tau) = T_\lambda \left(A\xi(\tau) + Bu_\lambda(\tau\, T_\lambda)\right),$$

and observe that $\hat{x}(\lambda, \tau) = x_\lambda(T_\lambda \tau)$ gives its unique 1-periodic solution. Since the map $(\lambda, \tau) \mapsto \hat{x}(\lambda, \tau)$ is continuous, the set $\hat{x}([0,1] \times [0,1])$ is connected and coincides with the set of images of the maps $x_\lambda(\cdot)$, for $\lambda \in [0,1]$.

6.4 Nonlinear perturbations

This section is devoted to studying the control process (6.1). The main result is Theorem 6.4.1 which states that, under reasonable assumptions, the uniqueness of control sets with nonvoid interior for (6.1) holds. As we mentioned the argument of the proof is inspired by that of Theorem 6.3.4 above, although the technical details are more subtle.

Consider the following nonlinear perturbation of a linear hyperbolic control system.

$$\dot{\xi}(t) = A\xi(t) + Bu(t) + \varepsilon F(u(t), \xi(t), \varepsilon), \tag{6.10a}$$

$$u(t) \in U. \tag{6.10b}$$

Throughout this section F will be assumed C^1 with $\|D_1 F(v,p,\varepsilon)\| \leq M_1$ and $\|D_2 F(v,p,\varepsilon)\| \leq M_2$ uniformly. Furthermore, let K_A denote the constant given by Theorem 6.3.1.

Lemma 6.4.1. *Let F and A be as above. Take $\varepsilon \in [-\varepsilon_0, \varepsilon_0]$ with*

$$\varepsilon_0 = \min\left\{1\ ,\ \frac{1}{2K_A M_2}\right\}.$$

Then for every $T > 0$, the differential equation (6.10a) has a unique T-periodic solution, $\xi(\cdot, u, \varepsilon)$, for $u \in \mathcal{U}_T$, and the map $\mathcal{U}_T \times [-\varepsilon_0, \varepsilon_0] \to C_T^1(\mathbf{R}^m)$ given by $(u, \varepsilon) \mapsto x(\cdot, u, \varepsilon)$ is continuous.

Furthermore, assuming in addition that U contains the origin of $\mathbf{R}^m$ in its interior, one has

$$\sup_{t\in[0,T]} |x(t,u,\varepsilon)| \leq 2c_U K_A\Big(\|B\| + M_1\Big), \qquad c_U := \max\{|v| : v \in U\}, \tag{6.11}$$

for every $u \in \mathcal{U}_T$ and $\varepsilon \in [-\varepsilon_0, \varepsilon_0]$.

Proof. Rewrite the equation (6.10a) in the form:

$$Lx - \bar{A}x - \bar{B}u - \varepsilon\bar{F}(u,x,\varepsilon) = 0 \tag{6.12}$$

where we put

$$\begin{array}{lll} L : C_T^1(\mathbf{R}^d) \to C_T(\mathbf{R}^d) & \text{with} & (Lx)(t) = \dot{x}(t), \\ \bar{A} : C_T^1(\mathbf{R}^d) \to C_T(\mathbf{R}^d) & \text{with} & (\bar{A}x)(t) = Ax(t), \\ \bar{B} : C_T(\mathbf{R}^m) \to C_T(\mathbf{R}^d) & \text{with} & (\bar{B}u)(t) = Bu(t), \\ \bar{F} : \mathcal{U}_T \times C_T^1(\mathbf{R}^d) \times \mathbf{R} \to C_T(\mathbf{R}^d) & \text{with} & \bar{F}(u,x,\varepsilon)(t) = F(u(t),x(t),\varepsilon). \end{array}$$

From Theorem 6.3.1 follows that there exists $K_A > 0$ (independent of $T > 0$) such that if $(L - \bar{A})\,x = y$ then $\|x\|_1 \leq K_A \|y\|_0$. In other words

$$\left\|(L-\bar{A})^{-1}\right\| \leq K_A.$$

Let $\Phi : \mathcal{U}_T \times C_T^1(\mathbf{R}^d) \times \mathbf{R} \to C_T^1(\mathbf{R}^d)$, be given by

$$\Phi(u,x,\varepsilon) = -\left(L-\bar{A}\right)^{-1}\left(\bar{B}u + \varepsilon\bar{F}(u,x,\varepsilon)\right).$$

Then equation (6.12) is equivalent to

$$\Phi(u,x,\varepsilon) = x. \tag{6.13}$$

Let us show that for $|\varepsilon| < (2K_A M_2)^{-1}$, equation (6.13) admits exactly one solution for every $u \in \mathcal{U}_T$. In fact, for $\varepsilon = 0$ this follows from the hyperbolicity of A, and for $|\varepsilon| < (2K_A M_2)^{-1}$ we have

$$\begin{aligned} \|\Phi(u,x_1,\varepsilon) - \Phi(u,x_2,\varepsilon)\|_1 &\leq |\varepsilon| \left\|(L-\bar{A})^{-1}\right\| \left\|\bar{F}(u,x_1,\varepsilon) - \bar{F}(u,x_2,\varepsilon)\right\|_0 \\ &\leq |\varepsilon|\, K_A M_2 \|x_1 - x_2\|_1 \leq \frac{1}{2}\|x_1 - x_2\|_1 . \end{aligned}$$

Hence, for $|\varepsilon| < (2K_A M_2)^{-1}$ and every $u \in \mathcal{U}_T$, $\Phi(u,\cdot,\varepsilon)$ is a contraction. Then, the Banach Contraction Theorem yields the existence of a unique fixed point which we denote by $x(\cdot, u, \varepsilon)$. Furthermore, for fixed $T > 0$, $x(\cdot, u, \varepsilon)$ depends continuously on $(u, \varepsilon) \in \mathcal{U}_T \times [-\varepsilon_0, \varepsilon_0]$ (see e.g. [8], Proposition 1.2).

To prove the last assertion, notice that for a fixed point x of $\Phi(u, \cdot, \varepsilon)$ one has

$$\begin{aligned}
\|x\|_1 &= \|\Phi(u, x, \varepsilon) - \Phi(0, 0, \varepsilon)\|_1 \\
&\leq \|\Phi(u, x, \varepsilon) - \Phi(u, 0, \varepsilon)\|_1 + \|\Phi(u, 0, \varepsilon) - \Phi(0, 0, \varepsilon)\|_1 \\
&\leq \frac{1}{2} \|x\|_1 + \|(L - \bar{A})^{-1}\| \left(c_U \|B\| + \varepsilon \left\|\bar{F}(u, 0, \varepsilon) - \bar{F}(0, 0, \varepsilon)\right\|_0 \right) \\
&\leq \frac{1}{2} \|x\|_1 + c_U K_A \big(\|B\| + M_1 \big).
\end{aligned}$$

which implies the inequality (6.11).

This lemma, combined with Proposition 6.2.2, yields a bound on the control sets with nonvoid interior.

Corollary 6.4.1. *Let A, B and F be as in Lemma 6.4.1, and assume that U contains the origin of $\mathbf{R}^m$ in its interior and take*

$$\varepsilon_0 = \min\left\{1 \,,\ \frac{1}{2K_A M_2} \,,\ c_*\right\},$$

c_* as in Proposition 6.2.2. Then every control set with nonvoid interior of (6.1) is contained in the closed $2c_U K_A \big(\|B\| + M_1 \big)$-ball of $\mathbf{R}^d$ centered at the origin.

Proof. Assume that there exist a point p, laying outside the $2c_U K_A \big(\|B\| + M_1 \big)$-ball centered at the origin, but belonging to the interior of a control set. Then, by Proposition 6.2.2, there exists a periodic solution of (6.10a) whose image contains p. This contradicts the inequality (6.11).

The assertion follows since the local accessibility ensures that, for a control set D with nonvoid interior, one has $D \subset \operatorname{cl}\operatorname{int} D$.

In what follows we denote by $\varphi_\varepsilon(\cdot, u, \xi)$ the solution of the Cauchy problem

$$\begin{cases} \dot{x}(t) = Ax(t) + Bu(t) + \varepsilon F\big(u(t), x(t), \varepsilon\big), \\ x(0) = \xi, \end{cases} \tag{6.14}$$

where $u \in \mathcal{U}_T$, $\xi \in \mathbf{R}^d$, and $\varepsilon \in \mathbf{R}$ are given.

We want to prove that, reducing ε_0 if necessary, given $T > 0$ the image of the periodic solution given by the lemma above is contained in the *interior* of a control set, provided that the pair (A, B) is controllable.

Let $\mathcal{E}$ be the Banach subspace of $C(\mathbf{R}, \mathbf{R}^m)$ given by

$$\mathcal{E} = \left\{ v \in C(\mathbf{R}, \mathbf{R}^m) : \operatorname{supp}(v) \subset [0,1] \right\}.$$

Let $\bar{u}$ with $\bar{u}(t) \in \operatorname{int} U$ for all $t \in \mathbf{R}$ be given. Define the open subset $\mathcal{V}_{\bar{u}}$ of $\mathcal{E}$ as follows:

$$\mathcal{V}_{\bar{u}} = \left\{ v \in \mathcal{E} : \bar{u}(t) + v(t) \in \operatorname{int} U, \text{ for all } t \in \mathbf{R} \right\}.$$

Given $\varepsilon \in \mathbf{R}$ and $p_0 \in \mathbf{R}^d$ define the map $\Theta_{\varepsilon,p_0} : \mathcal{V}_{\bar{u}} \to \mathbf{R}^d$ by

$$\Theta_{\varepsilon,p_0}(v) = \varphi_\varepsilon(1, \bar{u} + v, p_0).$$

Let $p_1 = \Theta_{\varepsilon,p_0}(0)$, we want to show that, under suitable assumptions on A, B and F, for ε small enough, there exists a neighborhood of p_1 which consists of images of Θ_{ε,p_0}.

Lemma 6.4.2. *Assume that the pair (A, B) is controllable and that F is C^1 with $\|D_1 F\| \le M_1$ and $\|D_2 F\| \le M_2$ uniformly. Then there exists $\varepsilon_0 > 0$ such that for $|\varepsilon| \le \varepsilon_0$, $p_0 \in \mathbf{R}^d$, $\bar{u} \in \operatorname{int}\mathcal{U}$, there exists a neighborhood V of 0 in $\mathcal{E}$, such that $\Theta_{\varepsilon,p_0}(0) = \varphi_\varepsilon(1, \bar{u}, p_0)$ lies in the interior of $\Theta_{\varepsilon,p_0}(V)$.*

Furthermore, one can actually choose ε_0 of the form

$$\varepsilon_0 = \min\left\{ 1\,,\ \frac{e^{-2\|A\|} r_{A,B}}{M_1 + M_2\big(\|B\| + M_1\big)e^{M_2}} \right\}, \tag{6.15}$$

where $r_{A,B} > 0$ depends only on A and B.

Proof. By the Surjective Mapping Theorem, it is enough to prove that there exists $\varepsilon_0 > 0$, independent of $\bar{u}$ such that $\Theta'_{\varepsilon,p_0}(0)$ is surjective for all $|\varepsilon| < \varepsilon_0$, and p_0.

For $\varepsilon = 0$ one can write explicitly

$$\Theta'_{0,p_0}(0)\omega = \int_0^1 e^{(1-s)A} B\omega(s)\; ds$$

Note that $\Theta'_{0,p_0}(0) : \mathcal{E} \to \mathbf{R}^d$ does depend neither on $\bar{u}$ nor on p_0, and, by the controllability assumption on (A, B), it is surjective. Let us put $\Theta'_{0,p_0}(0) = \Lambda$.

Since the surjective linear maps form an open subset of the space $L(\mathcal{E}, \mathbf{R}^d)$, we have that there exists $r_{A,B} > 0$ such that any $H \in L(\mathcal{E}, \mathbf{R}^d)$, which satisfies $\|H - \Lambda\| \le r_{A,B}$, is surjective.

Let us consider now $\varepsilon > 0$. Observe that $\Theta'_{\varepsilon,p_0}(0) = D_2\varphi(1, \bar{u}, p_0)$. For $\omega \in \mathcal{E}$ we put

$$\alpha(t) = D_2\varphi_\varepsilon(t, \bar{u}, p_0)\omega,$$
$$\beta(t) = D_2\varphi_0(t, \bar{u}, p_0)\omega.$$

We get

$$\alpha(t) = \int_0^t \Big[A\alpha(s) + B\omega(s) + \varepsilon D_1 F\big(\bar{u}(s), \varphi_\varepsilon(s, \bar{u}, p_0), \varepsilon\big)\omega(s) \\ + \varepsilon D_2 F\big(\bar{u}(s), \varphi_\varepsilon(s, \bar{u}, p_0), \varepsilon\big)\alpha(s)\Big]\, ds, \tag{6.16}$$

and analogously

$$\beta(t) = \int_0^t \Big[A\beta(s) + B\omega(s)\Big]\, ds. \tag{6.17}$$

Hence,

$$|\alpha(t)| \le \|\omega\| \left(\|B\| + \varepsilon M_1\right) + \int_0^t \left(\varepsilon M_2 + \|A\|\right) |\alpha(s)|\, ds$$

where M_1 and M_2 are upper bounds for $\|D_1 F\|$ and $\|D_1 F\|$ respectively. By the Gronwall inequality, we get the following estimate for $|\alpha|$

$$|\alpha(t)| \le \|\omega\| \left(\|B\| + \varepsilon M_1\right) e^{t(\|A\| + \varepsilon M_2)}. \tag{6.18}$$

Moreover, using (6.16) and (6.17),

$$|\alpha(t) - \beta(t)| \le \varepsilon \left(M_1 \|\omega\| + \int_0^1 M_2 |\alpha(s)|\, ds \right) \\ + \int_0^t \|A\| \, |\alpha(s) - \beta(s)|\, ds. \tag{6.19}$$

Plugging (6.18) into (6.19), and assuming $\varepsilon \le 1$, we get

$$\begin{aligned} |\alpha(t) - \beta(t)| &\le \varepsilon \|\omega\| \left(M_1 + M_2(\|B\| + \varepsilon M_1) e^{\|A\| + \varepsilon M_2} \right) \\ &\quad + \int_0^t \|A\| \, |\alpha(s) - \beta(s)|\, ds \\ &\le \varepsilon D \|\omega\| + \int_0^t \|A\| \, |\alpha(s) - \beta(s)|\, ds, \end{aligned} \tag{6.20}$$

where we have put

$$D = e^{\|A\|} \left(M_1 + M_2\big(\|B\| + M_1\big) e^{M_2} \right).$$

Note that the estimate (6.20) is independent of $\bar{u}$ and p_0. Applying the Gronwall inequality to (6.20),

$$\sup_{t \in [0,1]} |\alpha(t) - \beta(t)| \le \varepsilon D \|\omega\| e^{\|A\|}.$$

In other words, recalling the definitions of α and β, for

$$\varepsilon \leq \min\left\{1\,,\ \frac{e^{-2\|A\|}r_{A,B}}{M_1 + M_2\big(\|B\| + M_1\big)e^{M_2}}\right\},$$

we have

$$\big\|\Theta'_{\varepsilon,p_0}(0) - \Theta'_{0,p_0}(0)\big\| = \big\|\Theta'_{\varepsilon,p_0}(0) - \Lambda\big\| \leq r_{A,B},$$

independently of $\bar{u}$ and p_0, which yields the surjectivity of $\Theta'_{\varepsilon,p_0}(0)$ for each $\bar{u}$.

Remark 6.4.1. Lemma 6.4.2 says that, if ε is small enough, then, given $\bar{u} \in \operatorname{int}\mathcal{U}$, it is possible to reach any point in a suitably small neighborhood of $\varphi_\varepsilon(1, \bar{u}, p_0)$ by varying the control function in a neighborhood of $\bar{u}$.

With only minor changes in the proof one can show that the set which can be reached in a given time $\tau \in (0,1]$ from any given point p_0 has nonempty interior. This property, often called strong accessibility, obviously implies local accessibility. Hence one can actually choose the constant c_* which appears in Proposition 6.2.2 equal to ε_0 in (6.15).

We need to extend the result of Lemma 6.4.2 to the case of a time $T > 1$.

For any $v \in \mathcal{E}$, we put $\tilde{v}(t) = v(t - T + 1)$ and define

$$\Psi_{T,\varepsilon,p_0}(v) = \varphi_\varepsilon(T, \tilde{v}, p_0) = \Theta_{\varepsilon,\varphi_\varepsilon(T-1,\bar{u},p_0)}(v).$$

Thus we immediately get

Corollary 6.4.2. *Assume that the pair (A, B) is controllable and that F is C^1 with $\|D_1 F\| \leq M_1$ and $\|D_2 F\| \leq M_2$ uniformly. Then, there exists $\varepsilon_0 > 0$ such that for $|\varepsilon| \leq \varepsilon_0$, $p_0 \in \mathbf{R}^d$, $\bar{u} \in \mathcal{U}$ and $T > 1$ given, there exists a neighborhood V of 0 in $\mathcal{E}$, such that* $\operatorname{int} V \subset \Psi_{T,\varepsilon,p_0}(V)$.

Furthermore, one can actually choose ε_0 of the form (6.15).

Proof. It follows from Lemma 6.4.2 applied to the function $\tilde{\bar{u}} : t \mapsto \bar{u}(t-T+1)$ and to the point $\Theta_{\varepsilon,\varphi_\varepsilon(T-1,\bar{u},p_0)}(p_0)$.

This corollary allows us to prove for (6.1) a result which is analogous to Lemma 6.3.7. From now on we will assume that

$$\varepsilon_0 = \min\left\{1\,,\ \frac{e^{-2\|A\|}r_{A,B}}{M_1 + M_2\big(\|B\| + M_1\big)e^{M_2}}\,,\ \frac{1}{2K_A M_2}\right\} \tag{6.21}$$

Lemma 6.4.3. *Let U have non empty interior. Assume that A is hyperbolic, that the pair (A, B) is controllable and that F is C^1 with $\|D_1 F\|$ and $\|D_2 F\|$ bounded. Then if $|\varepsilon| \leq \varepsilon_0$, given $T > 0$ and $\bar{u} \in \operatorname{int} \mathcal{U}_T$,* (6.10a) *has a unique T-periodic solution. Furthermore this solution is contained in the interior of a control set of (6.1).*

Proof. Observe that, given $T > 0$, a T-periodic function is also nT-periodic, $n \in \mathbf{N}$. Hence, without loss of generality, we can assume $T > 1$.

Lemma 6.4.1 yields the existence of a unique T-periodic solution of (6.10a) for $|\varepsilon| \leq \varepsilon_0$ and $\bar{u} \in \operatorname{int} \mathcal{U}_T$.

Fix $\bar{u} \in \operatorname{int} \mathcal{U}_T$, let p_0 be the starting point of the unique periodic T-periodic solution of (6.10a). From Corollary 6.4.2 it follows that there exists a neighborhood V of p_0 in $\mathbf{R}^d$ such that for any $q \in V$ there exists $w \in \operatorname{int} \mathcal{U}_T$ such that $q = \varphi_\varepsilon(T, w, p_0)$.

Considering the time reversed system and reducing V, if necessary, we can assume that within this set any point can be driven into any other point. Hence V is contained in the interior of a control set. To prove that the whole

$\varphi_\varepsilon([0, T], \bar{u}, p_0)$ is contained in the interior of a control set, we proceed as in the last part of the proof of Lemma 6.3.7.

A noteworthy consequence of Lemma 6.4.3 combined with Proposition 6.2.2 is the following.

Remark 6.4.2. Assume, in addition to the hypotheses of Lemma 6.4.3, that U contains 0 in its interior and that $F(0, 0, \varepsilon) \equiv 0$ for any $|\varepsilon| \leq \varepsilon_0$. Then the origin of $\mathbf{R}^d$ is contained in the interior of a control set. In fact, the origin can be regarded as a 1-periodic solution of (6.10a).

We are now in a position to prove the main result of this section.

Theorem 6.4.1. *Let U be convex with non empty interior. Assume that (A, B) in (6.1) is controllable and A is hyperbolic. Let F be C^1 with $\|D_1 F\|$ and $\|D_2 F\|$ bounded. Then the control process* (6.1) *admits exactly one control set D with nonvoid interior if $|\varepsilon| \leq \varepsilon_0$, ε_0 as in* (6.21).

Furthermore,

1. *for $|\varepsilon| \leq \varepsilon_0$ the control set D is contained in the $2c_U K_A(\|B\| + M_1)$-ball of $\mathbf{R}^d$ centered at the origin,*
2. *if $F(0, 0, \varepsilon) \equiv 0$ for any $|\varepsilon| \leq \varepsilon_0$, then the origin is contained in the interior of D.*

Proof. Let $T > 1$ and $\bar{u} \in \mathrm{int}\,\mathcal{U}_T$. Lemma 6.4.1 guarantees the existence of a T-periodic solution of (6.10a), whose image is, by Corollary 6.4.2, contained in the interior of a control set. This proves the existence of at least one control set.

Let us prove the uniqueness assertion. Assume by contradiction that for some $\varepsilon \in [-\varepsilon_0, \varepsilon_0]$ there exist two different control sets, say D_0 and D_1. Then, by Proposition 6.2.2, there exists $u_i \in \mathrm{int}\,\mathcal{U}_{T_i}$, $i \in \{0,1\}$, such that the corresponding T_i-periodic trajectory of (6.10a) is contained in the interior of D_i. As in the proof of Lemma 6.4.3 we can always assume that $T_i > 1$ for $i \in \{0,1\}$.

As in the proof of Theorem 6.3.4, put $T_\lambda = \lambda T_1 + (1-\lambda)T_0$ and define

$$u_\lambda(t) = \lambda u_1\left(\frac{T_1 t}{T_\lambda}\right) + (1-\lambda)u_0\left(\frac{T_0 t}{T_\lambda}\right).$$

Since U is assumed convex, $u_\lambda \in \mathrm{int}\,\mathcal{U}_{T_\lambda}$. By the choice of ε_0, the equation

$$\dot{x}(t) = Ax(t) + Bu_\lambda(t) + \varepsilon F(u_\lambda(t), x(t), \varepsilon),$$

admits a unique T_λ-periodic solution whose image is contained in the interior of a control set. By the argument used in the proof of Theorem 6.3.4 we get the existence of a continuum which joins D_0 and D_1 and whose points are all contained in the interior of a control set. This yields the desired contradiction.

The last two assertions follow from Corollary 6.4.1 and Remark 6.4.2.

Theorem 6.4.1 has the following remarkable consequence.

Corollary 6.4.3. *Assume (A,B) controllable. Let $G : \mathbf{R}^m \times \mathbf{R}^d \to \mathbf{R}^d$ be a C^1 function such that $\|D_1G(v,p)\| \le M_1$ and $\|D_2G(v,p)\| \le M_2$, for any $(v,p) \in \mathbf{R}^m \times \mathbf{R}^d$. If the bounds M_1 and M_2 for the partial derivatives are small enough, then the control process* (6.3) *admits a unique control set D with nonempty interior. Moreover D turns out to be bounded, and, if $G(0,0) = 0$, then D contains the origin of $\mathbf{R}^m$ in its interior.*

Proof. If M_1 and M_2 are small enough then $\varepsilon_0 = 1$ in formula (6.21). Hence the assertion follows directly from Theorem 6.4.1.

Acknowledgement:

This paper has been written when the second author was visiting Universität Augsburg as a NCN fellow in the TMR program. M. Spadini wishes to thank the Nonlinear Control Network and the personnel of the 'Institut für Mathematik' for their helpful assistance.

References

1. F. Colonius and W. Kliemann, *Limit behavior and genericity for nonlinear control systems*, J. Diff. Equations 109 (1994), pp. 8-41.

2. F. Colonius and W. Kliemann, *Continuous, smooth, and control techniques for stochastic dynamics*, in *Stochastic Dynamics* H. Crauel and M. Gundlach eds., Springer-Verlag, 1999.

3. F. Colonius and W. Kliemann, *The Dynamics of Control*, Birkhäuser 1999, to appear.

4. W. Kliemann, *Recurrence and invariant measures for degenerate diffusions*, Ann. Prob., 15 (1987), pp. 690-707.

5. S. Lang, *Real and Functional Analysis*, Graduate Texts in Mathematics 142, Springer-Verlag, New York, 1993.

6. V. Jurdjevic, Geometric Control Theory Cambridge Studies in Mathematics 51, Cambridge University Press, Cambridge, 1997.

7. E. Sontag *Mathematical Control Theory, 2nd edition*, Texts in Applied Mathematics 6, Springer-Velag, New York, 1998.

8. E. Zeidler, *Nonlinear Functional Analysis* vol.1, Springer-Verlag, New York, 1986.

7. Design of control Lyapunov functions for "Jurdjevic-Quinn" systems

Ludovic Faubourg and Jean-Baptiste Pomet

I.N.R.I.A
B.P. 93
06902 Sophia Antipolis cedex, France
`Ludovic.Faubourg@sophia.inria.fr`,
`Jean-Baptiste.Pomet@sophia.inria.fr`
Url: `www.inria.fr/miaou`

Summary.

This paper presents briefly a method to design explicit control Lyapunov functions for control systems that satisfy the so-called "Jurdjevic-Quinn conditions", i.e. posses an "energy-like" function that is naturally non-increasing for the un-forced system. The results with proof will appear in a future paper. The present note rather focuses on the method, and on its application to the model of a mechanical system, the translational oscillator with rotation actuator (TORA) (also known as RTAC).

7.1 Introduction

For differentiable dynamical systems (without control), a Lyapunov function is a convenient tool to analyze the asymptotic stability of an equilibrium. See [12] for instance. It is of course not the only one, and its main drawback is that there is no systematic way to find a Lyapunov function in general, even though converse Lyapunov theorems (see [12, 15]) tell us that existence of a Lyapunov function is equivalent to asymptotic stability.

For control systems and the stabilization problem, Lyapunov functions have also been used extensively. For linear systems, that have the nice property that Lyapunov functions may always be taken quadratic, optimizing a quadratic criteria or assigning a quadratic Lyapunov function are more or less synonymous (see for instance [4]). For nonlinear systems, the so-called "Lyapunov design" (see [2, 6]) consists in designing a Lyapunov function *together with* a control that makes it decrease, i.e. that assigns this function to be a Lyapunov function for the closed-loop system. Artstein's theorem [1] makes this method theoretically consistent by characterizing the functions that may be assigned to be Lyapunov functions via a suitable continuous feedback control. Let us recall it briefly : consider a control system

$$\dot{x} = f_0(x) + \sum_{k=1}^{m} u_k f_k(x) \tag{7.1}$$

with state $x \in \mathbb{R}^n$ and control $(u_1, \dots, u_m) = u \in \mathbb{R}^m$, the f_k's being smooth vector fields in $\mathbb{R}^n$.

Theorem 1 (Artstein's theorem) *A differentiable function V that is positive definite (zero at the origin, positive elsewhere) and infinite at infinity can be assigned to be a Lyapunov function for the closed-loop system via a continuous feedback in (7.1) if and only if*

1. *it is a* control Lyapunov function *(CLF) :*

$$\forall x \in \mathbb{R}^n \backslash \{0\}, \qquad \left.\begin{array}{c} L_{f_1}V(x) = 0 \\ \vdots \\ L_{f_m}V(x) = 0 \end{array}\right\} \Longrightarrow L_{f_0}V(x) < 0 \ , \tag{7.2}$$

2. *it satisfies the so called* small control property *(SCP) : for any $\varepsilon > 0$, there exists a $\delta > 0$ such that, for $x \in \mathbb{R}^n$,*

$$\left.\begin{array}{c} x \neq 0 \\ \|x\| < \delta \end{array}\right\} \exists u \left\{\begin{array}{l} \|u\| < \varepsilon \\ L_{f_0 + \sum_{k=1}^{m} u_k f_k} V(x) < 0 \ . \end{array}\right. \tag{7.3}$$

This is only an existence result, but Sontag gave in [20] an explicit formula that gives a systematic way to obtain a stabilizing control corresponding to a given CLF. On practical examples, there are of course many other possibilities.

Other methods than building a control Lyapunov function exist to design stabilizing control laws. However, it is well know (see for instance [18, 19, 6]) that a control Lyapunov function, when available, is a very convenient tool to analyze stability, and its robustness to perturbations, or even to modify the design to enhance robustness or performances. For these reasons, it is interesting to obtain control Lyapunov functions for systems that may be stabilized by other methods. This is the topic of the present paper, at least for the situation where stabilization has been obtained using the "Jurdjevic-Quinn" method, also called "damping control".

Let us call (7.2)-(7.3) Artstein's equation, and draw a parallel with optimal control and Hamilton-Jacobi-Bellman's equation . Optimal control (for instance minimum time) is a quantitative problem, whose solution is unique, and often very difficult to describe. Asymptotic stabilization, a qualitative problem, whose solution is highly non-unique, the requirement being somehow weaker. Artstein's equations play, for stabilization, the role of Hamilton-Jacobi-Bellman's equation for optimal control : the dynamic programming principle asserts that a solution to the HJB's equation yields (at least when it is differentiable !) the optimal synthesis, and in the same way, Artstein's theorem allows one to derive a stabilizing control law from a solution to Artstein's equations, with some universal formulas available. However it is known that the solutions to HJB's equation are in general non-smooth, and the analysis of this equation has given rise to a lot of mathematical developments to handle this problem. Artstein's equations are much less constrained, and for this reason, they have smooth solutions, far from being unique : for instance one may freely change V outside the origin and the points where its derivative along the control vector fields is zero, at least if the changes are not big enough (in C^1 topology) for the derivative of the deformed function along the control vector fields to vanish. We do not give a bibliography on HJB's equation and dynamic programming, but the reader may find in [11, 9, 18, 19] some investigations on the link between optimality, control Lyapunov functions and robustness. We shall not elaborate more along these line, but we do think that Artstein's equations deserve a deeper analysis. This paper is a contribution to this analysis in a particular case.

This paper is organized as follows. In section 7.2, we present the type of systems we are interested in, and review very briefly the popular method to stabilize them (without obtaining a CLF). The method is outlined in section 7.3 and gives some general results. We then apply, in section 7.4 the method to a mechanical system used as a benchmark in [19], see also [3]. Note that this model does not satisfy the assumptions theoretically needed in section 7.3, but the method still yields a global CLF.

7.2 Jurdjevic-Quinn systems

It was noticed in [13], that a first integral for the drift vector field, plus some controllability conditions allow one to derive smooth asymptotically stabilizing control laws. This method has been generalized in [10, 2, 17] for instance. This is now popular, under the name "Jurdjevic-Quinn method" or "damping control".

Let us be more specific. Consider the affine control system (7.1). The following conditions are a bit more restrictive than these in [17], more general that these in other papers, but the idea is basically the same. To state them, let us define, for a positive and radially unbounded function V_0, the set $\mathcal{W}_L(V_0)$:

$$\mathcal{W}_L(V_0) = \left\{ \begin{array}{c} x \in I\!\!R^n, \ L_{f_0} V_0(x) = L_{ad^i_{f_0}(f_k)} V_0(x) = 0 \\ k = 1 \ldots m; \ i = 1 \ldots L \end{array} \right\} \tag{7.4}$$

Assumption A ((weak) Jurdjevic-Quinn conditions)

1. *The vector field f_0 has only one equilibrium :*

$$f_0(x) = 0 \iff x = 0 \ . \tag{7.5}$$

2. *A function $V_0 : \mathbb{R}^n \to \mathbb{R}$ is known and has the following three properties :*

   a) it is positive definite and radially unbounded,

   b) it satisfies :

$$\forall x \in \mathbb{R}^n,\ L_{f_0} V_0(x) \ \leq \ 0 \ , \tag{7.6}$$

   c) there is an integer L such that

$$\mathcal{W}_L(V_0) \ = \ \{0\} \ . \tag{7.7}$$

Proposition 1 *Under assumption A, the smooth control law $u_k = -Lf_k V_0$ asymptotically stabilizes the origin for system (7.1).*

This proposition is also correct replacing assumption A by the following one, that obviously implies A, and corresponds to the most common situation :

Assumption A′ *Same as assumption A, but instead of (7.6), V_0 is assumed to be a first integral of the vector field f_0 :*

$$\forall x \in \mathbb{R}^n,\ L_{f_0} V_0(x) \ = \ 0 \ . \tag{7.8}$$

Let us make our discussion in the simpler case of assumption A′. The proof is based on the fact that that these u_k yield a non-positive $\dot{V}_0$:

$$\dot{V}_0 \ \leq \ - \sum_{k=1}^{m} (L_{f_k} V_0)^2 \ .$$

This is enough to prove asymptotic stability through LaSalle's invariance principle (see [16, 12]). However, under assumption A′, $\dot{V}_0$ is zero exactly at the points where all the functions $Lf_k V_0$ vanish, and no control can do better at these points because u has no effect on $\dot{V}_0$. This V_0 is therefore *not* a CLF for system (7.1) : in Artstein's equation (7.2), the strict inequality is replaced by an equality. Although this "weak" Lyapunov function is as good as a "strict" one for the purpose of proving asymptotic stability, the fact that V_0 is not strictly decreasing along the solutions, does not give a "margin" that can be exploited for robustness analysis or enhancement.

7.3 Reshaping Lyapunov functions

Let us present a way to construct a CLF using the known function V_0 described above.

7.3.1 The idea

It is tempting to consider V_0 as "almost" a CLF, and to design a "deformation" V_λ, with λ a real parameter, such that
- for all λ, V_λ is positive definite and radially unbounded,
- for $\lambda = 0$, V_λ is the above V_0, and
- for positive value of λ, V_λ is a CLF, i.e. instead of $= 0$, one gets the strict inequation (7.2), ideally more and more negative as λ grows.

This idea was already presented in [7]. Let us sum up the obtained results, announced in [7] in a more restrictive form, and presented into details in the forthcoming [8].

The type of deformation considered is the very natural

$$V_\lambda = V_0 \circ \phi_\lambda^G \ , \tag{7.9}$$

where G is a complete vector field, and ϕ_λ^G stands for the flow at time λ of G, i.e. :

$$\frac{\partial}{\partial \lambda}(\phi_\lambda^G(x)) \ = \ G(\phi_\lambda^G(x)) \ , \quad \phi_0^G(x) \ = \ x \ . \tag{7.10}$$

Since ϕ_λ^G is a diffeomorphism for all λ, V_λ is positive definite and radially unbounded for all λ. Also, the condition $G(0) = 0$ is sufficient to ensure $V_\lambda(0) = 0$.

Of course, V_λ being a CLF for $\lambda > 0$ requires further conditions on G. The least is to require that $Lf_0 V_\lambda(x)$ be a decreasing function of λ, for small values of λ and at points where $Lf_k V_0(x) = 0$. It turns out that the following formula holds, for all x under assumption A$'$, and at points x where $L_{f_0} V_0$ vanishes under assumption A :

$$\left.\frac{d}{d\lambda}\right|_{\lambda=0} (L_{f_0} V_\lambda^G)(x) = L_{f_0} L_G V_0(x) \tag{7.11}$$

(this is established from standard differential calculus). With in mind the concern of finding explicit formulae, using the flow of a vector field is not the best choice. In the (usual) case where ϕ_λ^G cannot be explicitly computed, we may use instead

$$W_\lambda = V_0 \circ (Id + \lambda G) \ , \tag{7.12}$$

or any W_λ that approaches V_λ up to order 1 at $\lambda = 0$ (in (7.12), this is obtained by using the first order approximation of the flow). Note that (7.9) is an intrinsic definition of V_λ whereas the above (7.12) is coordinate dependent : for instance, in coordinates where G would be a "constant" vector field (impossible near an equilibrium !), the formula (7.12) would be equivalent to (7.12).

7.3.2 The results

The following results can be found in [7] in the particular case when $L_{f_0}V_0$ is zero (assumption A′). They will appear, with proofs, in a future paper [8]. For the definitions of homogeneity and dilations, and related properties, see for instance [14]. We do not insist here on this property.

Theorem 2 *Suppose that the control affine system (7.1) satisfies the assumption A and furthermore, that the vector fields $f_0, \ldots, f_m$ are* homogeneous *with respect to a certain dilation, f_0 having a degree strictly larger than the others, and that the function V_0 is also homogeneous with respect to the same dilation. If G is a homogeneous vector field of degree 0 which satisfies, for all x in $\mathbb{R}^n\backslash\{0\}$,*

$$\left.\begin{array}{l} L_{f_0}V_0(x) = 0 \\ L_{f_1}V_0(x) = 0 \\ \quad\vdots \\ L_{f_m}V_0(x) = 0 \end{array}\right\} \Longrightarrow L_{f_0}L_GV_0(x) < 0 \tag{7.13}$$

Then there exists a positive real number λ_0 such that for all λ that satisfies $0 < \lambda < \lambda_0$, V_λ and W_λ are homogeneous CLFs of the same degree as V_0, satisfying the small control property. This would also be true for any formula for W_λ instead of (7.12), provided that the derivative with respect to λ at $\lambda = 0$ is the same, and W_λ is homogeneous.

This result is completed by the following, that gives a "universal" formula for a vector field G, homogeneous of degree 0, that satisfies condition (7.13).

Theorem 3 *Suppose that the control affine system (7.1) satisfies the assumption A and that the vector fields $f_0, \ldots, f_m$ and the function V_0 satisfy the homogeneity assumptions of theorem 2. Define the vector field G by :*

$$G = \sum_{i=0}^{L-1}\sum_{k=1}^{m} \lambda_{i,k}\, ad^i_{f_0} f_k \tag{7.14}$$

with $\lambda_{i,k}$ ($i = 0, \ldots, L-1; k = 1, \ldots, m$) some functions defined by

$$\begin{cases} \lambda_{i,k} = \sum_{j=i}^{L-1} (-1)^{j-i+1} \dfrac{L_{ad_{f_0}^{(2j-i+1)}(f_k)} V_0}{(2V_0)^{\alpha_{j,k}}} \\ \alpha_{j,k} = \frac{(2j+1)c_0 + 2c_k + d}{2} \end{cases} . \tag{7.15}$$

This vector field G is homogeneous of degree zero and satisfies the conditions of theorem 2.

7.3.3 Some remarks on using this method to obtain CLFs

To sum up, the proposed method for designing a strict Lyapunov function consists in :
- finding a vector field G that satisfies the condition (7.13),
- computing V_λ according to (7.9) or W_λ according to the explicit formula (7.12), or another one that gives the same $\partial W_\lambda / \partial \lambda$ at $\lambda = 0$.

This yields a function that, if the homogeneity assumptions are satisfied, is guaranteed to be a (global) control Lyapunov function for λ positive and small enough.

On the choice of G : Theorem 3 gives a "universal" formula to obtain a good G, but there are many other choices, and it is important to have some choice (see next remarks). It is usually a good idea, following (7.14), to write G as a linear combination of the vector fields $L_{f_0}^j f_k$, and to write the (linear differential) relations on the coefficients implied by (7.13). A thorough discussion will appear in [8]. This is illustrated on an example in section 7.4.

How large can λ be ? The theorems only say that V_λ or W_λ will be a CLF for *small enough* positive values of λ. It is of course important that λ can can be "reasonably large", since if it is not the case, $\dot{V}_\lambda$ or $\dot{W}_\lambda$ will he negative but very small at the points where the control has no effect on them. One needs to compute on each example to see how big λ can be chosen, and how good a Lyapunov function one obtains, but that usually depends on the choice of G (hence it is important to have some choice) and even on the formula one chooses for $\dot{W}_\lambda$.

If the homogeneity conditions are not met, this method still yields some positive definite functions V_λ or W_λ. The theorems however do not apply, and on some examples, these functions may fail to be CLFs (problems around the origin, and at infinity), or fail to satisfy the small control property (automatically satisfied in the homogeneous case). However, the experience shows that they often work as CLFs although the theorems do not apply. Again, having some choice on G is crucial here. The following example is an illustration.

7.4 Example: The TORA system

Using our ideas to find some control Lyapunov function for this system was suggested by Rodolphe Sepulchre, from Université de Liège.

Let us consider the mechanical system called TORA (Translational Oscillator with Rotating Actuator) in [19] and RTAC (Rotational/Translational proof-mass Actuator) in [3]. It consists of a platform connected to a fixed frame of reference by a linear spring. The platform can oscillate without friction in the horizontal plane. On the platform, an eccentric rotating mass is actuated by a DC motor. The control of this rotating motion is used to dampen the translational oscillations of the platform. A precise description is given in [5]. After normalization (the time is also normalized), the dimensionless variables (x_1, x_2, x_3, x_4) may be used, where x_1 and x_2 are proportional to the translational displacement (from the equilibrium position) and the velocity of the platform, x_3 and x_4 are the angle (from a direction perpendicular to the spring) and the angular velocity of the eccentric mass. The following equations are obtained, where ε is a parameter, depending on the mass of the cart, the mass, length and inertia of the eccentric mass, and the stiffness of the string. A typical value for ε is 0.1. Note that the picture below is in a horizontal plane, so that gravity is ignored.

$$\begin{aligned}
\dot{x}_1 &= x_2 \\
\dot{x}_2 &= \frac{-x_1 + \varepsilon {x_4}^2 \sin x_3}{1 - \varepsilon^2 \cos^2 x_3} + \frac{-\varepsilon \cos x_3}{1 - \varepsilon^2 \cos^2 x_3} u \\
\dot{x}_3 &= x_4 \\
\dot{x}_4 &= \frac{\varepsilon \cos x_3 (x_1 - \varepsilon {x_4}^2 \sin x_3) + u}{1 - \varepsilon^2 \cos^2 x_3}
\end{aligned}$$

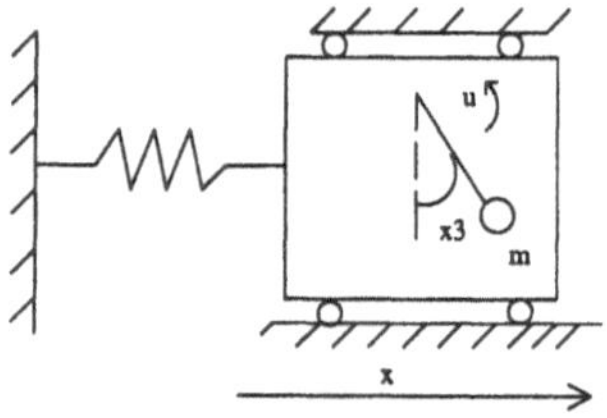

In [19], this system is extensively used as an illustrative example, and the papers in [3] also propose various control methods. The following is an illustration of our methods on this example to derive a (global) CLF. We do not propose new or "better" control laws.

Note that the coordinates of the center of mass are $(x_1 + \varepsilon \sin x_3, -\varepsilon \cos x_3)$. It is natural to use also the variables z_1 and z_2, that are the horizontal component of the position and velocity of the center of mass :

$$z_1 = x_1 + \varepsilon \sin x_3 \,, \quad z_2 = x_2 + \varepsilon x_4 \cos x_3 \,.$$

As noticed in [19] or some articles in [3], setting $z_3 = x_3$ and $z_4 = x_4$, the state equations take the following simpler form in the coordinates (z_1, z_2, z_3, z_4) :

$$\begin{aligned}
\dot{z}_1 &= z_2 \\
\dot{z}_2 &= -z_1 + \varepsilon \sin z_3 \\
\dot{z}_3 &= z_4 \\
\dot{z}_4 &= \frac{1}{1 - \varepsilon^2 \cos^2 z_3} [\varepsilon \cos z_3 (z_1 - \varepsilon(1 + {z_4}^2) \sin z_3) + u] \,.
\end{aligned} \tag{7.16}$$

It is also convenient in some occurrences to keep x_1 :

$$\begin{aligned}
\dot{x}_1 &= z_2 - \varepsilon z_4 \cos z_3 \\
\dot{z}_2 &= -x_1 \\
\dot{z}_3 &= z_4 \\
\dot{z}_4 &= \frac{1}{1 - \varepsilon^2 \cos^2 z_3} [\varepsilon \cos z_3 (x_1 - \varepsilon {z_4}^2 \sin z_3) + u]
\end{aligned} \tag{7.17}$$

The potential energy of the spring is (in dimensionless units) $\frac{1}{2} x_1^2$ and the kinetic energy is $\frac{1}{2} z_2^2 + \frac{1}{2} z_4^2 (1 + \mu \sin^2 z_3)$ with μ some coefficient. The total is naturally constant along the solutions for $u = 0$. However, it is not proper (infinite at infinity), and it may be zero for any value of z_3. It is therefore reasonable to "strengthen" it by adding a term that is proper with respect to z_3, like z_3^2. In fact, following [19], we modify the energy and use the following definite positive and radially unbounded function V_0 :

$$V_0(z) = \frac{1}{2} x_1^2 + \frac{1}{2} {z_2}^2 + \frac{k_1}{2} {z_3}^2 + \frac{1}{2} {z_4}^2 \tag{7.18}$$

$$= \frac{1}{2} (z_1 - \varepsilon \sin z_3)^2 + \frac{1}{2} z_2^2 + \frac{k_1}{2} {z_3}^2 + \frac{1}{2} {z_4}^2 \tag{7.19}$$

with k_1 some positive constant. Using the following preliminary feedback

$$\begin{aligned}
u = &-\varepsilon \cos z_3 (x_1 - \varepsilon {z_4}^2 \sin z_3) \\
&+ (1 - \varepsilon^2 \cos^2 z_3)(\varepsilon x_1 \cos z_3 - k_1 z_3 + v) \,,
\end{aligned} \tag{7.20}$$

we obtain a control system

$$\dot{x} \;=\; f_0(x) \;+\; v\, f_1(x)$$

where f_0 and f_1 are, in the coordinates (x_1, z_2, z_3, z_4),

$$f_0 = \begin{bmatrix} z_2 - \varepsilon z_4 \cos z_3 \\ -x_1 \\ z_4 \\ \varepsilon x_1 \cos z_3 - k_1 z_3 \end{bmatrix} \qquad f_1 = \begin{bmatrix} 0 \\ 0 \\ 0 \\ 1 \end{bmatrix} \tag{7.21}$$

A simple computation yields to the following equations :

$$\begin{aligned}
L_{f_0} V_0(z) &= 0 \\
L_{f_1} V_0(z) &= z_4 \\
L_{f_0} L_{f_1} V_0(z) &= \varepsilon \cos z_3 (z_1 - \varepsilon \sin z_3) - k_1 z_3 \\
L_{f_0}^2 L_{f_1} V_0(z) &= \varepsilon z_2 \cos z_3 + z_4(-\varepsilon z_1 \sin z_3 - \varepsilon^2 \cos 2 z_3 - k_1) \\
L_{f_0}^3 L_{f_1} V_0(z) &= \varepsilon \cos z_3 (-z_1 + \varepsilon \sin z_3) + z_4 h(z_1, z_2, z_3, z_4)
\end{aligned} \tag{7.22}$$

Hence this system satisfies the Jurdjevic-Quinn conditions (assumption A, and even A′). Precisely, (7.22) implies that $\mathcal{W}_3(V_0) = \{0\}$.

In [19] the authors are able to design a continuous feedback law which makes the origin of the system asymptotically stable by applying some "damping control" (see Proposition 1 above), and them modifying these control laws to get more robustness, but asymptotic stability is obtained via a weak Lyapunov function. Unfortunately this system does not meet the homogeneous assumptions of our theorems. Yet in this particular case we will apply successfully our method to obtain a control Lyapunov function.

The first step consists in finding vector field G that meets the condition (7.13). Let us illustrate the second remark in section 7.3.3, and try to find "all" the good G's instead of using formula (7.14)-(7.15). It is not difficult to see that the second and first component of G in the coordinates (z_1, z_2, z_3, z_4) should vanish when $\cos z_3$ vanishes, so let us take G of the form :

$$G = \begin{pmatrix} \varepsilon \cos z_3 g_1 \\ \varepsilon \cos z_3 g_2 \\ g_3 \\ g_4 \end{pmatrix} = \varepsilon \cos z_3 \left(g_1 \frac{\partial}{\partial z_1} + g_2 \frac{\partial}{\partial z_2} \right) + g_3 \frac{\partial}{\partial z_3} + g_4 \frac{\partial}{\partial z_4}$$

with g_1, g_2, g_3, and g_4 some functions. We have

$$[f_0, G] = \begin{pmatrix} \varepsilon \cos z_3 \left(f_0 g_1 - g_2 \right) - \varepsilon z_4 \sin z_3\, g_1 \\ \varepsilon \cos z_3 \left(f_0 g_2 + g_1 - g_3 \right) - \varepsilon z_4 \sin z_3\, g_2 \\ f_0 g_3 - g_4 \\ f_0 g_4 - \varepsilon \cos z_3 g_1 + \left(k_1 + \varepsilon z_1 \sin z_3 + \varepsilon^2 \cos(2 z_3) \right) g_3 \end{pmatrix}$$

and hence, since $L_{f_0} L_G V_0 = L_{[f_0, G]} V_0$,

$$\begin{aligned} L_{f_0} L_G V_0 = {} & (z_1 - \varepsilon \sin z_3) \left(\varepsilon \cos z_3 \left(f_0 g_1 - g_2 \right) - \varepsilon z_4 \sin z_3\, g_1 \right) \\ & + z_2 \left(\varepsilon \cos z_3 \left(f_0 g_2 + g_1 - g_3 \right) - \varepsilon z_4 \sin z_3\, g_2 \right) \\ & + \left(k_1 z_3 - \varepsilon \cos z_3 (z_1 - \varepsilon \sin z_3) \right) \left(f_0 g_3 - g_4 \right) \\ & + z_4 \left(\cdots \cdots \right) \end{aligned}$$

Finally, we have the following equality at points where $z_4 = 0$:

$$L_{f_0} L_G V_0 = -\varepsilon \cos z_3 \left(x_1 \alpha + z_2 \beta \right) - \left(k_1 z_3 - \varepsilon x_1 \cos z_3 \right) \gamma \tag{7.23}$$

with α, β and γ defined by

$$g_2 - L_{f_0} g_1 = \alpha \tag{7.24}$$

$$g_3 - L_{f_0} g_2 - g_1 = \beta \tag{7.25}$$

$$g_4 - L_{f_0} g_3 = \gamma\,. \tag{7.26}$$

Since $L_{f_0} V_0$ is identically zero and $L_{f_1} V_0$ is equal to z_4, equation (7.13) requires that $L_{f_0} L_G V_0$ be negative when $z_4 = 0$ and $z \neq 0$. All the solutions may be constructed in the following way : first find functions α, β and γ such that, for $z_4 = 0$, $\varepsilon \cos z_3 (x_1 \alpha + z_2 \beta) + (k_1 z_3 - \varepsilon x_1 \cos z_3) \gamma$ is positive, and then, solve (7.24)-(7.26) for the functions g_i (in fact, g_1 may be chosen arbitrary and defines g_2, g_3, g_4).

It is convenient to try to have $g_2 = 0$. This is possible if

$$\begin{aligned} g_1 &= \varepsilon \cos z_3 \, \rho(z_2) \\ \alpha &= \varepsilon \cos z_3 \, \rho'(z_2) x_1 + \varepsilon z_4 \, \rho(z_2) \sin z_3 \end{aligned}$$

where ρ is a function of z_2 such that

$$z_2 \neq 0 \quad \Rightarrow \quad z_2 \, \rho(z_2) > 0 \tag{7.27}$$

(we might take $\rho(z_2) = z_2$ here, but it will be necessary to have ρ growing slower). Then, one may chose

$$\begin{aligned} \beta &= \varepsilon \cos z_3 \, \rho(z_2) \, , \\ \gamma &= k_1 z_3 - \varepsilon x_1 \cos z_3 \, , \end{aligned} \tag{7.28}$$

to make (7.23) negative when $z_4 = 0$. This yields, from (7.24)-(7.26),

$$\begin{aligned} g_2 &= 0 \, , \quad g_3 = 2\varepsilon \cos z_3 \, \rho(z_2) \\ g_4 &= -\varepsilon x_1 \, (1 + 2\rho'(z_2)) \cos z_3 + k_1 z_3 - 2\varepsilon z_4 \, \rho(z_2) \sin z_3 \end{aligned} \tag{7.29}$$

and hence the vector field G from formula (7.4). ¿From now on, it is simpler to express the computations *in the coordinates* (x_1, z_2, z_3, z_4). In these coordinates, G reads :

$$G(x_1, z_2, z_3, z_4) = \begin{bmatrix} -\varepsilon^2 \rho(z_2) \cos^2 z_3 \\ 0 \\ 2\varepsilon \rho(z_2) \cos z_3 \\ -\varepsilon x_1 \, (1 + 2\rho'(z_2)) \cos z_3 + k_1 z_3 - 2\varepsilon z_4 \, \rho(z_2) \sin z_3 \end{bmatrix}$$

and we have, from (7.18) and (7.12),

$$\begin{aligned} W_\lambda = \; & \frac{1}{2} \left(x_1 - \lambda \varepsilon^2 \rho(z_2) \cos^2 z_3 \right)^2 + \frac{1}{2} \, {z_2}^2 + \frac{k_1}{2} \left(z_3 + 2\lambda\varepsilon\rho(z_2) \cos z_3 \right)^2 \\ & + \frac{1}{2} \left((1 - 2\lambda\varepsilon\rho(z_2) \sin z_3) z_4 - \lambda \varepsilon x_1 (1 + 2\rho'(z_2)) \cos z_3 + \lambda k_1 \, z_3 \right)^2 \end{aligned}$$

Recall that $L_{f_1} W_\lambda$ is simply $\partial W_\lambda / \partial z_4$. Hence

$$L_{f_1} W_\lambda = (1 - 2\lambda\varepsilon\rho(z_2) \sin z_3)^2 \, \xi_4 \, , \tag{7.30}$$

$$\text{with } \; \xi_4 = z_4 + \lambda \, \frac{k_1 z_3 - \varepsilon x_1 (1 + 2\rho'(z_2)) \cos z_3}{1 - 2\lambda\varepsilon\rho(z_2) \sin z_3} \, . \tag{7.31}$$

This vanishes either when $\xi_4 = 0$ or when $1 - 2\lambda\varepsilon\rho(z_2)\sin z_3 = 0$. It will be necessary to take $\rho(z_2)$ bounded and λ such that $2\lambda\varepsilon\rho(z_2) < 1$. Provided this is satisfied, a careful computation yields :

$$
\begin{aligned}
\dot{W}_\lambda = -\lambda & \left[k_1 z_3 - \varepsilon \cos z_3 \left(x_1 - \lambda(k_1 + \frac{\varepsilon^2}{2}\cos^2 z_3)\rho(z_2)\right)\right]^2 \\
& -\lambda\varepsilon^2 \cos^2 z_3\, \rho'(z_2) \left[x_1 - \lambda\frac{\varepsilon^2}{2}\cos^2 z_3 \rho(z_2)\right]^2 \\
& -\lambda\varepsilon^2 \cos^2 z_3\, \rho(z_2) \Bigg[z_2 - \lambda^2\rho(z_2) \\
& \qquad \left(\frac{\varepsilon^4}{4}(1+\rho'(z_2))\cos^4 z_3 + k_1\varepsilon^2\cos^2 z_3 + k_1^2\right)\Bigg] \\
& + \xi_4 \Bigg[\lambda R(\lambda, \varepsilon, x_1, z_2, z_3, z_4) + k_1 z_3 - \varepsilon x_1 \cos z_3 \\
& \qquad \frac{(1-2\lambda\varepsilon\rho(z_2)\sin z_3)^2}{1-\varepsilon^2\cos^2 z_3}\left(\varepsilon\cos z_3(x_1 - \varepsilon {z_4}^2 \sin z_3) + u\right)\Bigg]
\end{aligned}
\qquad (7.32)
$$

where R is some function, whose expression is somewhat lengthy but can be easily handled with a computer algebra system.

The three first terms depend only on x_1, z_2, z_3, and they are a negative definite function of these three variables provided $\rho'(z_2)$ is everywhere positive, bounded as well as ρ, say for instance $|\rho| < 1$ and $0 < |\rho'| < 1$, and λ is such that

$$
\lambda^2 \left(\frac{\varepsilon^4}{2} + k_1\varepsilon^2 + k_1^2\right) < 1 . \qquad (7.33)
$$

Note that the requirements on ρ are met for instance with $\rho(z_2) = \frac{2}{\pi}\mathrm{Arctan}\, z_2$, and that, $k_1 = 1$ and $\varepsilon = 0.1$, λ only has to be taken slightly less than 1.

If that is satisfied, then clearly W_λ is a control Lyapunov function, and it satisfies the small control property since equation (7.32) allows one to derive very explicitly a continuous stabilizing control by making the last term non-positive, and negative when $\xi_4 \neq 0$.

7.5 Conclusion

Section 7.3 gives a method to construct some functions from a known "energy-like" function (or "weak" Lyapunov function). We proved (Theorems 2 and 3) that these are control Lyapunov functions under some homogeneity assumptions. However, even for non homogeneous systems, this method can often be used, as illustrated here on the TORA system, to obtain global CLFs. We therefore have good hope that this gives in general a powerful guide to find CLFs from energy-like functions in conservative systems.

In fact the methodology should be refined to be proved to work in more general situations.

Acknowledgments

It is a pleasure to thank Rodolphe Sepulchre, from Université de Liège, for his interest in these ideas, and for suggesting the example in section 7.4.

References

1. Z. Artstein, Stabilization with relaxed control, *Nonlinear Analysis TMA*, 7 (1983), pp. 1163–1173.

2. A. Bacciotti, Local stabilizability of nonlinear control systems, vol. 8 of Series on advances in mathematics for applied sciences, World Scientific, Singapore, River Edge, London, 1992.

3. D. S. Bernstein, ed., Special Issue: A Nonlinear Benchmark Problem, *Int. J. of Robust & Nonlinear Cont.*, 8 (1998), No 4-5.

4. R. W. Brockett, Finite Dimensional Linear Systems, John Wiley and sons, New York, London, Sydney, Toronto, 1970.

5. R. T. Bupp, D. S. Bernstein, and V. T. Coppola, A benchmark problem for nonlinear control design, *Int. J. Robust & Nonlinear Cont.*, 8 (1998), pp. 307–310.

6. J.-M. Coron, L. Praly, and A. R. Teel, Feedback stabilization of nonlinear system : Sufficient conditions and Lyapunov and input-output techniques, in *Trends in Control, a European Perspective*, A. Isidori, ed., Springer-Verlag, 1995.

7. L. Faubourg and J.-B. Pomet, Strict control Lyapunov functions for homogeneous Jurdjevic-Quinn type systems, in *Nonlinear Control Systems Design Symposium (NOLCOS'98)*, H. Huijberts, H. Nijmeijer, A. van der Schaft, and J. Scherpen, eds., IFAC, July 1998, pp. 823–829.

8. L. Faubourg and J.-B. Pomet, Under preparation, 1999.

9. R. A. Freeman and P. V. Kokotovic, Inverse optimality in robust stabilization., *SIAM J. on Control and Optim.*, 34 (1996), pp. 1365–1391.

10. J.-P. Gauthier, Structure des Systèmes non-linéaires, Éditions du CNRS, Paris, 1984.

11. S. T. Glad, Robustness of nonlinear state feedback - a survey, *Automatica*, 23 (1987), pp. 425–435.

12. W. Hahn, Stability of Motion, vol. 138 of Grundlehren der Mathematischen Wissenschaften, Springer-Verlag, Berlin, New-York, 1967.

13. V. Jurjevic and J. P. Quinn, Controllability and stability, *J. of Diff. Equations*, 28 (1978), pp. 381–389.

14. M. Kawski, Homogeneous stabilizing feedback laws, *Control Th. and Adv. Technol.*, 6 (1990), pp. 497–516.

15. J. Kurzweil, On the inversion of Ljapunov's second theorem on stability of motion, A.M.S. Translations, ser. II, 24 (1956), pp. 19–77.

16. J.-P. LaSalle, Stability theory for ordinary differential equations, *J. of Diff. Equations*, 4 (1968), pp. 57–65.

17. R. Outbib and G. Sallet, Stabilizability of the angular velocity of a rigid body revisited, Syst. & Control Lett., 18 (1992), pp. 93–98.

18. L. Praly and Y. Wang, Stabilization in spite of matched un-modeled dynamics and equivalent definition of input-to-state stability.,*Math. of Control, Signals & Systems*, 9 (1996), pp. 1–33.

19. R. Sépulchre, M. Janković, and P. V. Kokotović, Constructive Nonlinear Control, Comm. and Control Engineering, Springer-Verlag, 1997.

20. E. D. Sontag, Feedback stabilization of nonlinear systems, in Robust control of linear systems and nonlinear control (vol. 2 of proceedings of MTNS'89), M. A. Kaashoek, J. H. van Schuppen, and A. Ran, eds., Basel-Boston, 1990, Birkhäuser, pp. 61–81.

8. Bifurcation analysis of a power factor precompensator

*Francisco Gordillo**, *Gerardo Escobar*** and *Javier Aracil**

*Escuela Superior de Ingenieros
Universidad de Sevilla
Camino de los Descubrimientos s/n
41092 Sevilla, Spain
gordillo@esi.us.es
aracil@esi.us.es

**Laboratoire des Signaux et Systèmes, C.N.R.S.
SUPELEC
91192 Gif-sur-Yvette, France
escobar@lss.supelec.fr

Summary.

This chapter presents a bifurcation analysis of a power converter addressed in power electronics literature as power factor precompensator (PFP) for which a feedback linearization scheme is considered. This analysis clarifies qualitatively the different behaviours that appear when the desired output voltage is swept form a permissible low value to a very big value that violates an equilibrium existence condition. This condition appears whenever the parasitic resistive effects in the circuit are considered. The analysis is performed using first harmonic balance which gives a close approximation to the qualitative behavior of this kind of circuits and in most of the cases it predicts and characterizes periodical behaviors.

8.1 Introduction

Control systems community is becoming aware of the complexity and richness of behavior that nonlinear systems can show ([1], [2], [3], [4], [5], [11], [12], [14]). These behaviors include the oscillations, which are out of the scope of the linear systems realm. Traditionally control engineers have been concerned about the local performance around the operating point where the system can be approximated by a linear model. However, the presence of nonlinearities should not be neglected in order to account for global phenomena as limit cycles or even chaos.

Nonlinear systems show two main differences in their behavior with respect to the one of linear systems: (1) they can have multiple equilibria; and (2) they can have limit sets more complex than the limit points, that is, limit cycles and chaotic attractors ([13], [15]). These attractors are organized in attraction basins, giving rise to landscapes far more complex than the ones of the linear systems, that are generically reduced to a single equilibrium point. To study these problems the tools supplied by the qualitative theory of nonlinear dynamical systems, and mainly bifurcation theory, could be used ([9], [10]). Bifurcation theory is concerned with the qualitative changes in the state space configuration as a result of the birth or of the disappearance of limit sets, and their corresponding basins.

In this paper a class of switched power converters commonly addressed as power factor precompensators is studied. The controller is taken from [7] when an adaptive scheme is adopted. The complexity of this system suggests to simplify it in order to perform the stability analysis. In this paper we consider only the non-adaptive case. Apart from this, the controller takes the same expression of [7].

We will be dealing with a forced system where the frequency of the driving variable is much larger than the one of the output variable. To overcome this difficulty a dynamical, first-harmonic balance is performed. The idea of the method, which is proposed in [8], is to substitute each state variable by a truncated Fourier series:

$$x_i(t) \approx a_{i0}(t) + a_{i1}(t)\sin \omega t + a_{i2}(t)\cos \omega t$$

where ω is the oscillation frequency of the variables (to be determined in some cases) and the behavior of $a_{i0}(t), a_{i1}(t)$ and $a_{i2}(t)$ is much slower than ωt, that is, $\dot{a}_{i0}(t), \dot{a}_{i1}(t), \dot{a}_{i2}(t) << \omega$. With this substitution the equations of the system are re-written and harmonics higher than one are neglected. The validity of the approximation depends on the low-pass filter characteristics of the system, which attenuates high frequency components. The great advantage of the method is that, even if it is approximate it easily yields conclusions which can serve as a guide to perform simulations or a more rigorous study. Moreover,, the method is not only local, as happens with linearization, but it catches some global behavior characteristics.

Studying the equilibrium points of the approximate model, a prediction for the limit cycle of the original system is obtained. Furthermore, some interesting conclusions can be drawn for an ideal controller with a feedback linearization philosophy, which show the complexity of the resultant system.

The rest of the paper is organized as follows. In Sect. 8.2 the problem formulation is stated. In Sect. 8.3 the dynamical first harmonic balance is performed while in Sect. 8.4 the results of the approximated model are presented. The paper closes with some conclusions.

8.2 Problem formulation

Consider the full bridge boost type power factor precompesator whose diagram is given in Fig. 8.1. The model describing the average behavior of this system is given by

$$L\dot{x}_1 = -ux_2 - rx_1 + v_i \tag{8.1}$$

$$C\dot{x}_2 = ux_1 - \frac{1}{R}x_2 \tag{8.2}$$

where $v_i = E \sin \omega t$ is the voltage of the AC-line source, x_1 is the input inductor current which is desired to be in phase with v_i, x_2 is the output capacitor voltage to be maintained, in average, in a desired, constant value V_d, R is the output resistance, r is a resistance that concentrates the parasitic effects and $u \in [-1, 1]$ is the control signal which represents the duty ratio of a PWM scheme.

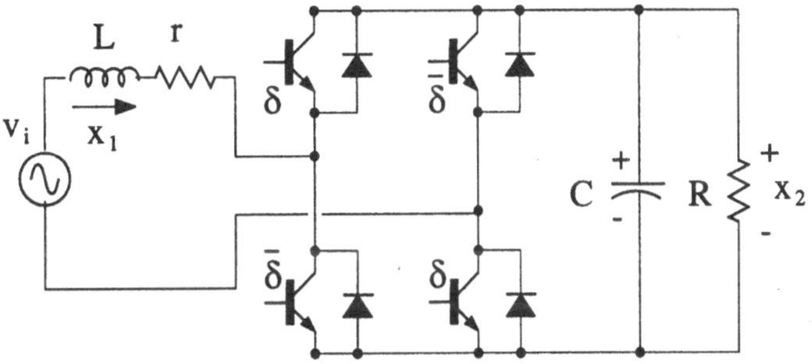

Fig. 8.1. Full bridge PFP boost circuit

The control objectives in this kind of circuits are

- to guarantee that x_1 is in phase with v_i in order to ensure a power factor close unity.
- to drive the bias component of the output voltage x_2 towards a constant desired level V_d.

It is well known that the system thus described turns out to be of nonminimum phase when the capacitor voltage, the variable of interest, is considered as the output [7]. For this reason efforts are directed to control the inductor current in order to indirectly regulate the output voltage. Hence, by defining a tracking problem on the signal x_1 towards a sinusoidal signal reference x_1^* with a constant amplitude I_d and in phase with the input voltage v_i, the twofold problem described above can be accomplished, where the value of I_d, yet to be determined, is such that the bias component of x_2 reaches the desired constant reference V_d. Furthermore, as pointed out in [7], $V_d > E$, that is the circuit can only provide voltages bigger than the amplitude of the sinusoidal source.

We remark that our goal is not to propose a new controller but to study the behavior of the system in closed loop with a given controller. We have taken one of the control schemes reported in [7] which is referred as feedback linearizing controller and is described by the following expression

$$u = \text{sat}(u_c) \tag{8.3}$$

$$u_c = \frac{1}{x_2}[v_i - rx_1 - L\dot{x}_1^* + R_1(x_1 - x_1^*)] \tag{8.4}$$

where u_c is the non-saturated output of the controller, u is the actual input to the system saturated by means of the function sat(.) and x_1^* represents the desired evolution for x_1. This signal must be computed as a function of the desired output voltage V_d and can be obtained by the following expression [7]

$$x_1^* = I_d \sin \omega t \tag{8.5}$$

where I_d comes from the solution of the following second order equation

$$I_d(E - rI_d) = \frac{2V_d^2}{R} \tag{8.6}$$

which has the following two solutions

$$I_d^+ = \frac{E + \sqrt{E^2 - \frac{8V_d^2 r}{R}}}{2r}$$

$$I_d^- = \frac{E - \sqrt{E^2 - \frac{8V_d^2 r}{R}}}{2r} \tag{8.7}$$

At this point we should make the following two assumptions

- The output capacitor voltage x_2 is accurately described by its bias component.
- The inductor current x_1 is accurately described by its first harmonic component.

This assumptions are very close to the circuit response in practice. And besides, they allow us to propose a harmonic balance approximation to study the behavior of the system.

In [7] the following proposition has been established which give a necessary and sufficient condition for the solvability of the problem. The condition gives an upper bound on the amplification gain of the PFP, which becomes arbitrarily large for small parasitic resistance r, but may become restrictive in the case of large r. Notice that this condition actually guarantees the solution of equation (8.6).

Proposition 8.2.1. *Consider the system (8.1), (8.2). A necessary and sufficient condition for the existence of a steady state regime, with the averaged values of x_2 and x_1 reaching V_d and $I_d sin\omega t$, respectively, is that the amplification gain satisfies the upper bound*

$$V_d < E\sqrt{\frac{R}{8r}} \tag{8.8}$$

The motivation for the present study is, on the one hand, to understand the meaning of the two solutions that appear for I_d in Eq. (8.6) provided that (8.8) is fulfilled. And, on the other hand, we would like to study, qualitatively, the changes in the system behavior when V_d varies.

8.3 Dynamical harmonic balance

Direct substitution of (8.3–8.4) into (8.1–8.2) yields the closed-loop system

$$L\dot{x}_1 = -x_2 \text{sat}\left(\frac{1}{x_2}\left[v_i - rx_1 - L\dot{x}_1^* + R_1(x_1 - x_1^*)\right]\right) - \tag{8.9}$$

$$rx_1 + v_i \tag{8.10}$$

$$C\dot{x}_2 = x_1 \text{sat}\left(\frac{1}{x_2}\left[v_i - rx_1 - L\dot{x}_1^* + R_1(x_1 - x_1^*)\right]\right) - \tag{8.11}$$

$$rx_1 + v_i - \frac{1}{R}x_2 \tag{8.12}$$

From the previous section, the reference signal x_1^* and its time derivative used in these expressions are computed as

$$x_1^* = I_d \sin\omega t$$
$$\dot{x}_1^* = I_d\omega \cos\omega t$$

where I_d will be defined below.

In order to perform a dynamical, first-harmonic balance the state variables x_1 and x_2 are expressed as

$$x_1(t) = a_{10}(t) + a_{11}(t)\sin\omega t + a_{12}(t)\cos\omega t \tag{8.13}$$

$$x_2(t) = a_{20}(t) \tag{8.14}$$

where a_{10}, a_{11}, a_{12} and a_{20} are assumed to vary much slower than ωt, that is $\dot{a}_{10}, \dot{a}_{11}, \dot{a}_{12}, \dot{a_{20}} << \omega$. Harmonics with order higher than one of x_1 and all the harmonics of x_2 have been neglected. The reason is that the dynamics of x_2 are slow compared against ωt (neglecting the ripple) while x_1 behaves as a modulated sinusoidal with frequency ω. Notice also that the frequency of the first harmonic of x_1 is assumed to be the same of v_i.

In [7] the following expression for I_d has been used, which allows an easier analysis and implementation of the controller without neglecting the parasitic resistance r.

$$I_d = \frac{2V_d^2}{R(E - ra_{11})}$$

In practice, a_{11} is obtained approximately by means of a rectifier and a filter. Nevertheless, in this paper it is assumed that this harmonic can be exactly obtained and the results will be given for this ideal case.

Due to the fact that a_{10} vanishes quickly and that a_{11}, a_{12} and a_{20} vary slowly, it can be assumed that $u_c(t)$ is approximately equal to a sinusoidal signal

$$u_c(t) \approx \frac{1}{a_{20}}\left[\left(E - ra_{11} + R_1 a_{11} - \frac{2V_d^2}{R(E - ra_{11})}\right)\sin\omega t + \right.$$
$$\left.\left((R_1 - r)a_{12} - \frac{2LV_d^2\omega}{R(E - ra_{11})}\right)\cos\omega t\right] \tag{8.15}$$

Therefore, $u_c(t)$ is approximately a sinusoidal signal whose amplitude U_c is

$$U_c = \frac{1}{a_{20}}\sqrt{(E - ra_{11} + R_1 a_{11} - I_d)^2 + ((R_1 - r)a_{12} - L\omega I_d)^2}$$

Now, the saturated control signal $u(t)$ can be approximated by means of the describing function γ of $u_c(t)$

$$u(t) = \gamma(U_c)u_c(t) \tag{8.16}$$

with

$$\gamma(U_c) = \begin{cases} 1 & \text{if } U_c \in [-1, 1] \\ \frac{2}{\pi}\left(\frac{1}{U_c}\sqrt{1 - 1/U_c^2} + \arcsin(1/U_c)\right) & \text{otherwise} \end{cases}$$

Now, we are ready to perform the harmonic balance. Substituting Eqs. (8.13) (8.14), (8.15) and (8.16) in Eq. (8.1) yields

$$\begin{aligned} &L(\dot{a_{10}} + \dot{a}_{11}\sin\omega t + a_{11}\omega\cos\omega t + \dot{a}_{12}\cos\omega t - a_{12}\omega\sin\omega t) = \\ &(a_{10} + a_{11}\sin\omega t + a_{12}\cos\omega t)[r(\gamma - 1)) - \gamma R_1] - \\ &\gamma L\frac{2V_d^2}{R(E - ra_{11})}\omega\cos\omega t + \gamma R_1\frac{2V_d^2}{R(E - ra_{11})}\sin\omega t + \\ &E\sin\omega t(1 - \gamma) \end{aligned} \tag{8.17}$$

Equating bias, sine and cosine coefficients, the following equations are obtained

$$L\dot{a}_{10} = -\gamma R_1 a_{10} - (1-\gamma) r a_{10} \tag{8.18}$$

$$L\dot{a}_{11} = L a_{12}\omega - \gamma R_1 a_{11} - (1-\gamma) r a_{11} + \gamma R_1 \frac{2V_d^2}{R(E - r a_{11})} + E(1-\gamma) \tag{8.19}$$

$$L\dot{a}_{12} = -L a_{11}\omega - \gamma R_1 a_{12} - (1-\gamma) r a_{12} + \gamma \frac{2LV_d^2}{R(E - r a_{11})}\omega \tag{8.20}$$

With respect to Eq. (8.2), substituting Eqs. (8.13) (8.14), (8.16) and (8.15) in Eq. (8.2) we get

$$C\dot{a}_{20} = \frac{\gamma}{a_{20}} \left[\left(E - r a_{11} + R_1 a_{11} - \frac{2V_d^2 R_1}{R(E - r a_{11})} \right) \sin\omega t + \left(R_1 a_{12} - \frac{2LV_d^2}{R(E - r a_{11})} \right) \cos\omega t \right] (a_{10} + a_{11}\sin\omega t + a_{12}\cos\omega t) - \frac{1}{R} a_{20}$$

Ignoring harmonics of order higher than one and equating bias, sine and cosine coefficients we obtain

$$C\dot{a}_{20} = \frac{\gamma a_{12}}{2a_{20}} \left(R_1 a_{12} - \frac{2LV_d^2\omega}{R(E - r a_{11})} \right) - \frac{a_{20}}{R} + \frac{\gamma a_{11}}{2a_{20}} \left(E - r a_{11} + R_1 a_{11} - \frac{2V_d^2 R_1}{R(E - r a_{11})} \right) \tag{8.21}$$

$$0 = \frac{\gamma a_{10}}{a_{20}} \left(R_1 a_{12} - \frac{2LV_d^2\omega}{R(E - r a_{11})} \right) \tag{8.22}$$

$$0 = \frac{\gamma a_{10}}{a_{20}} \left(E - r a_{11} + R_1 a_{11} - \frac{2V_d^2 R_1}{R(E - r a_{11})} \right) \tag{8.23}$$

Equations (8.22) and (8.23) are not differential due to the fact that the first harmonic terms of x_2 have not been considered. These equation are fulfilled when $a_{10} = 0$ which is achieved quickly by assumption. Equation (8.18) corroborates this assumption.

Therefore, the behavior of a_{10}, a_{11}, a_{12} and a_{20} is approximated by the dynamical system (8.18–8.21). It can be seen that the original, 2-dimensional, non-autonomous model can be approximated by a 4-dimensional, autonomous model out of which some qualitative conclusions can be drawn. In fact, the equilibrium points of system (8.18–8.21) will represent limit cycles of system (8.1–8.2).

8.4 Analysis of the approximated model

The behavior of the system can be studied for different values of V_d. In [7], the impossibility of achieving an output voltage greater than the bound given in (8.8) is presented. Furthermore, there are two solutions of Eq. (8.6) when V_d fulfills this bound. Now, we are ready to clarify these facts.

In order to obtain the equilibria of the system (8.18–8.21) the time derivatives should be equated to zero. In this way, an algebraic system of four equation with four unknowns is obtained

$$0 = -\gamma R_1 \overline{a}_{10} - (1-\gamma) r \overline{a}_{10} \tag{8.24}$$

$$0 = L\overline{a}_{12}\omega - \gamma R_1 \overline{a}_{11} - (1-\gamma) r \overline{a}_{11} + \gamma R_1 \frac{2V_d^2}{R(E - ra_{11})} + E(1-\gamma) \tag{8.25}$$

$$0 = -L\overline{a}_{11}\omega - \gamma R_1 \overline{a}_{12} - (1-\gamma) r \overline{a}_{12} + \gamma \omega L \frac{2V_d^2}{R(E - ra_{11})} \tag{8.26}$$

$$0 = \frac{\gamma \overline{a}_{12}}{2\overline{a}_{20}} \left(R_1 \overline{a}_{12} - \omega L \frac{2V_d^2}{R(E - ra_{11})} \right) - \frac{\overline{a}_{20}}{R} + \frac{\gamma \overline{a}_{11}}{2\overline{a}_{20}} \left(E - r\overline{a}_{11} + R_1 \overline{a}_{11} - R_1 \frac{2V_d^2}{R(E - ra_{11})} \right) \tag{8.27}$$

Non-saturated case. In order to clarify the analysis, the case when u is not saturated is considered first ($\gamma = 1$). In this case, Eqs. (8.24–8.27) give

$$0 = -R_1 \overline{a}_{10} \tag{8.28}$$

$$0 = L\overline{a}_{12}\omega - R_1 \overline{a}_{11} - R_1 I_d \tag{8.29}$$

$$0 = -L\overline{a}_{11}\omega - R_1 \overline{a}_{12} + \omega L I_d \tag{8.30}$$

$$0 = \frac{\overline{a}_{12}}{2\overline{a}_{20}} (R_1 \overline{a}_{12} - \omega L I_d) - \frac{\overline{a}_{20}}{R} + \frac{\overline{a}_{11}}{2\overline{a}_{20}} (E - r\overline{a}_{11} + R_1 \overline{a}_{11} - R_1 I_d) \tag{8.31}$$

Equation (8.28) gives $\overline{a}_{10} = 0$, while Eqs. (8.29) and (8.30) yields

$$0 = \overline{a}_{12} \tag{8.32}$$

$$0 = -\overline{a}_{11} RE + \overline{a}_{11}^2 Rr + 2V_d^2 \tag{8.33}$$

Comparing (8.33) and (8.6) it can be stated that, in the equilibrium $a_{11} = I_d$. When condition (8.8) is fulfilled I_d has two solutions as was pointed out before, and then $\overline{a}_{11}$ has two solutions as well. For both values, Eq. (8.31) yields $\overline{a}_{20} = \pm V_d$. In summary, there are four candidate equilibrium points in the state space $(\overline{a}_{10}, \overline{a}_{11}, \overline{a}_{12}, \overline{a}_{20})$

$$E_1 = (0, I_d^+, 0, V_d)$$
$$E_2 = (0, I_d^-, 0, V_d)$$
$$E_3 = (0, I_d^+, 0, -V_d)$$
$$E_4 = (0, I_d^-, 0, -V_d)$$

It can be verified that, for these four points, $U_c < 1$ and therefore, these points will also be equilibria of the system (8.18–8.21).

We remark that the equilibria E_3 and E_4 are not physically meaningful because the PFP circuit studied here is designed in order to provide a positive output voltage, and moreover, a positive voltage strictly bigger than the amplitude of the sinusoidal source, i.e., $0 < E < \overline{a}_{20}$. Thus, in the sequel, equilibria E_3 and E_4 are not considered.

To study the local stability of the equilibria E_1 and E_2 the eigenvalues of the Jacobian J matrix of system (8.18–8.21) are computed. For example, using the parameters of Tab. (8.1) with V_d equal to 210Volts. $I_d^+ = 72.07$ and

Parameter	Value
r	2.2
E	$115\sqrt{2}$
ω	$2\pi 60$
R	300
L	10×10^{-3}
C	2200×10^{-6}
R_1	10

Table 8.1. Parameter values

$I_d^- = 1.85$. The eigenvalues of the Jacobian matrix for each equilibrium are:

Equilibrium	Eigenvalues	Stability
E_1	-1000 -3.03 -1137.98 38005.83	*Unstable*
E_2	-1000 -3.03 $-987.14 \pm 371.89j$	*Stable*

where evidently E_2 is the unique stable equilibrium point. Besides, since the Jacobian for E_1 has one eigenvalue with positive real part, this equilibrium is a saddle point. It can be verified that E_1 and E_2 have the same stability for other values of V_d. As V_d increases I_d^- approaches I_d^+ so E_1 approaches E_2. For $V_d = V_d^{max}$ both equilibria coalesce. For $V_d > V_d^{max}$ the equilibria disappear. This is indeed the case of the well-known *saddle-node bifurcation* whose bifurcation diagram is sketched in Fig 8.2.

In the real model (8.1–8.2) the equilibria correspond to limit cycles and this bifurcation will be a saddle-node bifurcation of periodic orbits. This bifur-

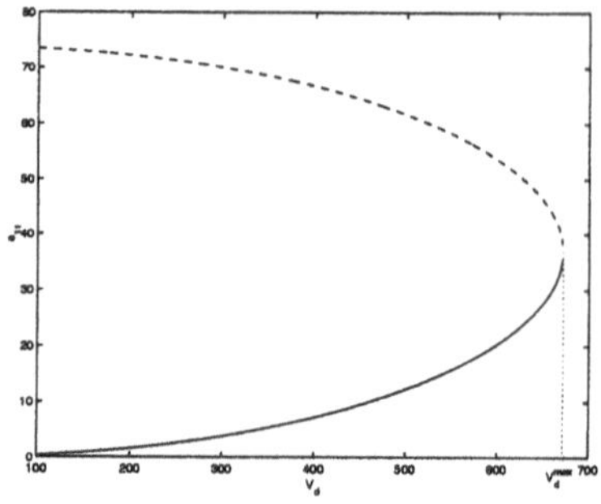

Fig. 8.2. Bifurcation diagram of the saddle-node bifurcation

cation happens when a stable limit cycle coalesces with an unstable limit cycle. As described in Fig. 8.3, before the bifurcation (portrait *a*) there exits a stable limit cycle, surrounded by an unstable one. The bifurcation is produced when the stable limit cycle grows and/or the unstable one decreases (portrait *b*) until both coalesce (portrait *c*) and disappear (portrait *d*); then the system becomes unstable. The transition between the portraits *a*, *b*, *c* and *d* is associated to the variation of the bifurcation parameter V_d.

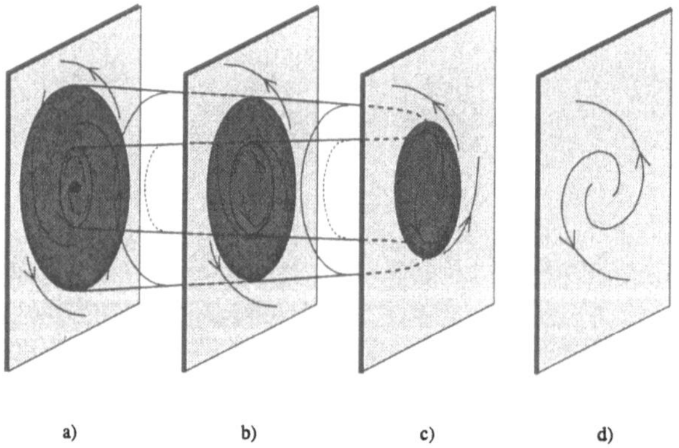

Fig. 8.3. Description of the saddle-node bifurcation of periodic orbits in four phase portraits.

Saturated case. When the control signal saturates the expressions are much more involved. Nevertheless, the bifurcation diagram can still be obtained by means of numerical continuation algorithms, like the program AUTO [6] which has been used here. The resultant diagrams are shown in Fig. 8.4. It can be seen that the lower part of Fig. 8.4a is equal to the one of Fig. 8.2 and, therefore, the saddle-node bifurcation of limit cycles holds. However, the diagram presents a new fold in the upper part which means that for some values

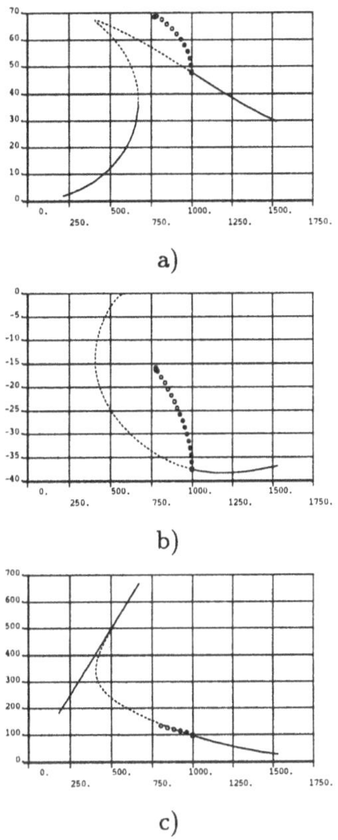

Fig. 8.4. Bifurcation diagram obtained with a continuation program. The x axis represents V_d while the y axis represents: a) a_{11} b) a_{12} c) a_{20}

of V_d there will exist three equilibria of the approximated system. Furthermore, for higher values of V_d a Hopf bifurcation occurs and the equilibrium becomes again stable. Nevertheless, this equilibrium is not desired since a_{12} does not reach zero, nor a_{20} goes to V_d. The resultant behaviour will be even more complicated, most of all in the original model where each equilibrium represents a limit cycle.

The main conclusion of this diagram is that, in spite of the fact that the equilibrium (or limit cycle for the original model) corresponding to I_d^- is the only stable one, the system is locally but not globally stable. The desired output voltage V_d should not be larger than V_d^{max} otherwise the system becomes unstable. Moreover, is not recommended to chose the reference voltage V_d close to V_d^{max}, since there is a big risk for the trajectories to leave the attractive region with a small disturbance due to the fact that the other equilibrium point (the one corresponding to I_d^+) is near the desired one. The last is put in evidence in the simulations presented below.

In Figs. 8.5 and 8.6 two simulations of the original system (8.1–8.2) are presented. In both simulations the parameters take values from Tab. 8.1 for which $V_d^{max} = 671.5$Volts and we have set $V_d = 650$Volts. The only difference between both simulations is in the initial conditions. It can be seen that the behaviour of the system in Fig. 8.5 is satisfactory (x_2 tends towards the desired equilibrium and x_1 has the same phase as v_i) while in Fig. 8.6 is not. This reinforces the fact that the desired equilibrium is locally but not globally stable.

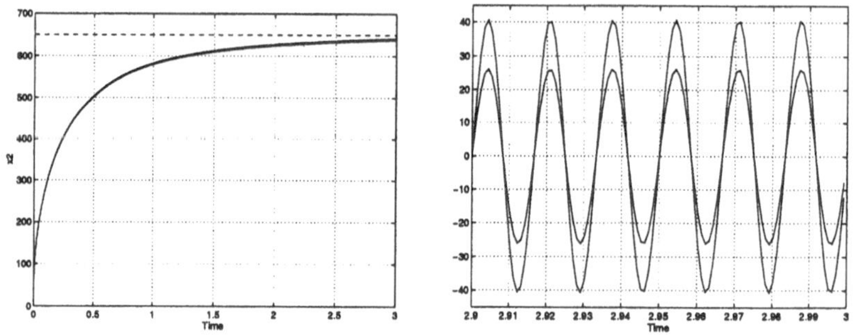

a b

Fig. 8.5. Simulation of the approximate model: a) Evolution of x_2 and b) Evolution of x_1 and $v_i/4$ in the last part of the simulation

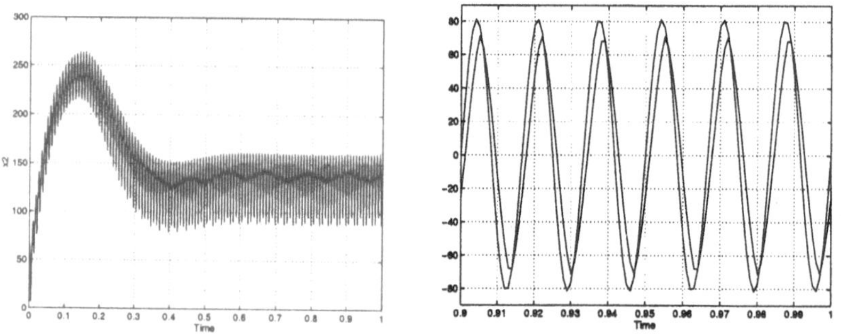

a b

Fig. 8.6. Simulation of the approximate model: a) Evolution of x_2 and b) Evolution of x_1 and $v_i/2$ in the last part of the simulation.

8.5 Conclusions

The ideal behaviour of a power factor precompensator with a feedback linearizing controller proposed in [7] has been studied. The adopted assumptions are:

- The inductor current x_1 can be approximated by a first order harmonic expansion with the same frequency as the input voltage.
- The dynamics of the output voltage x_2 are much slower than the dynamics of x_1.
- The bias component or zero-order harmonic coefficient of x_1 vanishes quickly.
- The first-order harmonic coefficient of x_1 which is in phase with the input voltage is measurable.

The first three assumptions are close to the real situation in practice. Besides, the saturation of the controller has been taken into account.

The analysis has been performed by means of a dynamical first-order harmonic balance which allows us to convert the problem of studying the limit cycles of the original system to the analysis of the equilibria of a new system. The results of the harmonic balance predict, first of all, a saddle-node bifurcation of periodic orbits when the desired output voltage grows. The validity of this prediction has been verified by means of simulations. The analysis also detects a branch of limit cycles for the new model which is a reflection of a complex behavior of the original system when the desired output voltage is very high. This phenomenon is out of the scope of this paper.

The main advantage of the chosen analysis method is that, in spite of its approximate nature, it easily yields good conclusions that have been verified by simulations. In this way, a global perspective of the qualitative behaviour modes of the system has been reached. This perspective allows a better understanding of the control problems of the power factor precompensator found in practice.

Acknowledgments

This work has been supported by the Spanish Ministry of Education and Culture under grant CICYT TAP 97-0553 and by the Consejo Nacional de Ciencia y Tecnología CONACYT of Mexico.

References

1. E.H. Abed, H.O. Wang and A. Tesi, 1996. Control of Bifurcations and Chaos, *The Control Handbook.* Levine W.S. (Ed.), 951-966. *IEEE Press.*

2. J. Alvarez and E. Curiel, 1997. Bifurcations and chaos in a linear control system with saturated input. *Int. J. of Bifurcation and Chaos.*, Vol. 7, No.10, 1811-1821.

3. J. Alvarez, E. Curiel and F. Verduzco, 1997. Complex dynamics in classical control systems. *Syst, and Control Let.*, Vol. 31, 277-285.

4. J. Aracil, K.J. Aström and D. Pagano, 1998. Global bifurcations in the Furuta Pendulum. *IFAC Nonlinear Control Systems Design*, in Enschede (The Netherlands), 37-41.

5. J. Aracil, F. Gordillo and T. Álamo, 1998. Global Stability Analysis of Second-Order Fuzzy Control Systems, *in R. Palm, D. Driankov and H. Hellendorn Eds. Advances in Fuzzy Control*, Springer-Verlag, 11-31.

6. E.J. Doedel, 1981. AUTO, a program for the automatic bifurcation analysis of autonomous systems. *Cong. Numer.*, 30, 265–384.

7. G. Escobar, D. Chevreau, R. Ortega and E. Mendes, 1999. An Adaptive Passivity-Based Controller for a Unity power Factor Rectifier Using a Full Bridge Boost Circuit: Stability Analysis and Experimental Results. Accepted for presentation at the European Control Conference, ECC'99, Karlsruhe, Germany.

8. A. Gelb and W.E. van der Velde, 1968. Multiple input describing function and nonlinear system design. *Mc Graw-Hill.*

9. J.K. Hale and H. Koçak, 1991. Dynamics and bifurcations. *Springer-Verlag.*

10. Y.A. Kuznetsov, 1995. Elements of Applied Bifurcation Theory. *Springer-Verlag.*

11. J. Llibre and E. Ponce, 1996. Global first harmonic bifurcation diagram for odd piecewise linear control systems. *Dynamics and Stability of Systems*, Vol. 11, No. 1, 49-88.

12. J. Moiola and G. Chen, 1996. Hopf Bifurcation Analysis : a Frequency Domain Approach. *World Scientific.*

13. A.H. Nayfeh and B. Balachandran, 1995. Applied Nonlinear Dynamics. *Wiley.*

14. D. Pagano, E. Ponce and J. Aracil, Bifurcations Analysis of Time-Delay Control Systems with Saturation. Accepted for publication in *International Journal of Bifurcation and Chaos in Applied Sciences and Engineering.*

15. S.H. Strogatz,1995. Nonlinear Dynamics and Chaos. *Addison-Wesley.*

9. Stabilization by sampled and discrete feedback with positive sampling rate

Lars Grüne

Fachbereich Mathematik
J.W. Goethe-Universität, Postfach 11 19 32
60054 Frankfurt am Main, Germany
gruene@math.uni-frankfurt.de

Summary.

In this paper we discuss recent results on stabilization by means of discontinuous feedback using a sampled closed loop system. Special emphasis is put on requirements on the sampling rate needed in order to achieve stability of the sampled closed loop system. In particular we focus on the cases where stabilization is possible using a fixed positive sampling rate, i.e. where the intersampling times do not tend to zero. A complete characterization of these cases is given for systems with certain homogenity properties.

9.1 Introduction

The problem of static state feedback stabilization of control systems is one of the classical problems in mathematical control theory. Whereas for linear control systems a well known result states that if a system is asymptotically controllable then it also asymptotically stabilizable by a continuous static state feedback (in fact, even by a linear one), this property fails to hold for nonlinear systems. The well known work of Brockett [2] makes this statement mathematically precise, and the recent survey [24] gives a good introduction into the geometrical obstructions to continuous feedback stabilization.

Thus, looking for stabilizing static state feedback laws for many nonlinear systems it is inevitable to consider also discontinuous feedback laws. This, however, causes a number of problems both in the theoretical analysis (due to the possible lack of uniqueness of trajectories) as well as in the practical implementation. A reasonable solution concept for systems controlled by discontinuous feedbacks is the idea of sampling: For a given sequence of increasing times (the "sampling times") one evaluates the feedback law at each of these sampling times and uses the resulting control value as a (constant) control up to the next sampling time. Continuing iteratively, it is not difficult to see that the usual assumptions on the right hand side of the control system indeed guarantee existence and uniqueness for this sampled trajectory. A slightly more specific concept is the notion of discrete feedback introduced in [7]: Here also sampled trajectories are considered, but instead of using arbitrary sequences of sampling times, here the intersampling times are fixed in advance, possibly depending on the state. Thus the resulting closed loop system is essentially equivalent to a discrete time system.

The concept of sampling is known for quite a while and also used in the context of stabilization, see e.g. [11, 12, 22], but only recently it was observed that for general nonlinear systems asymptotic stabilizability by sampled feedback laws is equivalent to asymptotic controllability [4]. However, one has to be careful in the definition of the behaviour of sampled systems: Although it is immediate that for each sequence of sampling times we obtain a unique trajectory, the asymptotic behaviour of this trajectory may strongly depend on the choice of the sampling rate (i.e. the maximal time allowed between two discrete sampling times) The general equivalence result mentioned above, for instance, is only true if we consider sampling rates tending to 0. Thus, it may be interpreted either as a practical stability result for fixed positive sampling rate, or as "real" stability for all possible limit trajectories for vanishing sampling rates. These, however, will in general not be unique.

In the present paper, we discuss recent results on sampled and discrete stability where special emphasis is put on requirements on the sampling rate needed in order to achieve stability of the sampled closed loop system. In particular we formulate the stability properties under consideration always as stability with *positive sampling rate*, thus describing the system behaviour of individual sampled trajectories rather than limits of trajectories with vanishing sampling rates. Using this approach we attempt to give a suitable mathematical description for implementations of sampled feedback e.g. using some digital controller, in which arbitrary small sampling rates in general will not be realizable. In fact, the investigation of the effect of different sampling rates is interesting not only for discontinuous feedback laws, since in practice also continuous laws are often implemented in a sampled way using digital controllers, and hence essentially the same problems occur.

For general nonlinear systems, a complete characterization of stabilizability with positive sampling rate has not yet been developed. Such a characterization is, however, possible for nonlinear systems with certain homogenity properties, and will be presented and illustrated in this paper.

For simplicity, here we will only deal with global or semi-global phenomena, however, the concepts can be transferred also to the case where stabilizability is only possible from a proper subset of the state space. Concerning the proofs of the results to be presented, instead of giving all the technical details (for which we will refer to the appropriate literature) we restrict ourselves to the main arguments hoping that this allows the reader to get some insight into the problems without being bothered by too much technicalities.

9.2 Setup and definitions

We consider nonlinear control systems of the form

$$\dot{x}(t) = f(x(t), u(t)) \tag{9.1}$$

where $u(\cdot) \in \mathcal{U} := \{u : \mathbb{R} \to U, \text{ measurable and locally essentially bounded}\}$, $U \subseteq \mathbb{R}^m$, $0 \in U$, $f : \mathbb{R}^d \times U \to \mathbb{R}^d$, $f(0,0) = 0$ and f is supposed to be continuous in both variables and Lipschitz in x for each $u \in U$.

For all $t \geq 0$ for which the (unique) open loop trajectory of (9.1) exists for some initial $x_0 \in \mathbb{R}^d$, some control function $u(\cdot) \in \mathcal{U}$, and initial time $t_0 = 0$ we denote it by $x(t, x_0, u(\cdot))$.

In order to characterize asymptotic behaviour at the origin, recall that a function $\alpha : [0, \infty) \to [0, \infty)$ is called of class $\mathcal{K}$, if it satisfies $\alpha(0) = 0$ and is continuous and strictly increasing (and class $\mathcal{K}_\infty$ if it is unbounded), and a continuous function $\beta : [0, \infty)^2 \to [0, \infty)$ is called of class $\mathcal{KL}$, if it is of class $\mathcal{K}$ in the first argument and decreasing to zero in the second variable.

Using this definition we are now able to characterize asymptotic controllability.

Definition 9.2.1. *System (9.1) is called* asymptotically controllable *(to the origin) if there exists a class $\mathcal{KL}$ function β such that for each $x_0 \in \mathbb{R}^d$ there exists $u_{x_0}(\cdot) \in \mathcal{U}$ with*

$$\|x(t, x_0, u_{x_0}(\cdot))\| \leq \beta(\|x_0\|, t) \text{ for all } t \geq 0,$$

and it is called asymptotically controllable with finite controls *if it is asymptotically controllable and there exists an open set $N \ni 0$ and a constant $C > 0$ such that for all $x_0 \in N$ the control $u_{x_0}(\cdot)$ from above can be chosen with $\|u_{x_0}(\cdot)\|_\infty < C$.*

Note that sometimes the definition of asymptotic controllability already includes finite controls, e.g. in [4, 24]. Here we do not necessarily demand this technical property, since for certain results we can do without it.

An important tool in the stability analysis is the control Lyapunov function as given by the following definition.

Definition 9.2.2. *A continuous function $V : \mathbb{R}^d \to [0, \infty)$ is called a* control Lyapunov function, *if it is positive definite (i.e. $V(0) = 0$ iff $V = 0$), proper* *(i.e. $V(x) \to \infty$ as $\|x\| \to \infty$), and there exists a continuous and positive definite function $W : \mathbb{R}^d \to [0, \infty)$ such that for each bounded subset $G \subset \mathbb{R}^d$ there exists a compact subset $U_G \subset U$ with*

$$\min_{v \in \mathrm{co} f(x, U_G)} DV(x; v) \le -W(x) \quad \text{for all } x \in G.$$

Here $DV(x;v)$ denotes the lower directional derivative

$$DV(x; v) := \liminf_{t \searrow 0, v' \to v} \frac{1}{t} \left(V(x + tv') - V(x) \right),$$

$f(x, U_G) := \{f(x, u) \,|\, u \in U_G\}$, and $\mathrm{co} f(x, U_G)$ denotes the convex hull of $f(x, U_G)$.

It is a well known result in control theory that system (9.1) admits a control Lyapunov functions if and only if it is asymptotically controllable with finite controls.

Finally, we introduce the concepts of sampled and discrete feedback control.

Definition 9.2.3. *(i) A* sampled feedback law *is is a (possibly discontinuous) map $F : \mathbb{R}^d \to U$ with $\sup_{x \in K} \|F(x)\| < \infty$ for all compact $K \subset \mathbb{R}^d$ which is applied the following way:*
An infinite sequence $\pi = (t_i)_{i \in \mathbb{N}_0}$ of times satisfying

$$0 = t_0 < t_1 < t_2 < \ldots \quad \text{and} \quad t_i \to \infty \text{ as } i \to \infty$$

is called a sampling schedule. *The values*

$$t_i, \quad \Delta t_i := t_{i+1} - t_i, \quad \text{and} \quad d(\pi) := \sup_{i \in \mathbb{N}_0} \Delta t_i$$

are called the sampling times, intersampling times, *and* sampling rate, *respectively. For any sampling schedule π the corresponding* sampled *or* π-trajectory *$x_\pi(t, x_0, F)$ with initial value $x_0 \in \mathbb{R}^d$ at initial time $t_0 = 0$ is defined inductively by*

$$x_\pi(t, x_0, F) = x(t - t_i, x_i, F(x_i)), \quad \text{for all } t \in [t_i, t_{i+1}], i \in \mathbb{N}_0$$

where $x_i = x_\pi(t_i, x_0, F)$ and $x(t, x_i, F(x_i))$ denotes the (open loop) trajectory of (9.1) with constant control value $F(x_i)$ and initial value x_i.

(ii) A discrete feedback law *is a sampled feedback law together with a (possibly state dependent) time step $h(x) > 0$, $x \in \mathbb{R}^d$ with $\inf_{x \in K} h(x) > 0$ for each compact set $K \not\ni 0$, which for each initial value $x_0 \in \mathbb{R}^d$ is applied using sampling schedules π satisfying $\Delta t_i = h(x_i)$. We denote the corresponding trajectories by $x_h(t_i, x_0, F)$.*

Observe that uniqueness of the π-trajectories for sampled and discrete feedbacks (on their maximal intervals of existence) follows immediately from the definition also for discontinuous feedback maps F.

The sampling schedules specified in the definition of the discrete feedback are uniquely determined by the initial value. The name "discrete feedback" origins from the fact that the resulting sampled closed loop system is in one-to-one correspondence to the discrete time system given by $x_{i+1} = x(h(x_i), x_i, F(x_i))$. The discrete feedback concept is particularly useful when numerical methods involving discretization of trajectories are used for feedback design, since in this situation the time step h can correspond to some numerical discretization parameter, cp. [7].

9.3 Stability concepts for sampled systems

In this section we introduce and discuss appropriate (asymptotic) stability concepts for nonlinear control systems with sampled and discrete feedback. In contrast to the classical case, here we have an additional parameter, namely the sampling rate, which we take into account in our definition.

Definition 9.3.1. *We call the sampled closed loop system from Definition 9.2.3(i)*
(i) semi-globally practically stable with positive sampling rate, *if there exists a class $\mathcal{KL}$ function β such that for each open set $B \subset \mathbb{R}^n$ and each compact set $K \subset \mathbb{R}^n$ satisfying $0 \in B \subset K$ there exists $\Delta t > 0$ such that*

$$x_\pi(t, x_0, F) \notin B \;\Rightarrow\; \|x_\pi(t, x_0, F)\| \leq \beta(\|x_0\|, t)$$

for all $t \geq 0$, all $x_0 \in K$ and all π with $d(\pi) \leq \Delta t$,
(ii) semi-globally stable with positive sampling rate, *if (i) holds and the sampling rate $\Delta t > 0$ can be chosen independently of B,*
(iii) globally practically stable with positive sampling rate *if (i) holds and the sampling rate $\Delta t > 0$ can be chosen independently of K,*
(iv) globally stable with positive sampling rate *if (i) holds and the sampling rate $\Delta t > 0$ can be chosen independently of K and B.*

We call the stability in (i)–(iv) exponential *if the function β satisfies*

$$\beta(\|x_0\|, t) \leq Ce^{-\sigma t}\|x_0\|$$

for constants $C, \sigma > 0$ which may depend on K, and uniformly exponential *if $C, \sigma > 0$ can be chosen independently of K.*

Note that each of the concepts (ii)–(iv) implies (i) which is exactly the s-stability property as defined in [4], cf. also [24, Sections 3.1 and 5.1]. In particular, any of these concepts implies *global stability* for the (possibly nonunique) limiting trajectories as $h \to 0$. The difference "only" lies in the performance with positive sampling rate. From the applications point of view, however, this is an important issue, since e.g. for an implementation of a feedback using some digital controller arbitrary small sampling rates in general will not be realizable. Furthermore if the sampling rate tends to zero the resulting stability may be sensitive to measurement errors, if the feedback is based on a non-smooth control Lyapunov function, see [17, 24]. In contrast to this it is quite straightforward to see that for a fixed sampling rate the stability is in fact robust to small errors in the state measurement (small, of course, relative to the norm of the current state of the system) if there exists a corresponding Lipschitz continuous control Lyapunov function, cf. [24, Theorem E].

Analogously, we define the corresponding concepts for systems controlled by discrete feedback.

Definition 9.3.2. *We call the discrete feedback controlled system from Definition 9.2.3(ii)*
(i) semi-globally practically stable with positive sampling rate, *if there exists a class $\mathcal{KL}$ function β such that*

$$\|x_h(t, x_0, F)\| \leq \beta(\|x_0\|, t)$$

for all $x_0 \in \mathbb{R}^d$,
(ii) semi-globally stable with positive sampling rate, *if (i) holds and the time step h satisfies $\inf_{x \in K} h(x) > 0$ for all compact sets $K \subset \mathbb{R}^d$,*
(iii) globally practically stable with positive sampling rate *if (i) holds and the time step h satisfies $\inf_{x \notin B} h(x) > 0$ for all open sets $B \subset \mathbb{R}^d$ with $0 \in B$,*
(iv) globally stable with positive sampling rate *if (i) holds and the time step h satisfies $\inf_{x \in \mathbb{R}^d} h(x) > 0$.*

Again, we call the stability in (i)–(iv) exponential *if β satisfies*

$$\beta(\|x_0\|, t) \leq Ce^{-\sigma t}\|x_0\|$$

for constants $C, \sigma > 0$ which may depend on K, and uniformly exponential *if $C, \sigma > 0$ can be chosen independently of K.*

In fact, it is not difficult to see that the following implications hold.

Proposition 9.3.1. *Each of the sampled stability concepts from Definition 9.3.1(i)–(iv) implies the corresponding discrete stability concept from Definition 9.3.2(i)–(iv).*

Proof. We show the implication Definition 9.3.1(i) $\Rightarrow$ Definition 9.3.2(i), the other implications follow similarly.

Assume Definition 9.3.1(i) holds for some class $\mathcal{KL}$ function β. Consider a sequence of compact sets $(K_i)_{i\in\mathbb{N}}$ with $K_i \subset K_{i+1}$ and $\bigcup_{i\in\mathbb{N}} K_i = \mathbb{R}^d$, and a sequence of open sets $(B_i)_{i\in\mathbb{N}}$ with $B_{i+1} \subset B_i$ and $\bigcap_{i\in\mathbb{N}} B_i = \{0\}$, such that $B_1 \subset K_1$. For each pair K_i and B_i, $i \in \mathbb{N}$ denote by $\tau_i > 0$ the value Δt from the assumption. Now for each point $x \in \mathbb{R}^d$ we pick the minimal index $i(x) \in \mathbb{N}$ such that $x \in K_{i(x)} \setminus B_{i(x)}$ and define the time step h via $h(x) := \tau_{i(x)}$.

Then from the construction of h and the assumption it follows that

$$||x_h(t, x_0, F)|| \le \beta(||x_0||, 0) \text{ for all } t \ge 0. \tag{9.2}$$

Furthermore we can conclude that for each $i \in \mathbb{N}$ there exists times $t_i > 0$ and $T_i > 0$ with

$$x_h(t, x_0, F) \in B_{i+1} \text{ for all } x_0 \in B_i,\, t \ge t_i$$

and

$$x_h(t, x_0, F) \in K_{i-1} \text{ for all } x_0 \in K_i,\, t \ge T_i.$$

Using the assumption and these two properties by induction it follows that there exist times $s_i > 0$ such that

$$x_h(t, x_0, F) \in B_i \text{ for all } x_0 \in K_i \setminus K_{i-1},\, t \ge s_i.$$

which, together with (9.2) implies the existence of the desired class $\mathcal{KL}$ function (which, however, in general will not coincide with the original β.)

It is an open question whether the converse implications also hold. The only exception is the case of semi-global practical stability where the following (much stronger) theorem holds, whose main statement goes back to [4].

Theorem 9.3.1. *Consider the system (9.1). Then the following properties are equivalent*
(i) The system is asymptotically controllable with finite controls
(ii) There exists a feedback F such that the sampled closed loop system is semi-globally practically stable with positive sampling rate
(iii) There esists a feedback F and a time step h such that the discrete feedback controlled system system is semi-globally practically stable with positive sampling rate

Sketch of Proof. "(ii) $\Rightarrow$ (iii)" follows from Proposition 9.3.1, "(iii) $\Rightarrow$ (i)" is immediately clear.

We sketch the basic idea of the proof of "(i) $\Rightarrow$ (ii)", for a detailled proof see [4]. From [23] asymptotic controllability with finite controls implies the existence of a continuous control Lyapunov function V_0.

For a positive parameter $\beta > 0$ we consider the approximation of V_0 by the (quadratic) inf-convolution

$$V_\beta(x) = \inf_{y \in \mathbb{R}^d} \left\{ V_0(y) + \frac{\|x-y\|^2}{2\beta^2} \right\}$$

For each $x \in \mathbb{R}^d$ we denote by $y_\beta(x)$ a point realizing the minimum on the right hand side of this definition, and define

$$\zeta_\beta(x) := \frac{x - y_\beta(x)}{2\beta^2}.$$

Then a straightforward but technical calculation shows that with F defined by

$$\langle \zeta_\beta(x), f(x, F(x)) \rangle = \inf_{u \in U_G} \langle \zeta_\beta(x), f(x,u) \rangle$$

we obtain

$$V_\beta(x(\tau, x_0, F(x_0))) - V_\beta(x) \leq -\tau W(x_0) + \omega_\beta(x_0)\tau + C(x_0)\frac{\tau^2}{\beta^2} \tag{9.3}$$

where $\omega_\beta(x_0) \to 0$ as $\beta \to 0$, ω_β depends on β and on the modulus of continuity of V in x_0, and $C(x_0) > 0$ is a suitable constant essentially depending on $|f(x_0, F(x_0))|$ (in fact, behind this estimate lies the theory of proximal sub- and supergradients, see e.g. [3] for an exposition).

By a compactness argument now on each ring

$$R = \{x \in \mathbb{R}^d \,|\, 0 < \alpha_1 \leq \|x\| \leq \alpha_2\}$$

we can formulate inequality (9.3) uniformly for $x_0 \in R$, which for $\beta > 0$ and $\tau > 0$ sufficiently small implies that on R the function V_β is a control Lyapunov function which decreases along $x(t, x_0, F(x_0))$ for $t \in [0, \tau]$. Choosing a growing family of rings $R_i \subset R_{i+1}$ covering $\mathbb{R}^d \setminus \{0\}$ and carefully (and rather technically) "gluing" the feedback together on ∂R_i finally yields the assertion.

This result in fact states that a stabilizing sampled feedback can always be found under the assumption of asymptotic controllabilty, provided we allow vanishing sampling rates. The question we want to address in the remaining sections is whether one can give conditions under which (sampled or discrete) stability with some *fixed positive* sampling rate can be achieved. Looking at the Proof of Theorem 9.3.1, one sees that the regularity of V plays a crucial role in estimate (9.3) (via the function ω) and hence in the choice of the time step τ. Thus one might conjecture that certain regularity properties of the corresponding control Lyapunov function could serve as a sufficient condition. However, the example discussed in the next section shows that even the existence of a C^∞ control Lapunov function does not necessarily help.

9.4 A counterexample to stabilizability with positive sampling rate

In this section we briefly discuss an example where stability by discrete or sampled feedback with positive sampling rate is not possible. Consider the system

$$\dot{r} = r(\theta - u)^2 - r^2$$
$$\dot{\theta} = 1$$

written in polar coordinates $r \in [0, \infty)$, $\theta \in [0, 2\pi)$, with $U = \mathbb{R}$.

Obviously the (classical) feedback $F(r, \theta) = \theta$ stabilizes this system.

However, considering the ball $B_1 := \{(\theta, r) \,|\, \theta \in [0, 2\pi), r \in [0, 1)\}$ and fixing some arbitrary $h > 0$ it is easily seen that any trajectory with initial value $(\theta_0, r_0) \in B_1$ which stays in B_1 for $t \in [0, h]$ satisfies

$$\|r(t, r_0, u)\| \geq C_1 r_0 \quad \text{for all} \quad u \in U,\, t \in [0, h] \tag{9.4}$$

for suitable some $C_1 > 0$. Moreover, there exist constants $u_0 > 0$ and $C_2 > 0$ such that

$$\|r(t, r_0, u)\| \leq C_2 r_0 \quad \text{for all} \quad |u| < u_0,\, t \in [0, h] \tag{9.5}$$

and

$$\|r(t, r_0, u)\| \geq r_0 + tC_1 r_0 \quad \text{for all} \quad |u| \geq u_0,\, t \in [0, h]. \tag{9.6}$$

Thus for each $u \in U$ with $|u| < u_0$ from (9.4) and (9.5) we can conclude

$$\begin{aligned} r(h, (r_0, \theta_0), u) - r_0 &\geq \int_0^h (\theta_0 + \tau - u)^2 C_1 r_0 - C_2^2 r_0^2 d\tau \\ &= \left((\theta_0 - u)^2 h + (\theta_0 - u)h^2 + \frac{h^3}{3}\right) C_1 r_0 - hC_2^2 r_0^2 \\ &\geq \frac{h^3}{12} C_1 r_0 - hC_2^2 r_0^2 \end{aligned}$$

for all trajectories with $r(t, (r_0, \theta_0), u) \in B_1$ for all $t \in [0, h]$ where for the last inequality we used that the minimum in $u \in U$ is attained for $u = h/2 + \theta_0$.

From this estimate and inequality (9.6) we can finally conclude that any sampled closed loop trajectory with intersampling times $\Delta t_i \geq h$ with $(\theta_0, r_0) \in B_\varepsilon(0) := \{(\theta, r) \,|\, \theta \in [0, 2\pi), r \in [0, \varepsilon)\}$ leaves $B_\varepsilon(0)$ in finite time for each $\varepsilon < \min\{1, C_1 h^2/(12C_2^2)\}$, and consequently neither sampled nor discrete stability with positive sampling rate are possible.

We finally note that the function $V(r, \theta) = r^2$ is a C^∞ control Lyapunov function for this system, and that the vector field is C^∞, hence these regularity properties do not imply stabilizability with positive sampling rate.

9.5 Homogeneous systems

In this section we summarize results from [10] which show that for homogeneous systems the stabilizability properties with positive sampling rate can be fully determined just by looking at the degree of the system. Stabilization of fig4-sch systems has already been investigated by a number of authors, see e.g. [14, 15, 16, 19, 20, 21, 25].

Let us start by defining what we mean by a "homogeneous system". Here we slightly relax the Lipschitz condition on the vector field f and do only assume Lipschitz continuity in $x \in \mathbb{R}^d \setminus \{0\}$.

Definition 9.5.1. *We call system (9.1)* homogeneous *if there exist*

$$r_i > 0, \quad i = 1, \ldots, d, \quad s_j > 0, \quad j = 1, \ldots, m$$

and $\tau \in (-\min_i r_i, \infty)$ *such that*

$$f(\Lambda_\alpha x, \Delta_\alpha u) = \alpha^\tau \Lambda_\alpha f(x, u) \text{ for all } u \in U, \alpha \geq 0 \tag{9.7}$$

and $\{\Delta_\alpha u \,|\, u \in U\} \subset U$ *for all* $\alpha > 0$.

For compact $U \subset \mathbb{R}^m$ *we call system (9.1)* homogeneous-in-the-state *if there exist* $r_i > 0$, $i = 1, \ldots, d$ *and* $\tau \in (-\min_i r_i, \infty)$ *such that*

$$f(\Lambda_\alpha x, u) = \alpha^\tau \Lambda_\alpha f(x, u) \text{ for all } u \in U, \alpha \geq 0 \tag{9.8}$$

Here

$$\Lambda_\alpha = \begin{pmatrix} \alpha^{r_1} & 0 & \cdots & 0 \\ 0 & \ddots & \ddots & \vdots \\ \vdots & \ddots & \ddots & 0 \\ 0 & \ldots & 0 & \alpha^{r_d} \end{pmatrix} \quad \text{and} \quad \Delta_\alpha = \begin{pmatrix} \alpha^{s_1} & 0 & \cdots & 0 \\ 0 & \ddots & \ddots & \vdots \\ \vdots & \ddots & \ddots & 0 \\ 0 & \ldots & 0 & \alpha^{s_m} \end{pmatrix}$$

are called dilation matrices. *With* $k = \min_i r_i$ *we denote the* minimal power *(of the state dilation) and the value* $\tau \in (-k, \infty)$ *is called the* degree *of the system.*

The core idea for the construction the stabilizing feedback here lies in finding a homogeneous control Lyapunov function in order to apply the construction of the proof of Theorem 9.3.1. This will first be accomplished for systems homogeneous-in-the-state with a very simple structure, using similar ideas as utilized for semilinear systems in [7, 8, 9]. Assume

$$f(\alpha x, u) = \alpha f(x, u) \text{ for all } \alpha > 0, u \in U \tag{9.9}$$

In the notation of Definition 9.5.1 this system is homogeneous-in-the-state with degree $\tau = 0$ with respect to the so-called standard dilation $\Lambda_\alpha = \alpha I$.

We assume furthermore that $U \subset \mathbb{R}^m$ is compact. Defining the exponential growth rates

$$\lambda^t(x_0, u(\cdot)) := \frac{1}{t} \ln \frac{\|x(t, x_0, u(\cdot))\|}{\|x_0\|}$$

for each $x_0 \neq 0$ and each $u(\cdot) \in \mathcal{U}$ it is easily seen from the homogenity property that the system is asymptotically controllable if and only if there exist $T, \sigma > 0$ such that for each $x_0 \neq 0$ there exists $u_{x_0}(\cdot) \in \mathcal{U}$ with

$$\lambda^t(x_0, u_{x_0}(\cdot)) \leq -\sigma < 0 \tag{9.10}$$

for all $x_0 \neq 0$ and all all $t \geq T$, cp. [10, Propositions 3.2 and 3.3]. (The idea of considering exponential growth rates is strongly connected with — and in fact inspired by — the spectral theory developed in [5, 6].)

Another easy consequence of this homogenity property is the fact that the projection

$$s(t, s_0, u(\cdot)) := \frac{x(t, x_0, u(\cdot))}{\|x(t, x_0, u(\cdot))\|}, \quad s_0 = \frac{x_0}{\|x_0\|}$$

of (9.9) onto the unit sphere $\mathbb{S}^{d-1}$ is well defined. A simple application of the chain rule shows that s is the solution of

$$\dot{s}(t) = f_{\mathbb{S}}(s(t), u(t)), \quad f_{\mathbb{S}}(s, u) = f(s, u) - \langle s, f(s, u)\rangle s$$

and that for $s_0 = x_0/\|x_0\|$ the exponential growth rate λ^t satisfies

$$\lambda^t(x_0, u(\cdot)) = \lambda^t(s_0, u(\cdot)) = \frac{1}{t} \int_0^t q(s(\tau, s_0, u(\cdot)), u(\tau)) d\tau$$

with $q(s, u) = \langle s, f(s, u)\rangle$. Thus defining the discounted integral

$$J_\delta(s_0, u(\cdot)) := \int_0^\infty e^{-\delta\tau} q(s(\tau, s_0, u(\cdot)), u(\tau)) d\tau$$

and the corresponding optimal value function

$$v_\delta(s_0) := \inf_{u(\cdot) \in \mathcal{U}} J_\delta(s_0, u(\cdot))$$

from (9.10) and [9, Lemma 3.5(ii)] we obtain that if system (9.9) is asymptotically controllable then for each $\rho \in (0, \sigma)$ there exists $\delta_\rho > 0$ such that for all $\delta \in (0, \delta_\rho]$ and all $s_0 \in \mathbb{S}^{n-1}$ the inequality

$$\delta v_\delta(s_0) < -\rho$$

holds. Note that v_δ is Hölder continuous and bounded for each $\delta > 0$, cp. e.g. [1]. We now fix some $\rho \in (0, \sigma)$ and some $\delta \in (0, \delta_\rho]$ and define

$$V_0(x) := e^{2v_\delta(x/\|x\|)}\|x\|^2.$$

Using Bellman's Optimality Principle a straightforward (but tedious) computation shows that the function V_0 is a control Lyapunov function which is homogeneous with degree $\tau = 1$ with respect to the standard dilation and satisfies

$$\min_{v \in \text{co} f(x,U)} DV_0(x; v) \leq -2\rho V_0(x),$$

cp. [10, Lemma 4.1].

Now we use this function as the starting point in the proof of Theorem 9.3.1, and proceed analogously (for details see [10, Proposition 4.2]). Note that V_β inherits the homogenity properties of V_0, thus F can be chosen to be constant on rays of the form αx, $\alpha > 0$, $x \in \mathbb{R}^d$. Now we chose a ring R containing $\mathbb{S}^{d-1}$ and consider inequality (9.3) (with $W(x) = 2\rho V_0(x)$). Again by a compactness argument, from this inequality we obtain

$$V_\beta(x(\tau, x_0, F(x_0)) - V_\beta(x_0) \leq -\tau\rho V_0(x_0)$$

for some $\beta > 0$ and some $\tau_0 > 0$ sufficiently small, all $\tau \in [0, \tau_0]$ and all $x_0 \in \mathbb{S}^{d-1}$. Then homogenity immediately implies this inequality for all $x_0 \in \mathbb{R}^d$ and hence the resulting feedback law globally stabilizes system (9.9) with positive sampling rate, in fact even uniformly exponentially.

This result can be carried over to the general homogeneous systems from Definition 9.5.1, leading to the following theorem. Here the function $N(x)$ is given by

$$N(x) := \left(\sum_{i=1}^{d} x_i^{\frac{p}{r_i}}\right)^{\frac{1}{p}}$$

with $p = 2\prod_{i=1}^{d} r_i$.

Theorem 9.5.1. *Consider a homogeneous system according to Definition 9.5.1 with dilation matrices Λ_α and Δ_α, minimal power $k > 0$, and degree $\tau \in (-k, \infty)$, and assume asymptotic controllability.*

Then there exists a feedback law $F : \mathbb{R}^d \to U$ satisfying $F(x) \in \Delta_{N(x)} U_0$ for some compact $U_0 \subset U$ and $F(\Lambda_\alpha x) = \Delta_\alpha F(x)$ for all $x \in \mathbb{R}^d$ and all $\alpha \geq 0$ such that the corresponding sampled closed loop system is either
(i) semi-globally stable (if $\tau > 0$), or
(ii) globally uniformly exponentially stable (if $\tau = 0$), or
(iii) globally practically exponentially stable (if $\tau < 0$)
with fixed sampling rate.

The analogous result holds for systems homogeneous-in-the-state; here F satisfies $F(x) \in U$ and $F(\Lambda_\alpha x) = F(x)$ for all $x \in \mathbb{R}^d$ and all $\alpha \geq 0$.

Sketch of Proof. (See [10, Theorem 2.6 and 4.3] for a detailled proof.)
First observe that the function N satisfies $N(\Lambda_\alpha x) = \alpha N(x)$. Hence if for a homogeneous system we replace f by $f(x, \Delta_{N(x)}u)$ we obtain a system homogeneous-in-the-state. A straightforward application of the homogenity yields that this system is asymptotically controllable with control values in some compact set $U_0 \subset U$ if and only if the original homogeneous system is asymptotically controllable, see [10, Proposition 6.1]; conversely if F stabilizes the system homogeneous-in-the-state then $\Delta_{N(x)}F(x)$ stabilizes the original homogeneous system. Hence it suffices to show the theorem for systems homogeneous-in-the-state.

To this end consider the manifold $N^{-1}(1) := \{x \in \mathbb{R}^d \,|\, N(x) = 1\}$. Obviously the function $S(x) = x/\|x\|$ gives a diffeomorphism from $N^{-1}(1)$ to $\mathbb{S}^{d-1}$. Thus the function $\Psi(x) = N(x)^k S(P(x))$ with $P(x) = \Lambda_{N(x)}^{-1}x$ is a continuous cordinate transformation with continuous inverse (both are also differentiable except possibly at the origin), and replacing $f(x,u)$ by $D\Psi(\Psi^{-1}(x))f(\Psi^{-1}(x),u)$ we obtain a system which is homogeneous in the state with respect to the standard dilation and with degree $\gamma = \tau/k$. Replacing further $f(x,u)$ by $f(x,u)\|x\|^{-\gamma}$ — i.e. applying a time transformation — we end up with a system of type (9.9) for which the stabilizing feedback based on the control Lyapunov function V_β has been constructed above. Re-translating this to the general system we first have to remove the time transformation which essentially depends on the sign of degree of the system. This affects the sampling rates and thus leads to the three different cases (i), (ii) and (iii). Since the space transformation does not affect the stability properties of the sampled closed loop system we obtain the assertion.

Note that the numerical methods from [7] are easily transferred to the homogeneous case, thus they give a possibility to compute stabilizing discrete feedbacks numerically. See the next section for examples.

Observe that the stabilizing homogeneous feedback corresponds to a homogeneous control Lyapunov function obtained by applying the coordinate transformation Ψ^{-1} to V_β. This may be used to transfer these results to local results for systems approximated by homogeneous systems, similar to [13, 16, 18].

Furthermore, note that even if a homogeneous system admits a stabilizing continuous static state feedback law, a stabilizing *continuous and homogeneous* static state feedback for does not exist in general, cp. [21]. One way to overcome the non-homogenity is by using dynamic feedbacks, see [14], the above theorem in fact shows that discontinuous feedbacks provide another way.

If we assume Lipschitz continuity of the homogeneous system in the orogin we immediately obtain $\tau \geq 0$, and thus at least semi-global stabilizability. If we assume global Lipschitz continuity (i.e. the existence of a global Lipschitz constant) this implies $\tau = 0$ and thus even global stabilizability.

9.6 Examples

Let us now illustrate our results by two examples. The first example, given by the vector field

$$f(x,u) = \begin{pmatrix} x_1 + u \\ 3x_2 + x_1 u^2 \end{pmatrix} \tag{9.11}$$

for $x = (x_1, x_2)^T \in \mathbb{R}^2$, $u \in U = \mathbb{R}$, is taken from [21] where it has been shown that a stabilizing continuous and homogeneous feedback law cannot exist for this system. The vector field f is homogeneous with $\Lambda_\alpha = \mathrm{diag}(\alpha, \alpha^3)$ and

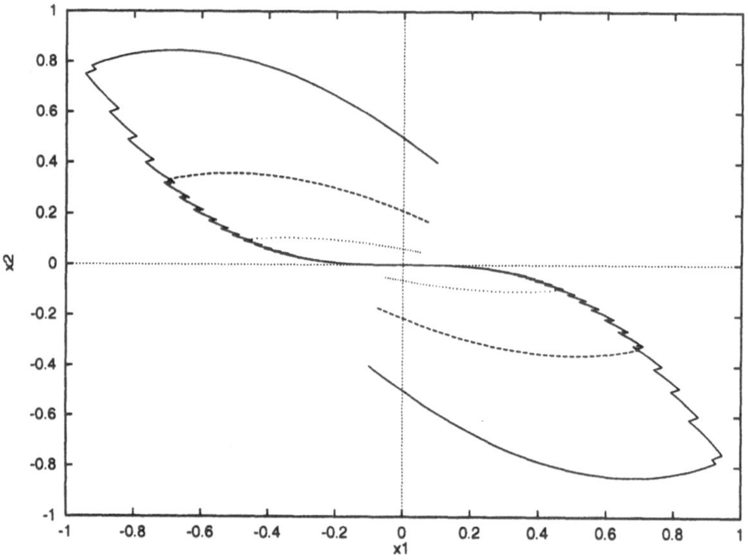

Fig. 9.1. Trajectories for stabilized system (9.11)

$\Delta_\alpha = \alpha$. Thus we obtain $N(x) = (x_1^6 + x_2^2)^{1/6}$. For system (9.11) a stabilizing discrete feedback has been computed numerically using the techniques from [7] extended to the general homogeneous case. Analyzing the switching curves of the numerical feedback in this case it was easy to derive the feedback

$$F(x) = \begin{cases} N(x), & x_1 \le -x_2^3 \\ -N(x), & x_1 > -x_2^3 \end{cases}$$

stabilizing the sampled system for all sufficiently small sampling rates. Figure 9.1 shows the corresponding (numerically simulated) sampled trajectories for some initial values, here the intersampling times have been chosen as $\Delta t_i = 0.01$ for all $i \in \mathbb{N}_0$.

The second example is the nonholonomic integrator given by Brockett [2] as an example for a system being asymptotically null controllable but not stabilizable by a continuous feedback law. In suitable coordinates (cf. [24], where also the physical meaning is discussed) it is given by the vector field

$$f(x,u) = \begin{pmatrix} u_1 \\ u_2 \\ x_1 u_2 \end{pmatrix} \tag{9.12}$$

for $x = (x_1, x_2, x_3)^T \in \mathbb{R}^3$, $u = (u_1, u_2)^T \in U = \mathbb{R}^2$. For this f we obtain homogenity with $\Lambda_\alpha = \mathrm{diag}(\alpha, \alpha, \alpha^2)$ and $\Delta_\alpha = \mathrm{diag}(\alpha, \alpha)$, hence $N(x) = (x_1^4 + x_2^4 + x_3^2)^{1/4}$. Again a stabilizing discrete feedback law has been computed numerically.

Also in this example it should be possible to derive an explicit formula from the numerical results. This is, however, considerably more complicated, since a number of switching surfaces have to be identified. Hence we directly used the numerically computed feedback for the simulation shown in the Figures 9.2–9.4 in different projections; the time step is $h \equiv 0.01$, the controlvalues were chosen as $U_0 = \{-1, 1\}$.

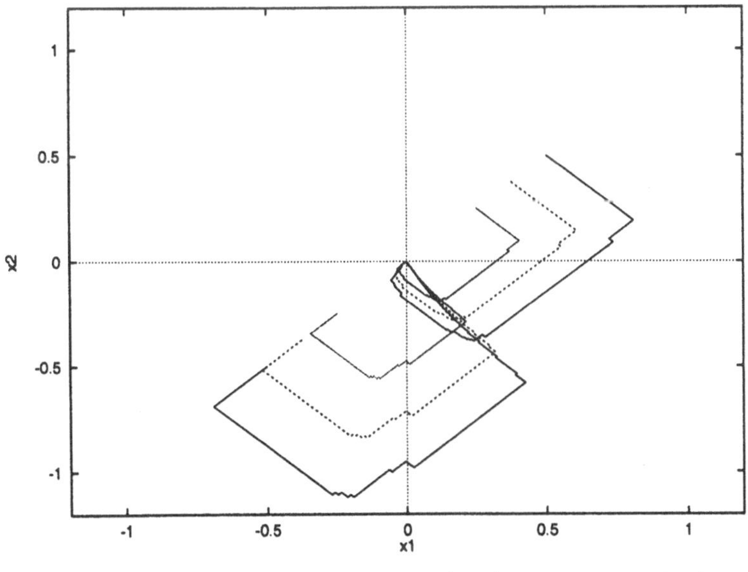

Fig. 9.2. Trajectories for stabilized system (9.12), projected to the (x_1, x_2) plane

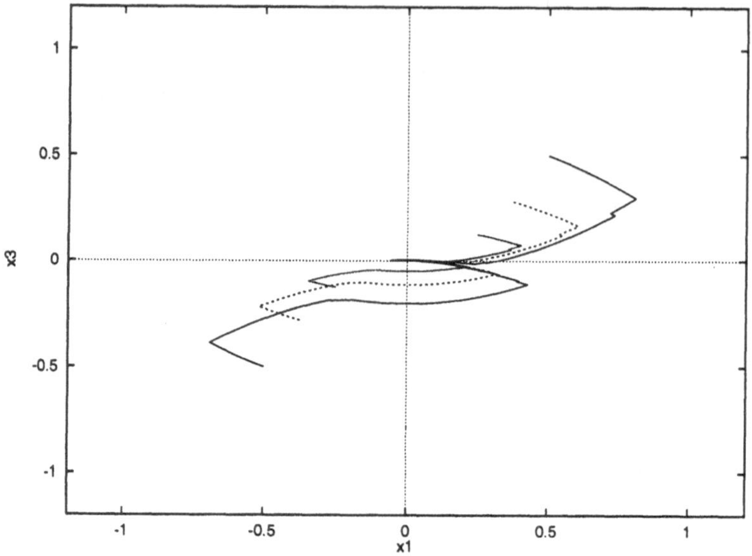

Fig. 9.3. Trajectories for stabilized system (9.12), projected to the (x_1, x_3) plane

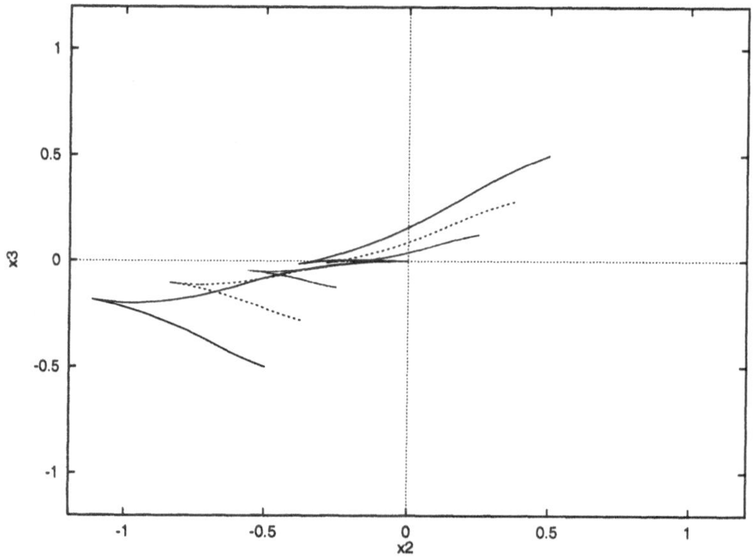

Fig. 9.4. Trajectories for stabilized system (9.12), projected to the (x_2, x_3) plane

Summary:

In this paper we discussed the stabilization of systems with sampled and discrete feedback. Whereas this is always possible provided the system under consideration is asymptotically controllable, in general it can only be achieved by using vanishing intersampling times close to the origin, or far away from it. This fact is illustrated by an example. For general vector fields conditions ensuring sampled or discrete stabilizability with positive sampling rate are still unknown. For homogeneous systems, however, this property can be completely characterized by the degree of homogenity of the system. Two examples of stabilized homogeneous systems illustrate this fact.

Acknowledgement. Parts of this paper have been written while the author was visiting the Dipartimento di Matematica of the Universitá di Roma "La Sapienza", Italy, supported by DFG-Grant GR1569/2-1.

References

1. M. Bardi and I. Capuzzo Dolcetta, Optimal Control and Viscosity Solutions of Hamilton-Jacobi-Bellman equations, Birkhäuser, Boston, 1997.

2. R. Brockett, Asymptotic stability and feedback stabilization, in Differential Geometric Control Theory, R. Brockett, R. Millman, and H. Sussmann, eds., Birkhäuser, Boston, 1983, pp. 181–191.

3. F. Clarke, Methods of Dynamic and Nonsmooth Optimization, vol. 22 of CBMS-NSF Regional Conferences Series in Applied Mathematics, SIAM, Philadelphia, 1989.

4. F. Clarke, Y. Ledyaev, E. Sontag, and A. Subbotin, Asymptotic controllability implies feedback stabilization, IEEE Trans. Autom. Control, 42 (1997), pp. 1394–1407.

5. F. Colonius and W. Kliemann, Maximal and minimal Lyapunov exponents of bilinear control systems, J. Differ. Equations, 101 (1993), pp. 232–275.

6. F. Colonius and W. Kliemann, The Dynamics of Control, Birkhäuser, to appear.

7. L. Grüne, Discrete feedback stabilization of semilinear control systems, ESAIM Control Optim. Calc. Var., 1 (1996), pp. 207–224.

8. L. Grüne, Numerical stabilization of bilinear control systems, SIAM J. Control Optim., 34 (1996), pp. 2024–2050.

9. L. Grüne, Asymptotic controllability and exponential stabilization of nonlinear control systems at singular points, SIAM J. Control Optim., 36 (1998), pp. 1585–1603.

10. L. Grüne, Homogeneous state feedback stabilization of homogeneous control systems, Preprint, Nonlinear Control Abstracts NCA-9-2-981203. Submitted.

11. H. Hermes, On stabilizing feedback attitude control, J. Optimization Theory Appl., 31 (1980), pp. 373–384.

12. H. Hermes, On the synthesis of stabilizing feedback control via Lie algebraic methods, SIAM J. Control Optim., 18 (1980), pp. 352–361.

13. H. Hermes, Nilpotent and high order approximations of vector field systems, SIAM Rev., 33 (1991), pp. 238–264.

14. H. Hermes, Homogeneous feedback control for homogeneous systems, System & Control Lett., 24 (1995), pp. 7–11.

15. A. Iggidr and J.-C. Vivalda, Global stabilization of homogeneous polynomial systems, Nonlinear Anal., 18 (1992), pp. 1181–1186.

16. M. Kawski, Homogeneous feedback stabilization, in New Trends in Systems Theory (Genova, 1990), Progr. Systems Control Theory, vol. 7, Birkhäuser, Boston, 1991, pp. 464–471.

17. Y. Ledyaev and E. Sontag, A Lyapunov characterization of robust stabilization, J. Nonlinear Anal. To appear.

18. L. Rosier, Homogeneous Liapunov function for continuous vector fields, System & Control Lett., 19 (1992), pp. 467–473.

19. E. Ryan, Universal stabilization of a class of nonlinear systems with homogeneous vector field, System & Control Lett., 26 (1995), pp. 177–184.

20. R. Sepulchre and D. Ayels, Homogeneous Lyapunov functions and necessary conditions for stabilization, Math. Control Signals Systems, 9 (1996), pp. 34–58.

21. R. Sepulchre and D. Ayels, Stabilizability does not imply homogeneous stabilizability for controllable homogeneous systems, SIAM J. Control Optim., 34 (1996), pp. 1798–1813.

22. E. Sontag, Nonlinear regulation: The piecewise linear approach, IEEE Trans. Autom. Control, AC-26 (1981), pp. 346–358.

23. E. Sontag, A Lyapunov like characterization of asymptotic controllability, SIAM J. Control Optim., 21(1983), pp. 462–471

24. E. Sontag, Stability and stabilization: Discontinuities and the effect of disturbances, in Proc. NATO Advanced Study Institute "Nonlinear Analysis, Differential Equations, and Control" (Montreal, Jul/Aug 1998), Kluwer, 1999.

25. J. Tsinias, Remarks on feedback stabilizability of homogeneous systems, Control Theory Adv. Tech., 6 (1990), pp. 533–542.

10. Linear controllers for tracking chained-form systems

Erjen Lefeber, Anders Robertsson** and Henk Nijmeijer*,****

*Faculty of Mathematical Science
Dep. of Systems, Signals, and Control
University of Twente, P.O. Box 217
7500 AE Enschede, The Netherlands
Phone +31 53 489 3459/489 3370 Fax +31 53 434 0733,
A.A.J.Lefeber@math.utwente.nl
H.Nijmeijer@math.utwente.nl

**Department of Automatic Control
Lund Institute of Technology
Lund University, P.O.Box 118
SE-221 00, Lund, Sweden
Phone +46 46 222 8790 Fax +46 46 138118,
Anders.Robertsson@control.lth.se

***Faculty of Mechanical Engineering
Eindhoven University of Technology, P.O. Box 513
5600 MB Eindhoven, The Netherlands

Summary.

In this paper we study the tracking problem for the class of nonholonomic systems in chained-form. In particular, with the first and the last state component of the chained-form as measurable output signals, we suggest a solution for the tracking problem using output feedback by combining a time-varying state feedback controller with an observer for the chained-form system. For the stability analysis of the "certainty equivalence type" of controller we use a cascaded systems approach. The resulting closed loop system is globally $\mathcal{K}$-exponentially stable.

10.1 Introduction

In recent years a lot of interest has been devoted to (mainly) stabilization and tracking of nonholonomic dynamic systems, see e.g. [1, 6, 8, 15, 17]. One of the reasons for the attention is the lack of a continuous static state feedback control since Brockett's necessary condition for smooth stabilization is not met, see [3]. The proposed solutions to this problem follow mainly two routes, namely discontinuous and/or time-varying control. For a good overview, see the survey paper [12] and the references therein.

It is well known that the kinematic model of several nonholonomic systems can be transformed into a *chained-form system*. The global tracking problem for chained-form systems has recently been addressed in [4, 6, 7, 8, 17, 20]. In this paper we consider the tracking problem for chained form systems by means of output feedback, where we consider as output the first and last state component of the chained-form. To our knowledge, this problem has only been addressed in [9] where a backstepping approach is used. Our results are based on the construction of a linear time varying state feedback controller in combination with an observer. However, the stability analysis and design are based on results for (time-varying) cascaded systems [18]. In the design we divide the chained-form into a cascade of two sub-systems which we can stabilize independently of each other, and furthermore a similar partition into cascaded systems can be done for the controller-observer combination, where the same stability results apply. Regarding the latter part, similar ideas were recently presented for the combination of high-gain controllers and high-gain observer for a class of triangular nonlinear systems [2], see also [13].

The organization of the paper is as follows. Section 10.2 contains some definitions, preliminary results and the problem formulation. Section 10.3 addresses the tracking problem based on time-varying state feedback and in section 10.4 we design an exponentially convergent observer for the chained-form system. In section 10.5 we combine the control law from section 10.3 with the observer from section 10.4 in a "certainty equivalence" sense. This yields a globally $\mathcal{K}$-exponentially stable closed loop system under the condition of persistently exciting reference trajectories. Finally, section 10.6 concludes the paper.

10.2 Preliminaries and problem formulation

In this section we introduce the definitions and theorems used in the remainder of this paper and formulate the problem under consideration. We start with some basic stability concepts in 10.2.1, present a result for cascaded systems in 10.2.2 and recall some basic results from linear systems theory in 10.2.3. We conclude this section with the problem formulation in 10.2.4.

10.2.1 Stability

To start with, we recall some basic concepts (see e.g. [11, 23]).

Definition 10.2.1. *A continuous function* $\alpha : [0, a) \to [0, \infty)$ *is said to belong to* class $\mathcal{K}$ *if it is strictly increasing and* $\alpha(0) = 0$.

Definition 10.2.2. *A continuous function* $\beta : [0, a) \times [0, \infty) \to [0, \infty)$ *is said to belong to* class $\mathcal{KL}$ *if, for each fixed s, the mapping* $\beta(r, s)$ *belongs to class* $\mathcal{K}$ *with respect to r and, for each fixed r, the mapping* $\beta(r, s)$ *is decreasing with respect to s and* $\beta(r, s) \to 0$ *as* $s \to \infty$.

Consider the system

$$\dot{x} = f(t, x), \quad f(t, 0) = 0 \quad \forall t \geq 0 \tag{10.1}$$

with $x \in I\!R^n$ and $f(t, x)$ piecewise continuous in t and locally Lipschitz in x.

Definition 10.2.3. *The system (10.1) is* uniformly stable *if for each* $\epsilon > 0$ *there is* $\delta = \delta(\epsilon) > 0$, *independent of* t_0, *such that*

$$||x(t_0)|| < \delta \Rightarrow ||x(t)|| < \epsilon, \quad \forall t \geq t_0 \geq 0.$$

Definition 10.2.4. *The system (10.1) is* globally uniformly asymptotically stable
(GUAS) *if it is uniformly stable and globally attractive, that is, there exists a class* $\mathcal{KL}$ *function* $\beta(\cdot, \cdot)$ *such that for every initial state* $x(t_0)$:

$$||x(t)|| \leq \beta(||x(t_0)||, t - t_0), \quad \forall t \geq t_0 \geq 0$$

Definition 10.2.5. *The system (10.1) is* globally exponentially stable (GES) *if there exist* $k > 0$ *and* $\gamma > 0$ *such that for any initial state*

$$||x(t)|| \leq ||x(t_0)|| k \exp[-\gamma(t - t_0)].$$

A slightly weaker notion of exponential stability is the following (cf. [21])

Definition 10.2.6. *We call the system (10.1)* globally $\mathcal{K}$-exponentially stable *if there exist* $\gamma > 0$ *and a class* $\mathcal{K}$ *function* $\kappa(\cdot)$ *such that*

$$||x(t)|| \leq \kappa(||x(t_0)||) \exp[-\gamma(t - t_0)] \tag{10.2}$$

Definition 10.2.7. *We call the (locally integrable) vector-valued function*

$$w(t) = [w_1(t), \ldots, w_n(t)]^T$$

persistently exciting *if there exist* $\delta, \varepsilon_1, \varepsilon_2 > 0$ *such that for all* $t > 0$:

$$\varepsilon_1 I \leq \int_t^{t+\delta} w(\tau) w(\tau)^T d\tau \leq \varepsilon_2 I$$

10.2.2 Cascaded systems

Consider the system

$$\begin{cases} \dot{z}_1 = f_1(t, z_1) + g(t, z_1, z_2)z_2 \\ \dot{z}_2 = f_2(t, z_2) \end{cases} \tag{10.3}$$

where $z_1 \in I\!R^n$, $z_2 \in I\!R^m$, $f_1(t, z_1)$ is continuously differentiable in (t, z_1) and $f_2(t, z_2)$, $g(t, z_1, z_2)$ are continuous in their arguments, and locally Lipschitz in z_2 and (z_1, z_2) respectively.

We can view the system (10.3) as the system

$$\Sigma_1 : \dot{z}_1 = f_1(t, z_1)$$

that is perturbed by the state of the system

$$\Sigma_2 : \dot{z}_2 = f_2(t, z_2).$$

When Σ_2 is asymptotically stable, we have that z_2 tends to zero, which means that the z_1 dynamics in (10.3) asymptotically reduces to Σ_1. Therefore, we can hope that asymptotic stability of both Σ_1 and Σ_2 implies asymptotic stability of (10.3).

Unfortunately, this is not true in general. However, from the proof presented in [18] it can be concluded that:

Theorem 10.2.1 (based on [18]). *The cascaded system (10.3) is GUAS if the following three assumptions hold:*

- assumption on Σ_1: *the system $\dot{z}_1 = f_1(t, z_1)$ is GUAS and there exists a continuously differentiable function $V(t, z_1) : I\!R_+ \times I\!R^n \to I\!R$ that satisfies*

$$W(z_1) \le V(t, z_1), \tag{10.4}$$

$$\frac{\partial V}{\partial t} + \frac{\partial V}{\partial z_1} \cdot f_1(t, z_1) \le 0, \quad \forall \|z_1\| \ge \eta, \tag{10.5}$$

$$\left\| \frac{\partial V}{\partial z_1} \right\| \|z_1\| \le cV(t, z_1), \quad \forall \|z_1\| \ge \eta, \tag{10.6}$$

  where $W(z_1)$ is a positive definite proper function and $c > 0$ and $\eta > 0$ are constants,

- assumption on the interconnection: *the function $g(t, z_1, z_2)$ satisfies for all $t \ge t_0$:*

$$\|g(t, z_1, z_2)\| \le \theta_1(\|z_2\|) + \theta_2(\|z_2\|)\|z_1\|,$$

  where $\theta_1, \theta_2 : I\!R_+ \to I\!R_+$ are continuous functions,

- assumption on Σ_2: *the system* $\dot{z}_2 = f_2(t, z_2)$ *is GUAS and for all* $t_0 \geq 0$*:*

$$\int_{t_0}^{\infty} \|z_2(t_0, t, z_2(t_0))\| dt \leq \kappa(\|z_2(t_0)\|),$$

where the function $\kappa(\cdot)$ *is a class* $\mathcal{K}$ *function,*

Remark 10.2.1. Notice that the assumption on Σ_1 is slightly weaker than the one presented in [18]. However, under the assumption mentioned above the result can still be shown to be true by (almost) exactly copying the proof presented in [18].

Lemma 10.2.1 (see [17]). *If in addition to the assumptions in Theorem 10.2.1 both* $\dot{z}_1 = f_1(t, z_1)$ *and* $\dot{z}_2 = f_2(t, z_2)$ *are globally* $\mathcal{K}$*-exponentially stable, then the cascaded system (10.3) is globally* $\mathcal{K}$*-exponentially stable.*

10.2.3 Linear time-varying systems

Consider the linear time-varying system

$$\begin{aligned} \dot{x}(t) &= A(t)x(t) + Bu(t) \\ y(t) &= Cx(t) \end{aligned} \tag{10.7}$$

and let $\Phi(t, t_0)$ denote the state-transition matrix for the system $\dot{x} = A(t)x$. We recall some results from linear control theory (cf. [10, 19]).

Definition 10.2.8. *The pair* $(A(t), B)$ *is* uniformly controllable *if there exist* $\delta, \varepsilon_1, \varepsilon_2 > 0$ *such that for all* $t > 0$*:*

$$\varepsilon_1 I \leq \int_t^{t+\delta} \Phi(t, \tau) B B^T \Phi^T(t, \tau) d\tau \leq \varepsilon_2 I$$

Definition 10.2.9. *The pair* $(A(t), C)$ *is* uniformly observable *if there exist* $\delta, \epsilon_1, \epsilon_2 > 0$ *such that for all* $t > 0$*:*

$$\epsilon_1 I \leq \int_{t-\delta}^{t} \Phi^T(\tau, t-\delta) C^T C \Phi(\tau, t-\delta) d\tau \leq \epsilon_2 I$$

From linear systems theory several methods are available to exponentially stabilize the linear time-varying system (10.7) via state or output feedback, in case the pairs $(A(t), B)$ and $(A(t), C)$ are uniformly controllable and observable respectively (cf. [19]):

Theorem 10.2.2. *Suppose that the system (10.7) is uniformly controllable and define for* $\alpha > 0$

$$W_\alpha(t, t+\delta) = \int_t^{t+\delta} 2e^{4\alpha(t-\tau)}\Phi(t,\tau)BB^T\Phi^T(t,\tau)d\tau \tag{10.8}$$

Then given any constant α *the state feedback* $u(t) = K_\alpha(t)x(t)$ *where*

$$K_\alpha(t) = -B^T W_\alpha^{-1}(t, t+\delta) \tag{10.9}$$

is such that the resulting closed-loop state equation is uniformly exponentially stable with rate α.

Theorem 10.2.3. *Suppose that the system (10.7) is uniformly controllable and uniformly observable and define for* $\alpha > 0$

$$M_\alpha(t-\delta, t) = \int_{t-\delta}^{t} 2e^{4\alpha(\tau-t)}\Phi^T(\tau, t-\delta)C^TC\Phi(\tau, t-\delta)d\tau$$

Then given $\alpha > 0$, *for any* $\eta > 0$ *the linear dynamic output feedback*

$$\begin{aligned}
u(t) &= K_{\alpha+\eta}(t)\hat{x}(t) \\
\dot{\hat{x}}(t) &= A(t)\hat{x}(t) + Bu(t) + H_{\alpha+\eta}(t)[y(t) - \hat{y}(t)], \qquad \hat{x}(t_0) = \hat{x}_0 \\
\hat{y}(t) &= C\hat{x}(t)
\end{aligned}$$

with feedback and observer gains

$$\begin{aligned}
K_{\alpha+\eta}(t) &= -B^T W_{\alpha+\eta}^{-1}(t, t+\delta) \\
H_{\alpha+\eta}(t) &= \left[\Phi^T(t-\delta, t)M_{\alpha+\eta}(t-\delta, t)\Phi(t-\delta, t)\right]^{-1} C^T
\end{aligned} \tag{10.10}$$

is such that the closed-loop state equation is uniformly exponentially stable with rate α.

10.2.4 Problem formulation

The class of chained-form nonholonomic systems we study in this paper is given by the following equations

$$\begin{aligned}
\dot{x}_1 &= u_1 \\
\dot{x}_2 &= u_2 \\
\dot{x}_3 &= x_2u_1 \\
&\vdots \\
\dot{x}_n &= x_{n-1}u_1
\end{aligned} \tag{10.11}$$

where $x = (x_1, \ldots, x_n)$ is the state, u_1 and u_2 are control inputs.

Consider the problem of tracking a reference trajectory (x_r, u_r) generated by the chained-form system:

$$
\begin{aligned}
\dot{x}_{1,r} &= u_{1,r} \\
\dot{x}_{2,r} &= u_{2,r} \\
\dot{x}_{3,r} &= x_{2,r}u_{1,r} \\
&\vdots \\
\dot{x}_{n,r} &= x_{n-1,r}u_{1,r}
\end{aligned}
\tag{10.12}
$$

where we assume $u_{1,r}(t)$ to $u_{2,r}(t)$ be continuous functions of time. This reference trajectory can be generated by any of the motion planning techniques available from the literature.

When we define the tracking error $x_e = x - x_r$ we obtain as tracking error dynamics

$$
\begin{aligned}
\dot{x}_{1,e} &= u_1 - u_{1,r} && = u_1 - u_{1,r} \\
\dot{x}_{2,e} &= u_2 - u_{2,r} && = u_2 - u_{2,r} \\
\dot{x}_{3,e} &= x_2u_1 - x_{2,r}u_{1,r} && = x_{2,e}u_{1,r} + x_2(u_1 - u_{1,r}) \\
&\vdots && \vdots \\
\dot{x}_{n,e} &= x_{n-1}u_1 - x_{n-1,r}u_{1,r} && = x_{n-1,e}u_{1,r} + x_{n-1}(u_1 - u_{1,r})
\end{aligned}
\tag{10.13}
$$

The state feedback tracking control problem then can be formulated as

Problem 10.2.1 (State feedback tracking control problem). Find appropriate state feedback laws u_1 and u_2 of the form

$$
u_1 = u_1(t, x, x_r, u_r) \quad \text{and} \quad u_2 = u_2(t, x, x_r, u_r) \tag{10.14}
$$

such that the closed-loop trajectories of (10.13,10.14) are globally uniformly asymtotically stable.

Consider the system (10.11) with output

$$
y = \begin{bmatrix} x_1 \\ x_n \end{bmatrix} \tag{10.15}
$$

then it is easy to show (see e.g. [1]) that the system (10.11) with output (10.15) is locally observable at any $x \in \mathbb{R}^n$. Clearly, this is the minimal number of state components we need to know for solving the output-feedbacl tracking problem.

Now we can formulate the output feedback tracking problem as

Problem 10.2.2 (Output feedback tracking control problem). Find appropriate control laws u_1 and u_2 of the form

$$u_1 = u_1(t, \hat{x}, y, x_r, u_r) \quad \text{and} \quad u_2 = u_2(t, \hat{x}, y, x_r, u_r) \tag{10.16}$$

where $\hat{x}$ is generated from an observer

$$\dot{\hat{x}} = f(t, \hat{x}, y, x_r, u_r) \tag{10.17}$$

such that the closed-loop trajectories of (10.13,10.16,10.17) are globally uniformly asymptotically stable.

10.3 The state feedback problem

The approach we use to solve our problem is based on the recently developed studies on cascaded systems [5, 14, 16, 18, 22], and that of Theorem 10.2.1 in particular, since it deals with time-varying systems.

We search for a subsystem which, with a stabilizing control law, can be written in the form $\dot{z}_2 = f_2(t, z_2)$ that is asymptotically stable. In the remaining dynamics we can then replace the appearance of z_2 by 0, leading to the system $\dot{z}_1 = f_1(t, z_1)$. If this system is asymptotically stable we might be able to conclude asymptotic stability of the overall system using Theorem 10.2.1.

Consider the tracking error dynamics (10.13). We can stabilize the $x_{1,e}$ dynamics by using the linear controller

$$u_1 = u_{1,r} - c_1 x_{1,e} \tag{10.18}$$

which yields GES for $x_{1,e}$, provided $c_1 > 0$.

If we now set $x_{1,e}$ equal to 0 in (10.13) we obtain

$$\begin{aligned} \dot{x}_{2,e} &= u_2 - u_{2,r} \\ \dot{x}_{3,e} &= x_{2,e} u_{1,r} \\ &\vdots \\ \dot{x}_{n,e} &= x_{n-1,e} u_{1,r} \end{aligned} \tag{10.19}$$

where we used (10.18).

Notice that the system (10.19) is a linear time-varying system:

$$\begin{bmatrix} \dot{x}_{2,e} \\ \dot{x}_{3,e} \\ \dot{x}_{4,e} \\ \vdots \\ \dot{x}_{n,e} \end{bmatrix} = \underbrace{\begin{bmatrix} 0 & \cdots & \cdots & \cdots & 0 \\ u_{1,r}(t) & \ddots & & & \vdots \\ 0 & u_{1,r}(t) & \ddots & & \vdots \\ \vdots & \ddots & \ddots & \ddots & \vdots \\ 0 & \cdots & 0 & u_{1,r}(t) & 0 \end{bmatrix}}_{A(t)} \begin{bmatrix} x_{2,e} \\ x_{3,e} \\ x_{4,e} \\ \vdots \\ x_{n,e} \end{bmatrix} + \underbrace{\begin{bmatrix} 1 \\ 0 \\ 0 \\ \vdots \\ 0 \end{bmatrix}}_{B} (u_2 - u_{2,r}) \tag{10.20}$$

that can be made exponentially stable by means of the controller $u(t) = K(t)x(t)$ provided the system (10.20) is uniformly controllable (cf. Theorem 10.2.2).

This observation leads to the following

Proposition 10.3.1. *Assume that the reference trajectory (x_r, u_r) satisfying (10.12) to be tracked by our chained form system is given. Define*

$$w_r(t,t_0) = \begin{bmatrix} 1 \\ \int_{t_0}^{t} u_{1,r}(\tau)d\tau \\ \left(\int_{t_0}^{t} u_{1,r}(\tau)d\tau\right)^2 \\ \vdots \\ \left(\int_{t_0}^{t} u_{1,r}(\tau)d\tau\right)^{n-2} \end{bmatrix} = \begin{bmatrix} 1 \\ x_{1,r}(t) - x_{1,r}(t_0) \\ (x_{1,r}(t) - x_{1,r}(t_0))^2 \\ \vdots \\ (x_{1,r}(t) - x_{1,r}(t_0))^{n-2} \end{bmatrix}$$

and assume that there exist $\delta, \varepsilon_1, \varepsilon_2 > 0$ such that for all $t > 0$:

$$\varepsilon_1 I \leq \int_t^{t+\delta} w_r(t,\tau)w_r(t,\tau)^T d\tau \leq \varepsilon_2 I. \tag{10.21}$$

Consider the system (10.13) in closed-loop with the linear controller

$$\begin{aligned} u_1 &= u_{1,r} - c_1 x_{1,e} \\ u_2 &= u_{2,r} + K(t) \begin{bmatrix} x_{2,e} \\ \vdots \\ x_{n,e} \end{bmatrix} \end{aligned} \tag{10.22}$$

where $c_1 > 0$ and $K(t)$ is given by

$$K(t) = -[1\ 0\ 0\ \ldots\ 0] \left[\int_t^{t+\delta} 2e^{4\alpha(t-\tau)} w_r(t,\tau)w_r(t,\tau)^T d\tau \right]^{-1} \tag{10.23}$$

with $\alpha > 0$. If $x_{2,r}(t), \ldots, x_{n-1,r}(t)$ are bounded then the closed-loop system (10.13, 10.22) is globally $\mathcal{K}$-exponentially stable.

Proof. We can see the closed-loop system (10.13,10.22) as a system of the form (10.3) where

$$\begin{aligned} z_1 &= [x_{2,e}, \dots, x_{n,e}]^T && (10.24)\\ z_2 &= x_{1,e} && (10.25)\\ f_1(t, z_1) &= (A(t) - BK(t))z_1 && (10.26)\\ f_2(t, z_2) &= -c_1 z_2 && (10.27)\\ g(t, z_1, z_2) &= -c_1[0, x_2, x_3, \dots, x_{n-1}]^T && (10.28) \end{aligned}$$

with

$$A(t) = \begin{bmatrix} 0 & \dots\dots & \dots & 0 \\ u_{1,r}(t) & \ddots & & \vdots \\ 0 & \ddots \ \ddots & & \vdots \\ \vdots & \ddots \ \ddots & \ddots & \vdots \\ 0 & \dots \ 0 & u_{1,r}(t) & 0 \end{bmatrix} \qquad B = \begin{bmatrix} 1 \\ 0 \\ \vdots \\ \vdots \\ 0 \end{bmatrix}$$

To be able to apply Theorem 10.2.1 we need to verify the three assumptions:

- assumption on Σ_1: Due to the assumption (10.21) on $u_{1,r}(t)$ we have that the system (10.20) is uniformly controllable (cf. Remark 10.3.2). Therefore, from Theorem 10.2.2 we know that $\dot{z}_1 = f_1(t, z_1)$ is GES and therefore GUAS. From converse Lyapunov theory (see e.g. [11]) the existence of a suitable V is guaranteed.
- assumption on connecting term: Since $x_{2,r}, \dots, x_{n-1,r}$ are bounded, we have

$$\begin{aligned} \|g(t, z_1, z_2)\| &\leq c_1 \left(\left\| \begin{bmatrix} 0 \\ x_{2,r} \\ \vdots \\ x_{n-1,r} \end{bmatrix} \right\| + \left\| \begin{bmatrix} 0 \\ x_{2,e} \\ \vdots \\ x_{n-1,e} \end{bmatrix} \right\| \right) && (10.29)\\ &\leq c_1 M + c_1 \|x\| && (10.30) \end{aligned}$$

- assumption on Σ_2: Follows from GES of $\dot{x}_2 = -c_1 x_2$.

Therefore, we can conclude GUAS from Theorem 10.2.1. Since both Σ_1 and Σ_2 are GES, Lemma 10.2.1 gives the desired result.

Remark 10.3.1. Notice that since

$$u_1(t) = u_{1,r}(t) - c_1 x_{1,e}(t_0) \exp(-c_1(t - t_0))$$

the condition (10.21) on $u_{1,r}(t)$ is satisfied if and only if a similar condition on $u_1(t)$ is satisfied (i.e. in which the r is omitted).

Therefore, we can also see the closed-loop system (10.13,10.22) as a system of the form (10.3) where

$$z_1 = [x_{2,e}, \dots, x_{n,e}]^T \tag{10.31}$$

$$z_2 = x_{1,e} \tag{10.32}$$

$$f_1(t, z_1) = (A(t) - BK(t))z_1 \tag{10.33}$$

$$f_2(t, z_2) = -c_1 z_2 \tag{10.34}$$

$$g(t, z_1, z_2) = -c_1[0, x_{2,r}, x_{3,r}, \dots, x_{n-1,r}]^T \tag{10.35}$$

with

$$A(t) = \begin{bmatrix} 0 & \dots\dots & \dots & 0 \\ u_1(t) & \ddots & & \vdots \\ 0 & \ddots & \ddots & \vdots \\ \vdots & \ddots & \ddots \quad \ddots & \vdots \\ 0 & \dots & 0 \quad u_1(t) & 0 \end{bmatrix} \qquad B = \begin{bmatrix} 1 \\ 0 \\ \vdots \\ \vdots \\ 0 \end{bmatrix}$$

Notice that we redefined $A(t)$ and that correspondingly the connecting term $g(t, z_1, z_2)$ changed. When we modify our controller accordingly, i.e. redefine $K(t)$ in (10.22) as

$$K(t) = -[1\ 0\ 0\ \dots\ 0]\left[\int_t^{t+\delta} 2e^{4\alpha(t-\tau)} w(t,\tau) w(t,\tau)^T d\tau\right]^{-1} \tag{10.36}$$

with $\alpha > 0$, where

$$w(t, t_0) = \begin{bmatrix} 1 \\ \int_{t_0}^t u_1(\tau)d\tau \\ \left(\int_{t_0}^t u_1(\tau)d\tau\right)^2 \\ \vdots \\ \left(\int_{t_0}^t u_1(\tau)d\tau\right)^{n-2} \end{bmatrix} = \begin{bmatrix} 1 \\ x_1(t) - x_1(t_0) \\ (x_1(t) - x_1(t_0))^2 \\ \vdots \\ (x_1(t) - x_1(t_0))^{n-2} \end{bmatrix}$$

we can copy the proof.

Moreover, since the connecting term $g(t, z_1, z_2)$ now can be bounded by a constant, we can claim not only global $\mathcal{K}$-exponential stability, but even GES. However, the disadvantage of (10.36) in comparison to (10.23) is that it depends on the state and therefore can not be determined a priori for a known reference trajectory in contrast to (10.23).

Remark 10.3.2. Notice that in general it is not easy to compute $\Phi(t, t_0)$. However, for the system (10.20) this turns out not to be too difficult, due to the nice and simple structure of the matrix $A(t)$. We find:

$$\Phi(t,t_0) = \begin{bmatrix} f_0(t,t_0) & 0 & \dots & 0 \\ f_1(t,t_0) & f_0(t,t_0) & \ddots & \vdots \\ \vdots & \ddots & \ddots & 0 \\ f_{n-2}(t,t_0) & \dots & f_1(t,t_0) & f_0(t,t_0) \end{bmatrix}$$

where

$$f_k(t,t_0) = \frac{1}{k!}\left[\int_{t_0}^{t} u_{1,r}(\sigma)d\sigma\right]^k = \frac{1}{k!}\left[x_{1,r}(t) - x_{1,r}(t_0)\right]^k$$

¿From this it is also straightforward to see that uniform controllability of the system (10.20) can also rephrased as persistency of excitation of the vector

$$\begin{bmatrix} f_0(t,t_0) \\ f_1(t,t_0) \\ \vdots \\ f_{n-2}(t,t_0) \end{bmatrix}$$

Remark 10.3.3. Notice that the persistency of excitation condition (10.21) is obviously met in case $\liminf_{t\to\infty} u_{1,r}(t) = \varepsilon > 0$, so that the results of [6, 7, 8, 17] are included in this result.

10.4 An observer

The observability property for chained-form systems was considered in [1], in which a (local) observer was proposed in case $u_1(t) = -c_1 x_1(t)$. In this section we propose a globally exponentially stable observer for the chained system under an observability condition which is related to the persistence of excitation with respect to the first component of the state.

Proposition 10.4.1. *Consider the chained-form system (10.11) with output (10.15). Define*

$$w(t,t_0) = \begin{bmatrix} 1 \\ \int_{t_0}^{t} u_1(\tau)d\tau \\ \left(\int_{t_0}^{t} u_1(\tau)d\tau\right)^2 \\ \vdots \\ \left(\int_{t_0}^{t} u_1(\tau)d\tau\right)^{n-2} \end{bmatrix} = \begin{bmatrix} 1 \\ x_1(t) - x_1(t_0) \\ (x_1(t) - x_1(t_0))^2 \\ \vdots \\ (x_1(t) - x_1(t_0))^{n-2} \end{bmatrix}$$

Assume that there exist $\delta, \varepsilon_1, \varepsilon_2 > 0$ such that for all $t > 0$:

$$\varepsilon_1 I \leq \int_t^{t+\delta} w(t,\tau)w(t,\tau)^T d\tau \leq \varepsilon_2 I.$$

Then the observer

$$\begin{bmatrix} \dot{\hat{x}}_2 \\ \dot{\hat{x}}_3 \\ \dot{\hat{x}}_4 \\ \vdots \\ \dot{\hat{x}}_n \end{bmatrix} = \begin{bmatrix} 0 & \dots & \dots & \dots & 0 \\ u_1 & \ddots & & & \vdots \\ 0 & u_1 & \ddots & & \vdots \\ \vdots & \ddots & \ddots & \ddots & \vdots \\ 0 & \dots & 0 & u_1 & 0 \end{bmatrix} \begin{bmatrix} \hat{x}_2 \\ \hat{x}_3 \\ \hat{x}_4 \\ \vdots \\ \hat{x}_n \end{bmatrix} + \begin{bmatrix} 1 \\ 0 \\ 0 \\ \vdots \\ 0 \end{bmatrix} u_2 + H(t)\tilde{x}_n$$

where $\tilde{x}_n = x_n - \hat{x}_n$ and

$$H(t) = \left[\Phi^T(t-\delta,t)M_\alpha(t-\delta,t)\Phi(t-\delta,t)\right]^{-1} C^T \quad (\alpha > 0)$$

guarantees that the observation error $\tilde{x} = x - \hat{x}$ converges to zero exponentially.

Proof. Because of the assumption on $u_1(t)$ we have a uniformly observable linear time-varying system. The result follows readily from standard linear theory (see e. g. [19]).

10.5 The output feedback problem

In section 3 we derived a state feedback controller for tracking a desired trajectory, whereas in section 4 we derived an observer for a system in chained-form. We can also combine these two results in a "certainty equivalence" sense:

Proposition 10.5.1. *For the reference trajectory x_r, u_r) satisfying (10.12) define*

$$w_r(t,t_0) = \begin{bmatrix} 1 \\ \displaystyle\int_{t_0}^t u_{1,r}(\tau)d\tau \\ \left(\displaystyle\int_{t_0}^t u_{1,r}(\tau)d\tau\right)^2 \\ \vdots \\ \left(\displaystyle\int_{t_0}^t u_{1,r}(\tau)d\tau\right)^{n-2} \end{bmatrix} = \begin{bmatrix} 1 \\ x_{1,r}(t) - x_{1,r}(t_0) \\ (x_{1,r}(t) - x_{1,r}(t_0))^2 \\ \vdots \\ (x_{1,r}(t) - x_{1,r}(t_0))^{n-2} \end{bmatrix}$$

and assume that there exist $\delta, \varepsilon_1, \varepsilon_2 > 0$ such that for all $t > 0$:

$$\varepsilon_1 I \leq \int_t^{t+\delta} w_r(t,\tau) w_r(t,\tau)^T d\tau \leq \varepsilon_2 I.$$

Consider the system (10.13) in closed-loop with the linear controller-observer-combination

$$\begin{aligned}
u_1 &= u_{1,r} - c_1 x_{1,e} \\
u_2 &= u_{2,r} + K(t) \begin{bmatrix} \hat{x}_{2,e} \\ \vdots \\ \hat{x}_{n,e} \end{bmatrix}
\end{aligned}$$

$$\begin{bmatrix} \dot{\hat{x}}_{2,e} \\ \dot{\hat{x}}_{3,e} \\ \dot{\hat{x}}_{4,e} \\ \vdots \\ \dot{\hat{x}}_{n,e} \end{bmatrix} = \begin{bmatrix} 0 & \dots & \dots & \dots & 0 \\ u_{1,r} & \ddots & & & \vdots \\ 0 & u_{1,r} & \ddots & & \vdots \\ \vdots & \ddots & \ddots & \ddots & \vdots \\ 0 & \dots & 0 & u_{1,r} & 0 \end{bmatrix} \begin{bmatrix} \hat{x}_{2,e} \\ \hat{x}_{3,e} \\ \hat{x}_{4,e} \\ \vdots \\ \hat{x}_{n,e} \end{bmatrix} + \begin{bmatrix} 1 \\ 0 \\ 0 \\ \vdots \\ 0 \end{bmatrix} u_2 + H(t)\tilde{x}_n \tag{10.37}$$

where $\tilde{x}_n = x_n - \hat{x}_n$, $c_1 > 0$ and $K(t)$ and $H(t)$ are given by

$$K(t) = -[1\ 0\ 0\ \dots\ 0] \left[\int_t^{t+\delta} 2e^{4\alpha(t-\tau)} w_r(t,\tau) w_r(t,\tau)^T d\tau \right]^{-1}$$

$$H(t) = \left[2e^{4\alpha(\tau-t)} w_r(\tau, t-\delta) w_r(\tau, t-\delta)^T d\tau\, \Phi(t-\delta, t) \right]^{-1} w_r(t, t-\delta)$$

with $\alpha > 0$. If $x_{2,r}, \dots, x_{n-1,r}$ are bounded then the closed-loop system (10.13,10.37) is globally $\mathcal{K}$-exponentially stable.

Proof. Similar to that of Proposition 10.3.1. Note that due to the assumption on $u_{1,r}$ we have both uniform controllability and uniform controllability. ¿From Theorem 10.2.3 we then know that the system

$$\begin{bmatrix} \dot{z}_1 \\ \dot{\hat{z}}_1 \end{bmatrix} = \begin{bmatrix} A(t) & -BK(t) \\ A(t) + H(t)C & -BK(t) - H(t)C \end{bmatrix} \begin{bmatrix} z_1 \\ \hat{z}_1 \end{bmatrix}$$

is globally exponentially stable.

Since we can write the closed-loop system (10.13,10.37) as

$$\begin{aligned}
\begin{bmatrix} \dot{z}_1 \\ \dot{\hat{z}}_1 \end{bmatrix} &= \begin{bmatrix} A(t) & -BK(t) \\ A(t) + H(t)C & -BK(t) - H(t)C \end{bmatrix} \begin{bmatrix} z_1 \\ \hat{z}_1 \end{bmatrix} + \begin{bmatrix} g(t, \begin{bmatrix} z_1 \\ \hat{z}_1 \end{bmatrix}, z_2) \\ 0 \end{bmatrix} z_2 \\
\dot{z}_2 &= -c_1 z_2
\end{aligned}$$

where

$$z_1 = [x_{2,e}, \dots, x_{n,e}]^T$$
$$z_2 = x_{1,e}$$
$$g(t, \begin{bmatrix} z_1 \\ \hat{z}_1 \end{bmatrix}, z_2) = -c_1[0, x_2, x_3, \dots, x_{n-1}]^T$$

The proof can be completed similar to that of Proposition 10.3.1.

10.6 Conclusions

In this paper we considered the tracking problem for nonholonomic systems in chained-form by means of output feedback. We combined a time-varying state feedback controller with an observer for the chained-form in a "certainty equivalence" way. The stability of the closed loop system is shown using results from time-varying cascaded systems. Under a condition of persistence of excitation, we have shown globally $\mathcal{K}$-exponential stability of the closed loop system.

References

1. A. Astolfi. Discontinuous output feedback control of nonholonomic chained systems. In *Proceedings of the 3rd European Control Conference*, pages 2626–2629, Rome, Italy, September 1995.

2. G. Besançon. State-affine systems and observer based control. In *Proceedings of the Fourth IFAC Symposium on Nonlinear Control Systems Design (NOLCOS'98)*, volume 2, pages 399–404, Enschede, The Netherlands, July 1998.

3. R.W. Brockett. Asymptotic stability and feedback stabilization. In R.W. Brockett, R.S. Millman, and H.J. Sussmann, editors, *Differential Geometric Control Theory*, pages 181–191. Birkhauser, Boston, MA, 1983.

4. M. Egerstedt, X. Hu, and A. Stotsky. Control of a car-like robot using a virtual vehicle approach. In *Proceedings of the 37th Conference on Decision and Control*, pages 1502–1507, Tampa, Floria, USA, December 1998.

5. M. Jankovic, R. Sepulchre, and P. Kokotovic. Constructive Lyapunov design of nonlinear cascades. *IEEE Transactions on Automatic Control*, 41(12):1723–1735, December 1996.

6. Z.-P. Jiang, E. Lefeber, and H. Nijmeijer. Stabilization and tracking of a nonholonomic mobile robot with saturating actuators. In *Proceedings of CONTROLO'98, 3rd Portugese Conference on Automatic Control*, volume 2, pages 315–320, Coimbra, Portugal, September 1998.

7. Z.-P. Jiang and H. Nijmeijer. Backstepping-based tracking control of nonholonomic chained systems. In *Proceedings of the 4th European Control Conference*, Brussels, Belgium, 1997. Paper 672 (TH-M A2).

8. Z.-P. Jiang and H. Nijmeijer. Tracking control of mobile robots: a case study in backstepping. *Automatica*, 33(7):1393–1399, 1997.

9. Z.-P. Jiang and H. Nijmeijer. Observer-controller design for nonholonomic systems. In H. Nijmeijer and T.I. Fossen, editors, *New Trends in Nonlinear Observer Design*, Lecture Notes in Control and Information Sciences. Springer Verlag, Londen, 1999.

10. T. Kailath. *Linear Systems*. Prentice-Hall, 1980.

11. H.K. Khalil. *Nonlinear Systems*. Prentice-Hall, Upper Saddle River, NJ USA, second edition, 1996.

12. I. Kolmanovsky and N.H. McClamroch. Developments in nonholonomic control problems. *IEEE Control Systems Magazine*, 16(6):20–36, December 1995.

13. A. Loría, E. Panteley, H. Nijmeijer, and T.I. Fossen. Robust adaptive control of passive systems with unknown disturbances. In *Proceedings of the Fourth IFAC Symposium on Nonlinear Control Systems Design (NOLCOS'98)*, volume 3, pages 866–871, Enschede, The Netherlands, July 1998.

14. F. Mazenc and L. Praly. Adding integrators, saturated controls and global asymptotic stibilization of feedforward systems. *IEEE Transactions on Automatic Control*, 41:1559–1579, 1996.

15. P. Morin and C. Samson. Application of backstepping techniques to the time-varying exponential stabilisation of chained form systems. *European Journal of Control*, 3:15–36, 1997.

16. R. Ortega. Passivity properties for stabilization of nonlinear cascaded systems. *Automatica*, 29:423–424, 1991.

17. E. Panteley, E. Lefeber, A. Loría, and H. Nijmeijer. Exponential tracking control of a mobile car using a cascaded approach. In *Proceedings of the IFAC Workshop on Motion Control*, pages 221–226, Grenoble, September 1998.

18. E. Panteley and A. Loría. On global uniform asymptotic stability of nonlinear time-varying systems in cascade. *Systems and Control Letters*, 33(2):131–138, February 1998.

19. W.J. Rugh. *Linear Systems Theory*. Prentice-Hall, 1996.

20. C. Samson and K. Ait-Abderrahim. Feedback control of a nonholonomic wheeled cart in cartesian space. In *Proceedings of the 1991 IEEE International Conference on Robotics and Automation*, pages 1136–1141, Sacramento, USA, 1991.

21. O.J. Sørdalen and O. Egeland. Exponential stabilization of nonholonomic

chained systems. *IEEE Transactions on Automatic Control*, 40(1):35–49, January 1995.

22. E. Sontag. Smooth stabilization implies coprime factorization. *IEEE Transactions on Automatic Control*, 34:435–443, 1989.

23. M. Vidyasagar. *Nonlinear Systems Analysis.* Prentice-Hall, Englewood Cliffs, New Jersey USA, second edition, 1993.

11. Asymptotic methods in stability analysis and control

Luc Moreau and Dirk Aeyels

SYSTeMS
Universiteit Gent
Technologiepark 9
B-9052 Zwijnaarde, Belgium
Tel: +32 9 264 56 55 Fax: +32 9 264 58 40
Luc.Moreau@rug.ac.be
Dirk.Aeyels@rug.ac.be

Summary.

Systems depending on a small parameter are considered, and the interplay between convergence results for trajectories and stability properties is investigated. Under a continuity assumption for solutions, practical and exponential stability results are obtained. The presented results are useful for constructive stabilization of control systems and robustness analysis.

11.1 Introduction

The aim of this chapter is to derive stability results for dynamical systems based on a trajectory-oriented approach.

Consider a system that depends on a small parameter $\varepsilon > 0$

$$\dot{x} = f^{\varepsilon}(t, x)$$

and a system

$$\dot{x} = g(t, x) ,$$

with the assumption that trajectories of $\dot{x} = f^{\varepsilon}(t, x)$ converge uniformly on compact time intervals to trajectories of $\dot{x} = g(t, x)$ as $\varepsilon \downarrow 0$. The following

[1] This chapter presents research results of the Belgian Programme on Interuniversity Poles of Attraction, initiated by the Belgian State, Prime Minister's Office for Science, Technology and Culture. The scientific responsibility rests with its authors.
[2] The first author is supported by BOF grant 011D0696 of the University of Ghent.

question relates convergence of trajectories with stability properties: if the origin is an asymptotically stable equilibrium point of $\dot{x} = g(t,x)$, what does this imply for $\dot{x} = f^\varepsilon(t,x)$? In this chapter, we offer various answers to this question.

The relevance of this for control theory is twofold. First, it provides a theoretical motivation for the stabilization paradigm where one constructs feedback laws that depend on a small parameter ε in such a way that trajectories of the closed-loop system converge uniformly on compact time intervals to trajectories of an asymptotically stable system as $\varepsilon \downarrow 0$; see, for example, [10, 5, 8]. Second, results obtained in this framework are relevant for a robustness analysis of control systems with respect to general perturbations that leave trajectories close to those of the idealized model.

Related stability results may be found, for example, in [12, 1, 4, 10]. The stability results from these references are obtained within the framework of Lyapunov theory. The present chapter offers an alternative approach for stability analysis, which is not Lyapunov-based. We believe that the ideas presented here may also prove useful in other contexts.

11.2 Preliminaries

The state space for all systems featuring in the present chapter is $\mathbf{R}^n$ with $n \in \mathbf{N}$. $\|\cdot\|$ denotes the Euclidean norm on $\mathbf{R}^n$.

Throughout the chapter we consider the following data: a system that depends on a parameter $\varepsilon \in (0, \varepsilon_0]$ ($\varepsilon_0 \in (0, \infty)$)

$$\dot{x} = f^\varepsilon(t,x) \tag{11.1}$$

$\mathbf{R} \times \mathbf{R}^n$ to $\mathbf{R}^n$, and a system

$$\dot{x} = g(t,x) \,. \tag{11.2}$$

$\mathbf{R}^n$. We make the following hypothesis.

Hypothesis 1 *For each ε, the function $f^\varepsilon : \mathbf{R} \times \mathbf{R}^n \to \mathbf{R}^n$ is continuous and the system $\dot{x} = f^\varepsilon(t,x)$ has the uniqueness property of solutions*[1]. *The function $g : \mathbf{R} \times \mathbf{R}^n \to \mathbf{R}^n$ is continuous and the system $\dot{x} = g(t,x)$ has the uniqueness property of solutions.*

[1] By this we mean the following. For every $(t_0, x_0) \in \mathbf{R} \times \mathbf{R}^n$, there is a solution ξ of this ordinary differential equation with $\xi(t_0) = x_0$ such that (i) the domain of ξ is an open interval containing t_0, and (ii) for any other solution $\bar{\xi}$ of this ordinary differential equation with $\bar{\xi}(t_0) = x_0$, (a) the domain of $\bar{\xi}$ is contained in the domain of ξ, and (b) $\bar{\xi}$ and ξ coincide on the common part of their domains. We call this solution ξ the *trajectory* of the system passing through state x_0 at time t_0.

Let $\phi^\varepsilon(t, t_0, x_0)$ be the trajectory of $\dot{x} = f^\varepsilon(t, x)$ passing through state x_0 at time t_0 evaluated at time t. The function $(t, t_0, x_0) \mapsto \phi^\varepsilon(t, t_0, x_0)$ is called the *flow* of this system. For each ε, the domain of ϕ^ε is open and ϕ^ε is continuous on its domain [2]. Similarly, the flow of $\dot{x} = g(t, x)$ is defined as the function $(t, t_0, x_0) \mapsto \psi(t, t_0, x_0)$ with $\psi(t, t_0, x_0)$ the trajectory of $\dot{x} = g(t, x)$ passing through state x_0 at time t_0 evaluated at time t. The domain of ψ is open and ψ is continuous on its domain.

Throughout the chapter, we assume that trajectories of (11.1) converge to trajectories of (11.2) in the following sense.

Hypothesis 2 *For every $T \in (0, \infty)$ and compact set $K \subset \mathbf{R}^n$ satisfying $\{(t, t_0, x_0) \in \mathbf{R} \times \mathbf{R} \times \mathbf{R}^n : t \in [t_0, t_0 + T],\ x_0 \in K\} \subset \mathrm{Dom}\,\psi$, for every $d \in (0, \infty)$, there exists $\varepsilon^* \in (0, \varepsilon_0]$ such that for all $t_0 \in \mathbf{R}$, for all $x_0 \in K$ and for all $\varepsilon \in (0, \varepsilon^*)$*

$$\begin{cases} \phi^\varepsilon(t, t_0, x_0) \text{ exists} \\ \|\phi^\varepsilon(t, t_0, x_0) - \psi(t, t_0, x_0)\| < d \end{cases} \quad \forall t \in [t_0,\, t_0 + T].$$

In other words, we require that trajectories of (11.1) converge uniformly on compact time intervals to trajectories of (11.2) as $\varepsilon \downarrow 0$, and furthermore we assume that this convergence is uniform with respect to t_0 and x_0 for $t_0 \in \mathbf{R}$ and x_0 belonging to compact sets. [7] or [8]. See these papers. It is important to notice the following: the assumed convergence is not stated in terms of vectorfields, but in terms of trajectories; we do not assume that f^ε converges pointwise to g as $\varepsilon \downarrow 0$. The following two examples illustrate this.

Example 11.2.1 (Fast time-varying systems). Given $f : \mathbf{R} \times \mathbf{R}^n \to \mathbf{R}^n : (t, x) \mapsto f(t, x)$ and $f_{av} : \mathbf{R}^n \to \mathbf{R}^n : x \mapsto f_{av}(x)$ satisfying the following three conditions: (i) f is continuous, $f(t, \cdot) : \mathbf{R}^n \to \mathbf{R}^n$ is locally Lipschitz uniformly with respect to t for $t \in \mathbf{R}$, and $f(\cdot, x) : \mathbf{R} \to \mathbf{R}^n$ is bounded uniformly with respect to x for x in compact subsets of $\mathbf{R}^n$; (ii) f_{av} is locally Lipschitz; and (iii) for each compact set $K \subset \mathbf{R}^n$, for each $\theta \in (0, \infty)$ and $i \in \{0\} \cup \mathbf{N}$

$$\int_{t_0+i\theta}^{t_0+(i+1)\theta} \left(f\left(\frac{s}{\varepsilon}, x\right) - f_{av}(x)\right) ds \to 0 \tag{11.3}$$

as $\varepsilon \downarrow 0$ uniformly with respect to t_0 and x for $t_0 \in \mathbf{R}$ and $x \in K$. System

$$\dot{x} = f\left(\frac{t}{\varepsilon}, x\right) \tag{11.4}$$

is called a fast time-varying system, and

$$\dot{x} = f_{av}(x) \tag{11.5}$$

the associated averaged system. These systems satisfy Hypothesis 1 —this is a standard result within the theory of ordinary differential equations— and Hypothesis 2 —this may be proven based on the Gronwall Lemma; cf. [9]. Consequently, all results obtained in the general framework of the present chapter apply in particular to fast time-varying systems (11.4) and their averaged (11.5).

Example 11.2.2 (Highly oscillatory systems). Given vectorfields $X_i : \mathbf{R}^n \to \mathbf{R}^n : x \to X_i(x)$ $(i \in \{1, 2, 3\})$ of class C^2. System

$$\dot{x} = X_1(x) + \frac{1}{\sqrt{\varepsilon}} \cos(\frac{t}{\varepsilon}) X_2(x) + \frac{1}{\sqrt{\varepsilon}} \sin(\frac{t}{\varepsilon}) X_3(x) \tag{11.6}$$

is called a highly oscillatory system, and

$$\dot{x} = X_1(x) + \frac{1}{2}[X_2, X_3](x) \tag{11.7}$$

the associated extended system. These systems satisfy Hypothesis 1 —this is a standard result within the theory of ordinary differential equations— and Hypothesis 2 —this may be proven based on partial integration and the Gronwall Lemma; cf. [9]. Consequently, all results obtained in the general framework of the present chapter apply in particular to highly oscillatory systems (11.6) and their extended system (11.7).

11.3 Practical stability

Consider systems $\dot{x} = f^\varepsilon(t, x)$ and $\dot{x} = g(t, x)$ introduced above satisfying Hypotheses 1 and 2. Assume that the origin is a locally uniformly asymptotically stable equilibrium point of $\dot{x} = g(t, x)$. It is well known that this does not imply that the origin is an asymptotically stable equilibrium point of $\dot{x} = f^\varepsilon(t, x)$ for ε sufficiently small. It seems however reasonable to expect that $\dot{x} = f^\varepsilon(t, x)$ inherits some weaker notion of stability. In the present section we identify such a weaker notion of stability, which we call *practical local uniform asymptotic stability*[2].

We start with the relevant stability definitions. First we recall the definition of local uniform asymptotic stability —see, for example, [3]— and then we introduce the notion of practical local uniform asymptotic stability.

[2] On the one hand, this terminology refers to the situation where $\dot{x} = f^\varepsilon(t, x)$ is the actual system and $\dot{x} = g(t, x)$ its studied idealization. In this case, LUAS for the idealized model implies PLUAS in practice; see Theorem 11.3.1. On the other hand, this terminology reflects that for many practical purposes, PLUAS is a satisfactory stability property.

Definition 11.3.1. *Let the origin be an equilibrium point of system $\dot{x} = g(t, x)$ introduced above, and assume that Hypothesis 1 is satisfied. This equilibrium point is called* locally uniformly asymptotically stable *(LUAS) if the following two conditions are both satisfied.*

1. *Uniform stability. For every $c_2 \in (0, \infty)$, there exists $c_1 \in (0, \infty)$ such that for all $t_0 \in \mathbf{R}$ and for all $x_0 \in \mathbf{R}^n$ with $\|x_0\| < c_1$*

$$\begin{cases} \psi(t, t_0, x_0) \text{ exists} & \forall t \in [t_0, \infty), \\ \|\psi(t, t_0, x_0)\| < c_2 & \forall t \in [t_0, \infty). \end{cases}$$

2. *Local uniform attractivity. There exists $c_1 \in (0, \infty)$ such that for all $c_2 \in (0, \infty)$, there exists $T \in (0, \infty)$ such that for all $t_0 \in \mathbf{R}$ and for all $x_0 \in \mathbf{R}^n$ with $\|x_0\| < c_1$*

$$\begin{cases} \psi(t, t_0, x_0) \text{ exists} & \forall t \in [t_0, \infty), \\ \|\psi(t, t_0, x_0)\| < c_2 & \forall t \in [t_0 + T, \infty). \end{cases}$$

Definition 11.3.2. *Consider the system $\dot{x} = f^{\varepsilon}(t, x)$ introduced above, and assume that Hypothesis 1 is satisfied. We call the origin of this system* practically locally uniformly asymptotically stable *(PLUAS) if the following two conditions are both satisfied.*

1. *For every $c_2 \in (0, \infty)$, there exist $c_1 \in (0, \infty)$ and $\hat{\varepsilon} \in (0, \varepsilon_0]$ such that for all $t_0 \in \mathbf{R}$, for all $x_0 \in \mathbf{R}^n$ with $\|x_0\| < c_1$ and for all $\varepsilon \in (0, \hat{\varepsilon})$*

$$\begin{cases} \phi^{\varepsilon}(t, t_0, x_0) \text{ exists} & \forall t \in [t_0, \infty), \\ \|\phi^{\varepsilon}(t, t_0, x_0)\| < c_2 & \forall t \in [t_0, \infty). \end{cases}$$

2. *There exists $c_1 \in (0, \infty)$ such that for all $c_2 \in (0, \infty)$, there exist $T \in (0, \infty)$ and $\hat{\varepsilon} \in (0, \varepsilon_0]$ such that for all $t_0 \in \mathbf{R}$, for all $x_0 \in \mathbf{R}^n$ with $\|x_0\| < c_1$ and for all $\varepsilon \in (0, \hat{\varepsilon})$*

$$\begin{cases} \phi^{\varepsilon}(t, t_0, x_0) \text{ exists} & \forall t \in [t_0, \infty), \\ \|\phi^{\varepsilon}(t, t_0, x_0)\| < c_2 & \forall t \in [t_0 + T, \infty). \end{cases}$$

It is instructive to have a closer look at the strong similarities between these two definitions. The notion of PLUAS may be interpreted as follows. Condition 1 of Definition 11.3.2 defines a practical version of uniform stability. Condition 2 of Definition 11.3.2 captures a practical notion of local uniform attractivity: all trajectories starting in some fixed ball end up in an arbitrarily small ball for appropriate – depending on the radius of this small ball – values of the parameter ε. Notice that the origin is *not* required to be an equilibrium point in Definition 11.3.2.

Theorem 11.3.1. *Given systems $\dot{x} = f^{\varepsilon}(t, x)$ and $\dot{x} = g(t, x)$ introduced above satisfying Hypotheses 1 and 2. If the origin is a LUAS equilibrium point of $\dot{x} = g(t, x)$, then the origin of $\dot{x} = f^{\varepsilon}(t, x)$ is PLUAS.*

Proof of Theorem 11.3.1 First of all, let $c \in (0, \infty)$ be the constant c_1 featuring in condition 2 of Definition 11.3.1. Then $\psi(t, t_0, x_0)$ exists for all $t \in [t_0, \infty)$, for all $t_0 \in \mathbf{R}$ and for all $x_0 \in \mathbf{R}^n$ with $\|x_0\| < c$. We successively prove that conditions 1 and 2 of Definition 11.3.2 are satisfied.

1. Take an arbitrary $c_2 \in (0, \infty)$ and let $b_2 \in (0, c_2)$. By the LUAS property of ψ – in particular, by uniform stability – there exists $c_1 \in (0, \infty)$, which we choose to be in $(0, c)$, such that

$$\|\psi(t, t_0, x_0)\| < b_2 \quad \forall t \in [t_0, \infty), \\ \forall t_0 \in \mathbf{R}, \ \forall x_0 \in \mathbf{R}^n \text{ with } \|x_0\| < c_1. \tag{11.8}$$

Let $b_1 \in (0, c_1)$. Since the equilibrium point $x = 0$ of ψ is locally uniformly attractive and $\{x \in \mathbf{R}^n : \|x\| < c_1 < c\}$ is contained in its region of attraction, there exists $T \in (0, \infty)$ such that

$$\|\psi(t, t_0, x_0)\| < b_1 \quad \forall t \in [t_0 + T, \infty), \\ \forall t_0 \in \mathbf{R}, \ \forall x_0 \in \mathbf{R}^n \text{ with } \|x_0\| < c_1. \tag{11.9}$$

(11.9) for ψ with $0 < b_1 < c_1 < c$, $0 < b_2 < c_2$ and $T > 0$. Let $d = \min\{c_1 - b_1, c_2 - b_2\}$. Notice that $\{(t, t_0, x_0) \in \mathbf{R} \times \mathbf{R} \times \mathbf{R}^n : t \in [t_0, t_0 + T], \ \|x\| \le c_1 < c\} \subset \operatorname{Dom} \psi$. Hence, invoking Hypothesis 2 – with $K = \{x \in \mathbf{R}^n : \|x\| \le c_1\}$ – yields the existence of $\widehat{\varepsilon} \in (0, \varepsilon_0]$ such that

$$\begin{cases} \phi^\varepsilon(t, t_0, x_0) \text{ exists} \\ \|\phi^\varepsilon(t, t_0, x_0) - \psi(t, t_0, x_0)\| < d \end{cases} \quad \forall t \in [t_0, t_0 + T], \\ \forall t_0 \in \mathbf{R}, \ \forall x_0 \in \mathbf{R}^n \text{ with } \|x_0\| \le c_1, \ \forall \varepsilon \in (0, \widehat{\varepsilon}). \tag{11.10}$$

Estimates (11.8), (11.9) and (11.10) together yield

$$\begin{cases} \phi^\varepsilon(t, t_0, x_0) \text{ exists} & \forall t \in [t_0, t_0 + T], \\ \|\phi^\varepsilon(t, t_0, x_0)\| < c_2 & \forall t \in [t_0, t_0 + T], \\ \|\phi^\varepsilon(t, t_0, x_0)\| < c_1 & \text{for } t = t_0 + T, \end{cases} \\ \forall t_0 \in \mathbf{R}, \ \forall x_0 \in \mathbf{R}^n \text{ with } \|x_0\| < c_1, \ \forall \varepsilon \in (0, \widehat{\varepsilon}). \tag{11.11}$$

An iterative application of this expression yields

$$\begin{cases} \phi^\varepsilon(t, t_0, x_0) \text{ exists} \\ \|\phi^\varepsilon(t, t_0, x_0)\| < c_2 \end{cases} \quad \forall t \in [t_0, \infty), \\ \forall t_0 \in \mathbf{R}, \ \forall x_0 \in \mathbf{R}^n \text{ with } \|x_0\| < c_1, \ \forall \varepsilon \in (0, \widehat{\varepsilon}), \tag{11.12}$$

which is the property we had to prove.

2. Let $c_1 \in (0, c)$. Take an arbitrary $c_2 \in (0, \infty)$. By practical uniform stability —condition 1 of Definition 11.3.2— proven above, there exist $c_3 \in (0, \infty)$ and $\varepsilon^* \in (0, \varepsilon_0]$ such that

$$\begin{cases} \phi^\varepsilon(t,t_0,x_0) \text{ exists} \\ \|\phi^\varepsilon(t,t_0,x_0)\| < c_2 \end{cases} \forall t \in [t_0,\, \infty),$$
$$\forall t_0 \in \mathbf{R},\ \forall x_0 \in \mathbf{R}^n \text{ with } \|x_0\| < c_3,\ \forall \varepsilon \in (0,\, \varepsilon^*). \tag{11.13}$$

Let $b_3 \in (0,\, c_3)$. Since the equilibrium point $x = 0$ of ψ is locally uniformly attractive and $\{x \in \mathbf{R}^n : \|x\| < c_1 < c\}$ is contained in its region of attraction, there exists $T \in (0,\, \infty)$ such that

$$\|\psi(t,t_0,x_0)\| < b_3 \quad \forall t \in [t_0 + T,\, \infty),$$
$$\forall t_0 \in \mathbf{R},\ \forall x_0 \in \mathbf{R}^n \text{ with } \|x_0\| < c_1. \tag{11.14}$$

Let $d = c_3 - b_3$. Notice that $\{(t,t_0,x_0) \in \mathbf{R} \times \mathbf{R} \times \mathbf{R}^n : t \in [t_0,\, t_0 + T],\ \|x\| \le c_1 < c\} \subset \operatorname{Dom}\psi$. Hence, invoking Hypothesis 2 – with $K = \{x \in \mathbf{R}^n : \|x\| \le c_1\}$ – yields the existence of $\varepsilon^\# \in (0,\, \varepsilon_0]$ such that

$$\begin{cases} \phi^\varepsilon(t,t_0,x_0) \text{ exists} \\ \|\phi^\varepsilon(t,t_0,x_0) - \psi(t,t_0,x_0)\| < d \end{cases} \forall t \in [t_0,\, t_0 + T],$$
$$\forall t_0 \in \mathbf{R},\ \forall x_0 \in \mathbf{R}^n \text{ with } \|x_0\| \le c_1,\ \forall \varepsilon \in (0,\, \varepsilon^\#). \tag{11.15}$$

Estimates (11.14) and (11.15) yield

$$\begin{cases} \phi^\varepsilon(t,t_0,x_0) \text{ exists} & \forall t \in [t_0,\, t_0 + T], \\ \|\phi^\varepsilon(t,t_0,x_0)\| < c_3 & \text{for } t = t_0 + T, \end{cases}$$
$$\forall t_0 \in \mathbf{R},\ \forall x_0 \in \mathbf{R}^n \text{ with } \|x_0\| < c_1,\ \forall \varepsilon \in (0,\, \varepsilon^\#). \tag{11.16}$$

This, together with (11.13), leads to

$$\begin{cases} \phi^\varepsilon(t,t_0,x_0) \text{ exists} & \forall t \in [t_0,\, \infty), \\ \|\phi^\varepsilon(t,t_0,x_0)\| < c_2 & \forall t \in [t_0 + T,\, \infty), \end{cases}$$
$$\forall t_0 \in \mathbf{R},\ \forall x_0 \in \mathbf{R}^n \text{ with } \|x_0\| < c_1,\ \forall \varepsilon \in (0,\, \widehat{\varepsilon}), \tag{11.17}$$

where $\widehat{\varepsilon} = \min\{\varepsilon^*, \varepsilon^\#\}$. This is the second property we had to prove; and thus the theorem is proven.

□

Example 11.3.1 (Fast time-varying systems). Consider again the fast time-varying system (11.4) and its averaged (11.5) introduced in Example 11.2.1. An application of Theorem 11.3.1 yields: if the origin is a LUAS equilibrium point of the averaged, then the origin of the original fast time-varying system is PLUAS and thus, in particular, trajectories of (11.4) starting in some fixed ball end up in an arbitrarily small ball provided system (11.4) is "sufficiently —depending on the considered neighborhoods— fast time-varying".

Example 11.3.2 (Highly oscillatory systems). Consider again the highly oscillatory system (11.6) and its extended system (11.7) introduced in Example

11.2.2. An application of Theorem 11.3.1 yields: if the origin is a LUAS equilibrium point of the extended system, then the origin of the original highly oscillatory system is PLUAS and thus, in particular, trajectories of (11.6) starting in some fixed ball end up in an arbitrarily small ball provided system (11.6) is "sufficiently —depending on the considered neighborhoods— highly oscillatory".

11.4 Convergence results on an infinite time scale

Convergence results for trajectories as in Hypothesis 2 may typically be proven based on the Gronwall Lemma; see Examples 11.2.1 and 11.2.2. This type of convergence is on a finite time scale; that is, $\phi^\varepsilon(t, t_0, x_0)$ converges to $\psi(t, t_0, x_0)$ as $\varepsilon \downarrow 0$ uniformly with respect to t for t belonging to *compact* time intervals. For analysis as well as for control purposes, it may be interesting to have convergence results on an infinite time scale; that is, results that state that $\phi^\varepsilon(t, t_0, x_0)$ converges to $\psi(t, t_0, x_0)$ as $\varepsilon \downarrow 0$ uniformly with respect to t for t belonging to *infinite* time intervals. This type of convergence can not be concluded from the Gronwall Lemma alone.

In the present section, we give a convenient way to obtain convergence results on an infinite time scale: as an application of Theorem 11.3.1, we show how in the presence of an asymptotically stable attractor, convergence results on a finite time scale as in Hypothesis 2 extend to an infinite time scale.

Theorem 11.4.1. *Given systems $\dot{x} = f^\varepsilon(t, x)$ and $\dot{x} = g(t, x)$ introduced above satisfying hypotheses 1 and 2. Consider an initial state $\overline{x}_0$ and let the origin be a LUAS equilibrium point of $\dot{x} = g(t, x)$ with $\overline{x}_0$ in its region of attraction; that is, for all $c \in (0, \infty)$, there exists $T \in (0, \infty)$ such that for all $t_0 \in \mathbf{R}$*

$$\begin{cases} \psi(t, t_0, \overline{x}_0) \text{ exists} & \forall t \in [t_0, \infty), \\ \|\psi(t, t_0, \overline{x}_0)\| < c & \forall t \in [t_0 + T, \infty). \end{cases}$$

Then, $\phi^\varepsilon(t, t_0, \overline{x}_0)$ converges to $\psi(t, t_0, \overline{x}_0)$ as $\varepsilon \downarrow 0$ uniformly with respect to t and t_0 for $t \in [t_0, \infty)$ and $t_0 \in \mathbf{R}$; that is, for every $d' \in (0, \infty)$, there exists $\varepsilon^* \in (0, \varepsilon_0]$ such that for all $t_0 \in \mathbf{R}$ and for all $\varepsilon \in (0, \varepsilon^*)$

$$\left\{ \begin{array}{l} \phi^\varepsilon(t, t_0, \overline{x}_0) \text{ exists} \\ \|\phi^\varepsilon(t, t_0, \overline{x}_0) - \psi(t, t_0, \overline{x}_0)\| < d' \end{array} \right. \quad \forall t \in [t_0, \infty).$$

Remark 11.4.1. A closely related result in this context is Theorem 4.2.1 (Eckhaus/Sanchez-Palencia) reported in [11]. The present theorem is a generalization of that result for the following reasons: (i) Theorem 11.4.1 introduced here extends Theorem 4.2.1 from [11], which is a specific averaging result, to the general context of the present chapter. This general context includes averaging as a special case, but also for example highly oscillatory systems. (ii) In Theorem 11.4.1 introduced here, the origin is assumed to be asymptotically stable, whereas Theorem 4.2.1 from [11] assumes exponential stability. Furthermore, in Theorem 4.2.1 from [11], additional technical assumptions are made.

Proof of Theorem 11.4.1 Take an arbitrary $d' \in (0, \infty)$. Since the origin is a LUAS equilibrium point of $\dot{x} = g(t, x)$, the origin of $\dot{x} = f^{\varepsilon}(t, x)$ is PLUAS by Theorem 11.3.1. Hence, there exist $c_1 \in (0, \infty)$ and $\hat{\varepsilon} \in (0, \varepsilon_0]$ such that

$$\begin{cases} \phi^{\varepsilon}(t, t_0, x_0) \text{ exists} \\ \|\phi^{\varepsilon}(t, t_0, x_0)\| < d'/2 \end{cases} \forall t \in [t_0, \infty),$$
$$\forall t_0 \in \mathbf{R},\ \forall x_0 \in \mathbf{R}^n \text{ with } \|x_0\| < c_1,\ \forall \varepsilon \in (0, \hat{\varepsilon}). \tag{11.18}$$

Notice that $c_1 \leq d'/2$. Since $\bar{x}_0$ is in the attraction region of the origin for $\dot{x} = g(t, x)$, there exists $T \in (0, \infty)$ such that

$$\begin{cases} \psi(t, t_0, \bar{x}_0) \quad \text{exists} & \forall t \in [t_0, \infty), \\ \|\psi(t, t_0, \bar{x}_0)\| < c_1/2 & \forall t \in [t_0 + T, \infty), \end{cases}$$
$$\forall t_0 \in \mathbf{R}. \tag{11.19}$$

Invoking Hypothesis 2 – with $K = \{\bar{x}_0\}$ and $d = c_1/2$ – yields the existence of $\varepsilon^{\#} \in (0, \varepsilon_0]$ such that

$$\begin{cases} \phi^{\varepsilon}(t, t_0, \bar{x}_0) \text{ exists} \\ \|\phi^{\varepsilon}(t, t_0, \bar{x}_0) - \psi(t, t_0, \bar{x}_0)\| < c_1/2 \end{cases} \forall t \in [t_0, t_0 + T],$$
$$\forall t_0 \in \mathbf{R},\ \forall \varepsilon \in (0, \varepsilon^{\#}). \tag{11.20}$$

Let $\varepsilon^* = \min\{\hat{\varepsilon}, \varepsilon^{\#}\}$. We show that the conclusion of the theorem follows from a suitable application of estimates (11.18), (11.19) and (11.20).

First, estimate (11.20) gives

$$\begin{cases} \phi^{\varepsilon}(t, t_0, \bar{x}_0) \text{ exists} \\ \|\phi^{\varepsilon}(t, t_0, \bar{x}_0) - \psi(t, t_0, \bar{x}_0)\| < d' \end{cases} \forall t \in [t_0, t_0 + T],$$
$$\forall t_0 \in \mathbf{R},\ \forall \varepsilon \in (0, \varepsilon^*) \tag{11.21}$$

since $c_1/2 < c_1 \leq d'/2 < d'$ and $\varepsilon^* \leq \varepsilon^{\#}$. Next, applying estimates (11.19) and (11.20) at time $t = t_0 + T$ yields $\|\phi^{\varepsilon}(t_0 + T, t_0, \bar{x}_0)\| < c_1$ for all $t_0 \in \mathbf{R}$ and for all $\varepsilon \in (0, \varepsilon^*)$. This, together with (11.18), yields

$$\begin{cases} \phi^{\varepsilon}(t, t_0, \bar{x}_0) \text{ exists} \\ \|\phi^{\varepsilon}(t, t_0, \bar{x}_0)\| < d'/2 \end{cases} \forall t \in [t_0 + T, \infty),$$
$$\forall t_0 \in \mathbf{R},\ \forall \varepsilon \in (0, \varepsilon^*), \tag{11.22}$$

and thus, by (11.19),

$$\begin{cases} \phi^{\varepsilon}(t, t_0, \overline{x}_0) \text{ exists} \\ \|\phi^{\varepsilon}(t, t_0, \overline{x}_0) - \psi(t, t_0, \overline{x}_0)\| < d'/2 + c_1/2 < d' \end{cases} \quad \forall t \in [t_0 + T, \infty),$$
$$\forall t_0 \in \mathbf{R},\ \forall \varepsilon \in (0, \varepsilon^*). \tag{11.23}$$

Estimates (11.21) and (11.23) prove the theorem. □

11.5 Homogeneous systems

Consider again systems $\dot{x} = f^{\varepsilon}(t, x)$ and $\dot{x} = g(t, x)$ introduced above satisfying Hypotheses 1 and 2. In Section 11.3 we have seen that, if the origin is a LUAS equilibrium point of $\dot{x} = g(t, x)$, it need not be a LUAS equilibrium point of $\dot{x} = f^{\varepsilon}(t, x)$ for ε sufficiently small. Instead we have identified a weaker stability property for $\dot{x} = f^{\varepsilon}(t, x)$: practical stability (PLUAS).

Although in general LUAS for $\dot{x} = g(t, x)$ does not imply LUAS for $\dot{x} = f^{\varepsilon}(t, x)$ for ε sufficiently small, there may still be particular situations where this implication holds after all. This is the subject of the present section. In addition to Hypotheses 1 and 2, we assume that the systems are *zero-order homogeneous*, and we prove that in this case, LUAS for $\dot{x} = g(t, x)$ implies LUAS for $\dot{x} = f^{\varepsilon}(t, x)$ for ε sufficiently small

This interplay between homogeneity, convergence of trajectories, and stability may further clarify the important role that homogeneity – together with averaging or the theory of highly oscillatory systems – plays in some recent stabilization schemes [10, 5].

assumption. First, homogeneity has recently proven to be a fruitful concept for solving stabilization problems. See for example [6] and [10]. Several classes of systems discussed in that reference fit into the framework framework of the present section. The present development may shed some new light on the results from these references. Second, in the context of homogeneous approximations, the results from the present section are useful for extensions to the non-homogeneous case.

For expository reasons, we restrict attention to zero-order homogeneity with respect to the standard dilation[3].

Hypothesis 3 *For each $\varepsilon \in (0, \varepsilon_0]$, $f^{\varepsilon}(t, \lambda x) = \lambda f^{\varepsilon}(t, x)$ for all $t \in \mathbf{R}$, $x \in \mathbf{R}^n$ and $\lambda \in (0, \infty)$.*

Theorem 11.5.1. *Given system $\dot{x} = f^{\varepsilon}(t, x)$ introduced above satisfying Hypotheses 1 and 3. If the origin of $\dot{x} = f^{\varepsilon}(t, x)$ is PLUAS, then there exist*

[3] The case of general dilations is treated in [7].

$\hat{\varepsilon} \in (0, \varepsilon_0]$, $\mu \in [1, \infty)$ and $\nu \in (0, \infty)$ such that for all $t_0 \in \mathbf{R}$, for all $x_0 \in \mathbf{R}^n$ and for all $\varepsilon \in (0, \hat{\varepsilon})$

$$\|\phi^\varepsilon(t, t_0, x_0)\| \leq \mu e^{-\nu(t-t_0)}\|x_0\| \quad \forall t \in [t_0, \infty).$$

Proof of Theorem 11.5.1 First of all, notice that Hypothesis 3 implies that $\mathrm{Dom}\,\phi^\varepsilon = \mathbf{R} \times \mathbf{R} \times \mathbf{R}^n$ for each ε. (The constraint imposed by Hypothesis 3 on the grow rate of $f^\varepsilon(t, x)$ as $\|x\| \to \infty$ excludes the possibility of finite escape times.)

The origin of $\dot{x} = f^\varepsilon(t, x)$ is assumed to be PLUAS. Hence there exist $c_1, c_2, c_3 \in (0, \infty)$ with $c_1 < c_2 < c_3$ and there exist $T \in (0, \infty)$ and $\varepsilon^* \in (0, \varepsilon_0]$ such that

$$\begin{cases} \|\phi^\varepsilon(t, t_0, x_0)\| \leq c_3 & \forall t \in [t_0, \infty), \\ \|\phi^\varepsilon(t, t_0, x_0)\| \leq c_1 & \forall t \in [t_0 + T, \infty), \end{cases}$$
$$\forall t_0 \in \mathbf{R},\ \forall x_0 \in \mathbf{R}^n \text{ with } \|x_0\| \leq c_2,\ \forall \varepsilon \in (0, \varepsilon^*). \quad (11.24)$$

Notice that we have used $\leq$ signs in the above formulation. This will be convenient for the following development. By Hypothesis 3, the flow ϕ^ε has the following scaling property:

$$\phi^\varepsilon(t, t_0, x_0) = \lambda \phi^\varepsilon(t, t_0, \frac{1}{\lambda} x_0) \quad \forall t \in \mathbf{R},$$
$$\forall t_0 \in \mathbf{R},\ \forall x_0 \in \mathbf{R}^n,\ \forall \varepsilon \in (0, \varepsilon_0],\ \forall \lambda \in (0, \infty). \quad (11.25)$$

In the remainder of the proof, we consider two cases: $x_0 = 0$ and $x_0 \neq 0$. First $x_0 = 0$. The scaling property (11.25) implies

$$\phi^\varepsilon(t, t_0, 0) = 0 \quad \forall t \in \mathbf{R},$$
$$\forall t_0 \in \mathbf{R},\ \forall \varepsilon \in (0, \varepsilon_0]. \quad (11.26)$$

Next $x_0 \neq 0$. The scaling property (11.25) implies

$$\phi^\varepsilon(t, t_0, x_0) = \frac{\|x_0\|}{c_2} \phi^\varepsilon(t, t_0, \frac{c_2}{\|x_0\|} x_0) \quad \forall t \in \mathbf{R},$$
$$\forall t_0 \in \mathbf{R},\ \forall x_0 \in \mathbf{R}^n \text{ with } x_0 \neq 0, \forall \varepsilon \in (0, \varepsilon_0]. \quad (11.27)$$

Since $\|\frac{c_2}{\|x_0\|} x_0\| = c_2$, estimate (11.24) applies and yields

$$\begin{cases} \|\phi^\varepsilon(t, t_0, x_0)\| \leq \frac{c_3}{c_2}\|x_0\| & \forall t \in [t_0, \infty), \\ \|\phi^\varepsilon(t, t_0, x_0)\| \leq \frac{c_1}{c_2}\|x_0\| & \forall t \in [t_0 + T, \infty), \end{cases}$$
$$\forall t_0 \in \mathbf{R},\ \forall x_0 \in \mathbf{R}^n \text{ with } x_0 \neq 0,\ \forall \varepsilon \in (0, \varepsilon^*). \quad (11.28)$$

In particular this estimate implies that $\|\phi^\varepsilon(t_0 + T, t_0, x_0)\| \leq \frac{c_1}{c_2}\|x_0\| < \|x_0\|$. Since $\phi^\varepsilon(t_0 + T, t_0, x_0) \neq 0$ by the uniqueness property of trajectories, a second application of estimate (11.28) then yields

$$\begin{cases} \|\phi^\varepsilon(t,t_0,x_0)\| \le \frac{c_3}{c_2}\frac{c_1}{c_2}\|x_0\| & \forall t \in [t_0+T,\,\infty), \\ \|\phi^\varepsilon(t,t_0,x_0)\| \le (\frac{c_1}{c_2})^2\|x_0\| & \forall t \in [t_0+2T,\,\infty), \end{cases}$$
$$\forall t_0 \in \mathbf{R},\ \forall x_0 \in \mathbf{R}^n \text{ with } x_0 \neq 0,\ \forall \varepsilon \in (0,\,\varepsilon^*). \quad (11.29)$$

This process may be repeated, and eventually it leads to

$$\|\phi^\varepsilon(t,t_0,x_0)\| \le \frac{c_3}{c_2}(\frac{c_1}{c_2})^i\|x_0\| \quad \forall t \in [t_0+iT,\,\infty),\ \forall i \in \{0\}\cup\mathbf{N},$$
$$\forall t_0 \in \mathbf{R},\ \forall x_0 \in \mathbf{R}^n \text{ with } x_0 \neq 0,\ \forall \varepsilon \in (0,\,\varepsilon^*). \quad (11.30)$$

Since $\frac{c_1}{c_2} < 1$, this implies

$$\|\phi^\varepsilon(t,t_0,x_0)\| \le \frac{c_3}{c_2}(\frac{c_1}{c_2})^{\frac{t-t_0}{T}-1}\|x_0\| \quad \forall t \in [t_0,\,\infty),$$
$$\forall t_0 \in \mathbf{R},\ \forall x_0 \in \mathbf{R}^n \text{ with } x_0 \neq 0,\ \forall \varepsilon \in (0,\,\varepsilon^*). \quad (11.31)$$

Since $\frac{c_3}{c_2}(\frac{c_1}{c_2})^{\frac{t-t_0}{T}-1} = \frac{c_3}{c_1}\exp(\frac{\ln c_1/c_2}{T}(t-t_0))$ with $\frac{c_3}{c_1} \ge 1$ and $\frac{\ln c_1/c_2}{T} < 0$ the theorem is proven by estimates (11.26) and (11.31). □

Corollary 1 *Given systems $\dot{x} = f^\varepsilon(t,x)$ and $\dot{x} = g(t,x)$ introduced above satisfying Hypotheses 1, 2 and 3. If the origin is a LUAS equilibrium point of $\dot{x} = g(t,x)$, then there exist $\widehat{\varepsilon} \in (0,\,\varepsilon_0]$, $\mu \in [1,\,\infty)$ and $\nu \in (0,\,\infty)$ such that for all $t_0 \in \mathbf{R}$, for all $x_0 \in \mathbf{R}^n$ and for all $\varepsilon \in (0,\,\widehat{\varepsilon})$*

$$\|\phi^\varepsilon(t,t_0,x_0)\| \le \mu e^{-\nu(t-t_0)}\|x_0\| \quad \forall t \in [t_0,\,\infty).$$

In words, this corollary states that, assuming Hypotheses 1, 2 and 3, if the origin is a LUAS equilibrium point of $\dot{x} = g(t,x)$, then the origin is a globally uniformly exponentially stable equilibrium point of $\dot{x} = f^\varepsilon(t,x)$ for ε sufficiently small; and moreover, the bounds μ and ν for the convergence do not depend on ε.

Summary:

This chapter has investigated the interplay between convergence results for trajectories and stability properties.

References

1. D. Aeyels and J. Peuteman, On exponential stability of nonlinear time-varying differential equations. Automatica 35:1091–1100, 1999

2. P. Hartman, Ordinary Differential Equations, 2nd edn. Birkhäuser, 1982.

3. H.K. Khalil, Nonlinear Systems, 2nd edn. Prentice-Hall, 1996.

4. R.T. M'Closkey, An averaging theorem for time-periodic degree zero homogeneous differential equations. Systems and Control Letters 32:179–183, 1997.

5. R.T. M'Closkey and P. Morin, Time-varying homogeneous feedback: design tools for the exponential stabilization of systems with drift. International Journal of Control 71:837–869, 1998.

6. R.T. M'Closkey and R.M. Murray, Nonholonomic Systems and Exponential Convergence: Some Analysis Tools. In: Proceedings of the 32th IEEE Conference on Decision and Control 1993 (CDC'93), 943–948, 1993.

7. L. Moreau and D. Aeyels, Stability for Homogeneous Flows Depending on a Small Parameter. In: Proceedings of IFAC Nonlinear Control Systems Design Symposium 1998 (NOLCOS'98), 488–493, 1998.

8. L. Moreau and D. Aeyels, Practical Stability for Systems Depending on a Small Parameter. In: Proceedings of the 37th IEEE Conference on Decision and Control 1998 (CDC'98), 1428–1433, 1998.

9. Moreau L., Aeyels D. (1999) Local approximations in stability analyis. In preparation

10. P. Morin, J.-B. Pomet and C. Samson, Design of Homogeneous Time-Varying Stabilizing Control Laws for Driftless Controllable Systems via Oscillatory Approximation of Lie Brackets in Closed-Loop. Accepted for publication in SIAM Journal on Control & Optimization.

11. J.A. Sanders and F. Verhulst, Averaging Methods in Nonlinear Dynamical Systems. Applied Mathematical Sciences, Vol. 59. Springer-Verlag, 1985.

12. A.R. Teel, J. Peuteman and D. Aeyels, Semi-global practical asymptotic stability and averaging. To appear in Systems and Control Letters

12. Robust point-stabilization of nonlinear affine control systems

Pascal Morin and Claude Samson

I.N.R.I.A
B.P. 93
06902 Sophia-Antipolis Cedex, France
pascal.morin@inria.fr
claude.samson@inria.fr

Summary.

Exponential stabilization of nonlinear driftless affine control systems is addressed with the concern of achieving robustness with respect to imperfect knowledge of the system's control vector fields. The present paper gives an overview of the results developed by the authors in [11], and provides new results on the robustness with respect to sampling of the control laws. Control design for a dynamic extension of the original system is also considered. This study is inspired by [1], where the same robustness issue was first addressed. It is further motivated by the fact, proven in [7], according to which no *continuous homogeneous* time-periodic state-feedback can be a robust exponential stabilizer in the sense considered here. *Hybrid* open-loop/feedback controllers, more precisely described as continuous time-periodic feedbacks associated with a specific dynamic extension of the original system, are considered instead.

12.1 Introduction

We consider an analytic driftless system on $\mathbf{R}^n$

$$(S_0) \ : \qquad \dot{x} = \sum_{i=1}^{m} f_i(x)u_i \, , \qquad (m < n), \tag{12.1}$$

locally controllable around the origin, i.e.

$$\text{Span}\{f(0) : f \in \text{Lie}(f_1, \dots, f_m)\} = \mathbf{R}^n \, , \tag{12.2}$$

and address the problem of constructing explicit feedback laws which (locally) exponentially stabilize, in some sense specified later, the origin $x = 0$ of the controlled system. A further requirement is that these feedbacks should also be exponential stabilizers for any "perturbed" system in the form

$$(S_\varepsilon) \ : \qquad \dot{x} = \sum_{i=1}^{m} (f_i(x) + h_i(\varepsilon, x)) u_i \, , \tag{12.3}$$

with h_i analytic in $\mathbf{R} \times \mathbf{R}^n$ and $h_i(0, x) = 0$, when $|\varepsilon|$ is small enough. In other words, given a *nominal* control system (S_0), we would like to find *nominal* feedback controls, derived on the basis of this nominal system, that preserve the property of exponential stability when they are applied to "neighboring" systems (S_ε).

Explicit *homogeneous* exponential (time-periodic) stabilizers $u(x, t)$ for systems (S_0) have been derived in various previous studies (see [8, 10], for example). However, as demonstrated in [7], none of these controls solves the robustness problem stated above in the sense that there always exists some $h_i(\varepsilon, .)$ for which the origin of the associated controlled system is not stable when $\varepsilon \neq 0$. This negative result strongly suggests that no continuous feedback $u(x, t)$, not necessarily homogeneous, can be a robust exponential stabilizer. However, it does not imply that the problem cannot be handled via an adequate dynamic extension of the original nominal system. As a matter of fact, and as explained below, the present study may already be seen as a step in this direction.

An alternative to continuous state feedback control consists in considering *hybrid* open-loop/feedback controls such as open-loop controls which are periodically updated from the measurement $x(kT)$, $k \in \mathbf{N}$, of the state at discrete time-instants. The idea of using this type of control to achieve asymptotic stabilization of the origin of the class of nonlinear driftless systems considered here is not new. This possibility has sometimes been presented as an extension of solutions obtained when addressing the open-loop steering problem, i.e. the problem of finding an open-loop control which steers the system from an initial state to another desired one (see [9, 12], for example). Hybrid continuous/discrete time exponential stabilizers for chained systems, which do not specifically rely on open-loop steering control, have also been proposed in [14]. However, [1] is to our knowledge the first study where the robustness problem stated above has been formulated in a similar fashion and where it has been shown that this problem can be solved by using a hybrid open-loop/feedback control. In fact, although this is not specified in the abovementioned reference, the proposed control does not "strictly" ensure asymptotic stability, in the usual sense of Lyapunov, of the origin of the perturbed systems (S_ε). In order to be more specific about this technical point, and also clarify the meaning of "periodically updated open-loop control applied to a time-continuous system $\dot{x} = f(x, u)$", it is useful to introduce the following *extended* control system:

$$\begin{cases} \dot{x} = f(x, u) \\ \dot{y} = (\sum_{k \in \mathbf{N}} \delta_{kT})(x - y_{-\alpha}) \qquad 0 < \alpha < T \,, \end{cases} \tag{12.4}$$

with T denoting the updating time-period of the control part which depends upon y, δ_{kT} the classical Dirac impulse at the time-instant kT, and $y_{-\alpha}$ the delay operator such that $y_{-\alpha}(t) = y(t - \alpha)$. The extra equation in y just indicates that $y(t)$ is constant and equal to $x(kT)$ on the time-interval $[kT, (k+1)T)$. Therefore, any control the expression of which, on the time-interval $[kT, (k+1)T)$, is a function of $x(kT)$ and t, may just be interpreted as a feedback control $u(y, t)$ for the corresponding extended system. From now on, we will adopt this point of view whenever referring to this type of control. As commonly done elsewhere, we will also say that a feedback control $u(x, y, t)$ is a (uniform) *exponential stabilizer* for the extended system (12.4) if there exist an open set $U \in \mathbf{R}^n \times \mathbf{R}^n$ containing the point $(0, 0)$, a positive real number γ, and a function β of class $\mathcal{K}$ such that:

$$\forall t \geq t_0 \geq 0 \,, \; \forall (x(t_0), y(t_0)) \in U \,,$$
$$|(x(t), y(t))| \leq \beta(|(x(t_0), y(t_0))|) exp(-\gamma(t - t_0))$$

with $(x(t), y(t))$ denoting any solution of the controlled system. In our opinion, the importance of the contribution in [1] comes from that it convincingly demonstrates the possibility of achieving *robust* (with respect to unmodeled dynamics, as defined earlier) *exponential stabilization* (stability being now taken in the *strict* sense of Lyapunov) of an extended control system $(\bar{S}_0)$, defined as the "nominal" system within the set of systems

$$(\bar{S}_\varepsilon) \; : \quad \begin{cases} \dot{x} = \sum_{i=1}^{m} (f_i(x) + h_i(\varepsilon, x)) u_i \\ \dot{y} = (\sum_{k \in \mathbf{N}} \delta_{kT})(x - y_{-\alpha}) \qquad 0 < \alpha < T \,, \end{cases} \tag{12.5}$$

via the use of a *continuous time-periodic feedback* $u(y, t)$. The exploration of this possibility has been carried further on in [11], and a large part of the present paper is devoted to recalling the main results proven in this reference. These include i) a theorem stating sufficient conditions under which a continuous time-periodic feedback $u(y, t)$ is a robust stabilizer (Section 12.2), ii) a general control design algorithm which applies to any controllable analytic (differentiability up to a certain order is in fact sufficient) driftless control system affine in the control (Section 12.3.1), and iii) a set of simpler stabilizers for the subclass of nilpotent chained systems, obtained by further exploiting the internal structure of these systems (Section 12.3.3). We also complement the aforementioned study with two new results. First, we prove a robustness result with respect to sampling of the control law (Section 12.3.1).

Then, we show how to derive new stabilizing control laws for a dynamic extension of the system (consisting in adding an integrator at each input level) (Section 12.4).

The following notation is used.

The identity function on $\mathbf{R}^n$ is denoted id, $|.|$ is the Euclidean norm, and the transpose of a row-vector $(x_1, \dots, x_n)$ is denoted as $(x_1, \dots, x_n)'$.

For any vector field X and smooth function f on $\mathbf{R}^n$, Xf denotes the Lie derivative of f along the vector field X. When $f = (f_1, \dots, f_n)'$ is a smooth map from $\mathbf{R}^n$ to itself, Xf denotes the map $(Xf_1, \dots, Xf_n)'$.

A square matrix A is called *discrete-stable* if all its eigenvalues are strictly inside the complex unit circle.

Given a continuous functions g, defined on some neighborhood of the origin in $\mathbf{R}^n$, we denote $o(g)$ (resp. $O(g)$) any function or map such that $\frac{|o(g)(x)|}{|g(x)|} \longrightarrow 0$ as $|x| \longrightarrow 0$ (resp. such that $\frac{|O(g)(x)|}{|g(x)|} \leq K$ in some neighborhood of the origin). When $g = |.|$, we write $o(x)$ (resp. $O(x)$) instead of $o(g)(x)$ (resp. $O(g)(x)$).

12.2 Sufficient conditions for exponential and robust stabilization

Prior to stating the main result of this section, we review some properties of Chen-Fliess series that will be used in the sequel. The exposition is based on [4, 17], and limited here to driftless systems.

A *m-valued multi-index* I is a vector $I = (i_1, \dots, i_k)$ with k denoting a strictly positive integer, and $i_1, \dots, i_k$, integers taken in the set $\{1, \dots, m\}$. We denote the length of I as $|I|$, i.e. $I = (i_1, \dots, i_k) \Longrightarrow |I| = k$.

Given piecewise continuous functions $u_1, \dots, u_m$ defined on some time-interval $[0, T]$, and a m-valued multi-index $I = (i_1, \dots, i_k)$, we define

$$\int_0^t u_I = \int_0^t \int_0^{t_k} \cdots \int_0^{t_2} u_{i_k}(t_k) u_{i_{k-1}}(t_{k-1}) \cdots u_{i_1}(t_1)\, dt_1 \cdots dt_k \,. \tag{12.6}$$

Given smooth vector fields $f_1, \dots, f_m$ on $\mathbf{R}^n$, and a m-valued multi-index $I = (i_1, \dots, i_k)$, we define the k-th order differential operator $f_I : C^\infty(\mathbf{R}^n; \mathbf{R}) \longrightarrow C^\infty(\mathbf{R}^n; \mathbf{R})$ by

$$f_I\, g = f_{i_1} f_{i_2} \cdots f_{i_k}\, g \,. \tag{12.7}$$

The following proposition is a classical result (see e.g. [17] for the proof).

Proposition 1 *[17] Consider the analytic system* (S_0) *and a compact set* $K \subset \mathbf{R}^n$. *There exists* $\mu > 0$ *such that for* $M, T \geq 0$ *verifying*

$$MT \leq \mu \,, \tag{12.8}$$

and for any control $u = (u_1, \dots, u_m)$ *piecewise continuous on* $[0,T]$ *and verifying*

$$|u(t)| \leq M \,, \quad \forall t \in [0,T] \,, \tag{12.9}$$

the solution $x(.)$ *of* (S_0), *with* $x_0 \triangleq x(0) \in K$, *satisfies*

$$x(t) = x_0 + \sum_I (f_I\, id)(x_0) \int_0^t u_I \,, \quad \forall t \in [0,T] \,. \tag{12.10}$$

Furthermore, the series in the right-hand side of (12.10) is uniformly absolutely convergent w.r.t. $t \in [0,T]$ *and* $x_0 \in K$.

Note that the sum in the right-hand side of equality (12.10) can be developed as

$$\sum_{k=1}^{\infty} \sum_{i_1,\dots,i_k=1}^{m} (f_{i_1} \cdots f_{i_k}\, id)(x_0) \int_0^t \int_0^{t_k} \cdots \int_0^{t_2} u_{i_k}(t_k) u_{i_{k-1}}(t_{k-1}) \cdots u_{i_1}(t_1)\, dt_1 \cdots dt_k \,.$$

Let us also remark that the condition (12.8), which relates the integration time-interval to the control size, is specific to driftless systems. For a system which contains a drift term, it is *a priori* not true that decreasing the size of the control inputs allows to increase the time-interval on which the expansion (12.10) is valid. The fact that this property holds for driftless systems can be viewed as a consequence of time-scaling invariance properties.

Our first result points out sufficient conditions under which exponential stabilization robust to unmodeled dynamics is granted.

Theorem 12.2.1. *[11] Consider an analytic locally controllable system* (S_0), *a neighborhood* U *of the origin in* $\mathbf{R}^n$, *and a function* $u : U \times \mathbf{R}^+ \longrightarrow \mathbf{R}^m$, $(x,t) \longmapsto u(x,t)$, *periodic of period* T *w.r.t.* t, *continuous w.r.t.* x *and piecewise continuous*[1] *w.r.t.* t. *Assume that*

1. *there exist* $\alpha, K > 0$ *such that* $|u(x,t)| \leq K|x|^\alpha$ *for all* $(x,t) \in U \times [0,T]$,
2. *the solution* $x(.)$ *of*

[1] In [11], u is assumed continuous w.r.t. t, but the proof is unchanged if u is only piecewise continuous.

$$\dot{x} = \sum_{i=1}^{m} f_i(x)u_i(x_0, t)\,, \quad x(0) = x_0 \in U\,, \tag{12.11}$$

satisfies $x(T) = Ax_0 + o(x_0)$ *with* A *a discrete-stable matrix,*

3. *for any multi-index* I *of length* $|I| \leq 1/\alpha$ *(this assumption is only needed when* $\alpha < 1$*),*

$$\int_0^T u_I(x) = O(x)\,. \tag{12.12}$$

Then, given a family of perturbed systems (S_ε)*, there exists* $\varepsilon_0 > 0$ *such that the origin of* $(\bar{S}_\varepsilon)$ *controlled by* $u(y,t)$ *is locally exponentially stable for any* $\varepsilon \in (-\varepsilon_0, \varepsilon_0)$ *.*

The conditions imposed in the theorem upon the control law can be satisfied in many ways. For instance, when the system (S_0) is known to be differentially flat [2], adequate control functions can be obtained by considering specifically tailored flatness-based solutions to the open-loop steering problem, as done for example in [1] in the case of chained systems. Although the control design approach and robustness analysis in [1] are very different from the ones developed in [11], the set of specific conditions derived in this reference imply that the assumptions of Theorem 12.2.1 are verified. This suggests that these assumptions are not unduly strong and also illustrates the fact that the domain of application of Theorem 12.2.1 extends to different control design techniques.

12.3 Control design

This section addresses the problem of constructing explicit controllers that meet the conditions of Theorem 12.2.1. Such controllers have to be exponential stabilizers for the extended system $(\bar{S}_0)$. A general design algorithm is first proposed. It takes advantage of known techniques based on the use of oscillatory open-loop controls in order to achieve net motion in any direction of the state space. Unfortunately (and unavoidably), the procedure also inherits the complexity of the abovementioned techniques, itself directly related to the process of selecting the "right" frequencies which facilitate motion monitoring in the state space. Unsurprisingly, the selection of these frequencies gets all the more involved that controllability of the system relies on high-order Lie brackets of the control vector fields. The control design can in fact be carried out from the expression of either the original system (S_0) or any locally controllable homogeneous approximation of (S_0). Indeed, working with an homogeneous approximation preserves the robustness of the feedback law provided that an extra condition is satisfied by the control law. This is stated more precisely further in the paper after recalling basic definitions and facts about homogeneous systems.

12.3.1 A general algorithm

We present in this section a general algorithm to construct robust and exponential stabilizers for (S_0). The algorithm uses previous results by Sussmann and Liu [16], and Liu [6]. It is also much related to the one developed in [10] for the construction of continuous time-periodic feedbacks $u(x,t)$ which exponentially stabilize the origin of a driftless system (S_0), but present the shortcoming of not being endowed with the type of robustness here considered.

In order to give a complete exposition of the algorithm, it is first useful to recall some notations from [18]. With the set of control vector fields $\{f_1, \dots, f_m\}$ we associate a set of *indeterminates* $X = \{X_1, \dots, X_m\}$. Brackets in $L(X)$, the **free** Lie algebra in the indeterminates $X_1, \dots, X_m$, will be denoted with the letter $\mathcal{B}$. To any such bracket, one can associate a *length* and a *set of indeterminates*. For instance, $\mathcal{B} = [X_1, [X_2, X_1]]$ has length three, and his set of indeterminates is $\{X_1, X_2, X_1\}$. To each element A in $L(X)$, one can also associate an element in the **control** Lie algebra $Lie(f)$ by means of the *evaluation operator* Ev. More precisely, $Ev(f)(A)$ is the vector field obtained by plugging in the f_j's for the X_j's in A. For instance, if $\mathcal{B} = [X_1, X_2]$, then $Ev(f)(\mathcal{B})$ is the vector field $[f_1, f_2]$.

Finally, we recall some definitions on subsets of $\mathbf{R}$ [16, 6].

Definition 1 *Let Ω be a finite subset of $\mathbf{R}$ and $|\Omega|$ denote the number of elements of Ω. The set Ω is said to be "Minimally Canceling" (in short, MC) if and only if :*

i) $\sum_{\omega \in \Omega} \omega = 0$

ii)this is the only zero sum with at most $|\Omega|$ terms taken in Ω with possible repetitions:

$$\left.\begin{array}{l} \sum_{\omega \in \Omega} \lambda_\omega \omega = 0 \\ \sum_{\omega \in \Omega} |\lambda_\omega| \leq |\Omega| \\ (\lambda_\omega)_{\omega \in \Omega} \in \mathbf{Z}^{|\Omega|} \end{array}\right\} \Longrightarrow \left\{\begin{array}{rl} (\lambda_\omega)_{\omega \in \Omega} = & (0, \dots, 0) \\ & or\ (1, \dots, 1) \\ & or\ (-1, \dots, -1) \end{array}\right. \tag{12.13}$$

Definition 2 *Let $(\Omega^\xi)_{\xi \in E}$ be a finite family of finite subsets Ω^ξ of $\mathbf{R}$. The family $(\Omega^\xi)_{\xi \in E}$ is said to be "independent with respect to p" if and only if :*

$$\left.\begin{array}{l}\displaystyle\sum_{\xi\in E}\sum_{\omega\in\Omega^\xi}\lambda_\omega\omega=0\\ \displaystyle\sum_{\xi\in E}\sum_{\omega\in\Omega^\xi}|\lambda_\omega|\le p\\ (\lambda_\omega)_{\omega\in\Omega^\xi,\xi\in E}\in\mathbf{Z}^{\Sigma|\Omega^\xi|}\end{array}\right\}\Longrightarrow\sum_{\omega\in\Omega^\xi}\lambda_\omega\omega=0\quad\forall\xi\in E \tag{12.14}$$

Algorithm

Step 1. Determine n vector fields $\tilde{f}_j$ $(j=1,\dots,n)$, obtained as Lie brackets of length $\ell(j)$ of the control vector fields f_i, and such that the matrix

$$\tilde{F}(x)\triangleq\left(\tilde{f}_1(x),\dots,\tilde{f}_n(x)\right) \tag{12.15}$$

is nonsingular at $x=0$.

Step 2. Determine a matrix G such that the matrix $(I_n+\tilde{F}(0)G)$ is discrete-stable (with I_n denoting the n-dimensional identity matrix), and define the linear feedback

$$a(x)=\frac{1}{T}Gx\,. \tag{12.16}$$

Step 3. By Step 1, there exists, for each $j=1,\dots,n$, a bracket $\mathcal{B}_j$ such that $\tilde{f}_j=Ev(f)(\mathcal{B}_j)$. Partition the set $\{\mathcal{B}_1,\dots,\mathcal{B}_n\}$ in *homogeneous components* $P_1,\dots,P_K$, i.e.

i) all brackets in a homogeneous component P_k have the same length $l(k)$, and the same set of indeterminates $\{X_{\tau_1^k},\dots,X_{\tau_{l(k)}^k}\}$.

ii) given two homogeneous components P_k and $P_{k'}$ (with $k\neq k'$), either $l(k)\neq l(k')$, or $\{X_{\tau_1^k},\dots,X_{\tau_{l(k)}^k}\}\neq\{X_{\tau_1^{k'}},\dots,X_{\tau_{l(k')}^{k'}}\}$.

Step 4. The last four steps can be conducted either in the control Lie algebra (c.l.a.) framework or in the framework of free Lie algebras (f.l.a.)[2].
c.l.a.: For every $k=1,\dots,K$, find permutations $\sigma_1,\dots,\sigma_{\underline{C}(k)}$ in $\mathcal{S}(l(k))$ such that the vector fields

$$[f_{\tau^k_{\sigma(1)}},[f_{\tau^k_{\sigma(2)}},[\dots,f_{\tau^k_{\sigma(l(k))}}]\dots]]\qquad(\sigma\in\{\sigma_1,\dots,\sigma_{\underline{C}(k)}\})$$

form a basis of the linear sub-space (over $\mathbf{R}$) of $Lie(f)$ spanned by the vector fields

[2] Respective advantages and drawbacks will be pointed out later.

$$[f_{T^k_{\sigma(1)}}, [f_{T^k_{\sigma(2)}}, [\dots, f_{T^k_{\sigma(l(k))}}]\dots]] \qquad (\sigma \in \mathcal{S}(l(k)))\,.$$

f.l.a.: For every $k = 1, \dots, K$, find permutations $\sigma_1, \dots, \sigma_{\overline{C}(k)}$ in $\mathcal{S}(l(k))$ such that the brackets

$$[X_{T^k_{\sigma(1)}}, [X_{T^k_{\sigma(2)}}, [\dots, X_{T^k_{\sigma(l(k))}}]\dots]] \qquad (\sigma \in \{\sigma_1, \dots, \sigma_{\overline{C}(k)}\})$$

form a basis of the linear sub-space (over $\mathbf{R}$) of $L(X)$ spanned by the brackets

$$[X_{T^k_{\sigma(1)}}, [X_{T^k_{\sigma(2)}}, [\dots, X_{T^k_{\sigma(l(k))}}]\dots]] \qquad (\sigma \in \mathcal{S}(l(k)))\,.$$

Step 5.
c.l.a.: For every $k \in \{1, \dots, K\}$ such that $l(k) \geq 2$, determine $C(k) \triangleq \underline{C}(k)$ MC sets $\Omega^{k,c} = \{\omega_1^{k,c}, \dots, \omega_{l(k)}^{k,c}\}$, with $c = 1, \dots, C(k)$, such that

i) the family of sets $(\Omega^{k,c})_{c=1,\dots,C(k)}^{k=1,\dots,K}$ is independent w.r.t. $\max_{k \in \{1,\dots,K\}} l(k)$

ii) all elements in these sets have a common divisor $\bar{\omega}$ $(= 2\pi/T)$, i.e.

$$\omega_i^{k,c}/\bar{\omega} \in \mathbf{Z}, \qquad \forall (k, c, i),$$

iii) the $C(k)$ elements $g^{k,c}$ $(c = 1, \dots, C(k))$ of $Lie(f)$ defined by

$$g^{k,c} = \sum_{\sigma \in \mathcal{S}(l(k))} \frac{[f_{T^k_{\sigma(1)}}, [f_{T^k_{\sigma(2)}}, [\dots, f_{T^k_{\sigma(l(k))}}]\dots]]}{\omega^{k,c}_{\sigma(1)}(\omega^{k,c}_{\sigma(1)} + \omega^{k,c}_{\sigma(2)}) \cdots (\omega^{k,c}_{\sigma(1)} + \dots + \omega^{k,c}_{\sigma(l(k)-1)})}$$

are independent (over $\mathbf{R}$).

For every $k \in \{1, \dots, K\}$ such that $l(k) = 1$, just set $\omega_1^{k,1} = 0$.

Each family of sets $\{\Omega^{k,c}\}_{c=1,\dots,C(k)}$ is used to associate the following sine and cosine functions with P_k

$$\alpha^{k,c}_{T^k_i}(t) = \begin{cases} \cos \omega_i^{k,c} t \;\; (i = 1) \\ \sin \omega_i^{k,c} t \;\; (i = 2, \dots, l(k))\,. \end{cases} \tag{12.17}$$

f.l.a.: Same as above, with $C(k) \triangleq \overline{C}(k)$ instead of $\underline{C}(k)$, each f_i replaced by X_i, and $Lie(f)$ replaced by $L(X)$.

Step 6.
c.l.a.: For each $k \in \{1, \dots, K\}$ and j such that $\mathcal{B}_j \in P_k$, determine coefficients $\mu_j^{k,c}$ $(c = 1, \dots, C(k))$ such that

$$\tilde{f}_j = \frac{(-1)^{l(k)-1}}{l(k)2^{l(k)-1}} \sum_{c=1}^{C(k)} \mu_j^{k,c} g^{k,c} . \tag{12.18}$$

f.l.a.: Same as above, with $\tilde{f}_j$ replaced by $\mathcal{B}_j$.

Step 7.
c.l.a. and **f.l.a.**: For each $k \in \{1, \ldots, K\}$, determine $l(k)C(k)$ state dependent functions $v_{\tau_i^k}^{k,c}$ which are $O(|x|^{\frac{1}{r(k)}})$, and such that

$$\prod_{i=1}^{l(k)} v_{\tau_i^k}^{k,c}(x) = \sum_{j:\mathcal{B}_j \in P_k} \mu_j^{k,c} a_j(x) \tag{12.19}$$

(a_j is the j-th component of a defined by (12.16)).

The following result concludes the description of the algorithm and points out the robustness properties associated with the resulting control in connection with Theorem 12.2.1.

Theorem 12.3.1. *Let*

$$u_i(x,t) = \begin{cases} \displaystyle\sum_{k=1}^{K} \sum_{c=1}^{C(k)} \sum_{p:\tau_p^k = i} \alpha_{\tau_p^k}^{k,c}(t) v_{\tau_p^k}^{k,c}(x) & \text{if } \exists (k,p) : \tau_p^k = i \\ 0 & \text{otherwise}. \end{cases} \tag{12.20}$$

with $C(k)$ equal to $\underline{C}(k)$ in the c.l.a. case, and to $\overline{C}(k)$ in the f.l.a. case. Then,

i) in both cases, u defined by (12.20) belongs to $C^0(\mathbf{R}^n \times \mathbf{R}^+; \mathbf{R}^m)$, is T-periodic w.r.t. t, and satisfies the three assumptions of Theorem 12.2.1.

ii) in the f.l.a. case, local asymptotic stability of the origin of the perturbed system $(\bar{S}_\varepsilon)$ is guaranteed for any ε such that $I_n + \tilde{F}_\varepsilon(0)G$ is discrete-stable, where $\tilde{F}_\varepsilon$ denotes the matrix-valued function obtained from (12.15) by replacing each $\tilde{f}_j = Ev(f)(\mathcal{B}_j)$ by $\tilde{f}_{j,\varepsilon} = Ev(f + h(\varepsilon, .))(\mathcal{B}_j)$.

Property *ii)* above summarizes the main advantage of working in the f.l.a. framework. In this case, *asymptotic stability of the origin of the controlled perturbed system is just equivalent to discrete-stability of the matrix $I_n + \tilde{F}_\varepsilon(0)G$.* This result is conceptually interesting because it is reminiscent of a well known robustness result associated with linear systems. On the other hand, the fact that the number $\underline{C}(k)$ is usually smaller than $\overline{C}(k)$ characterizes the main advantage of the c.l.a. framework over the f.l.a. one in terms of complexity of the control expression (12.20), as measured by the number of terms and time-periodic functions involved in this expression. Further explanations and comments about the algorithm are given in [11].

Now we show that robustness of the hybrid law (12.20) is conserved when sampling the control function at a large enough frequency.

Proposition 2 *Let u be defined by (12.20), and denote u_N (with $N \in \mathbf{N}$) the sampled function defined by*

$$\begin{aligned} &\forall k \in \mathbf{N}\,,\ \forall n = 0, \dots, N-1\,,\ \forall t \in \left[kT + \tfrac{nT}{N}, kT + \tfrac{(n+1)T}{N}\right)\,, \\ &\qquad u_N(x,t) = u(x, kT + nT/N)\,. \end{aligned} \tag{12.21}$$

Then, there exists $N_0 \in \mathbf{N}$ such that, for $N \geq N_0$, u_N is also a robust exponential stabilizer for (S_0).

Proof: The proof consists in showing that u_N satisfies the three assumptions of Theorem 12.2.1. It is clear from (12.21) that Assumption 1 is satisfied for u_N since, from Theorem 12.3.1, u satisfies Assumption 1. Let us now consider Assumption 2. Using the Chen-Fliess series, the solution $x(.)$ of (12.11) with u_N as control satisfies

$$\begin{aligned} x(T) &= x_0 + \sum_{|I| \leq 1/\alpha} (f_I\, id)(x_0) \int_0^T u_{N,I}(x_0) + o(x_0) \\ &= x_0 + \sum_{|I| \leq 1/\alpha} (f_I\, id)(x_0) \int_0^T u_I(x_0) + o(x_0) \\ &\quad + \sum_{|I| \leq 1/\alpha} (f_I\, id)(x_0) \left(\int_0^T u_{N,I}(x_0) - \int_0^T u_I(x_0) \right) \\ &= Ax_0 + o(x_0) \\ &\quad + \sum_{|I| \leq 1/\alpha} (f_I\, id)(x_0) \left(\int_0^T u_{N,I}(x_0) - \int_0^T u_I(x_0) \right)\,, \end{aligned} \tag{12.22}$$

where we have used the fact that u and u_N satisfy Assumption 1 with the same value of α, and the fact that u satisfies Assumption 2. Let us now consider each term

$$\int_0^T u_{N,I}(x_0) - \int_0^T u_I(x_0)$$

in (12.22). Using (12.20), we can rewrite this term as

$$\int_0^T u_{N,I}(x_0) - \int_0^T u_I(x_0) = \sum_q v_I^q(x_0) \left(\int_0^T \alpha_{N,I}^q - \int_0^T \alpha_I^q \right)\,. \tag{12.23}$$

This expression reads as follows. Each q denotes a family $(q_1, \dots, q_{|I|})$ with $q_i = (k_i, c_i)$, and

$$v_I^q(x_0) = v_{i_1}^{q_1}(x_0) \cdots v_{i_{|I|}}^{q_{|I|}}(x_0),$$

$$\int_0^T \alpha_{N,I}^q = \int_0^T \alpha_{N,i_I}^{q_{|I|}}(t_{|I|}) \int_0^{t_{|I|}} \dots \int_0^{t_2} \alpha_{N,i_1}^{q_1}(t_1) dt_1 \dots dt_{|I|},$$

and

$$\int_0^T \alpha_I^q = \int_0^T \alpha_{i_I}^{q_{|I|}}(t_{|I|}) \int_0^{t_{|I|}} \dots \int_0^{t_2} \alpha_{i_1}^{q_1}(t_1) dt_1 \dots dt_{|I|}.$$

Specifying further the (finite) set on which the sum in (12.23) is taken is not important. Note that the integrals in the right-hand side of (12.23) are iterated integrals of sine or cosine functions, and sampled sine or cosine functions, which are independent of x_0. To proceed with the proof, we need the following lemma.

Lemma 1 *Each term*

$$v_I^q(x_0)\left(\int_0^T \alpha_{N,I}^q - \int_0^T \alpha_I^q\right) \tag{12.24}$$

in (12.23), viewed as a function of x_0, satisfies one of the following properties

a) it is a $o(x_0)$,

b) it is a linear function of x_0,

c) it is identically zero for N large enough.

(Proof given farther)

Since each term

$$\int_0^T \alpha_{N,I}^q - \int_0^T \alpha_I^q$$

obviously tends to zero as N tends to infinity, we deduce from Lemma 1, that the term (12.24) is either a $o(x_0)$, or a term $A_N(I,q)x_0$ with $A_N(I,q)$ a matrix which tends to zero as N tends to infinity, or zero for N large enough. Therefore, from (12.22), (12.23), and using the facts that the f_i's are smooth, and that the number of multi-indices I such that $|I| \leq 1/\alpha$ is finite, there exists a matrix $B(N)$ which tends to zero as N tends to infinity, and such that

$$x(T) = Ax_0 + B(N)x_0 + o(x_0).$$

This clearly implies that Assumption 2 is satisfied for N large enough. Finally the satisfaction Assumption 3 is a direct consequence of (12.23), Lemma 1, and the fact that u satisfies this assumption. There remains to prove Lemma 1.

Proof of Lemma 1: Assuming that neither a) nor b) hold, we show that c) must be satisfied. The proof consists in expanding each sampled sine or cosine function as a Fourier series, in order to evaluate each term

$$\int_0^T \alpha_{N,I}^q \,. \tag{12.25}$$

First, we establish the following

Claim 1 *Let* $\{\omega_1, \dots, \omega_{|I|}\}$ *denote the set of frequencies associated with the functions* $\alpha_1, \dots, \alpha_{|I|}$ *in*

$$\int_0^T \alpha_I^q \,.$$

Then, for each M.C. set $\Omega^{k,c}$, $\{\omega_1, \dots, \omega_{|I|}\}$ *contains at most* $l(k)$ *elements which belong to* $\Omega^{k,c}$, *and does not contain* $\Omega^{k,c}$ *itself.*

We prove the claim by contradiction, and first assume that $\{\omega_1, \dots, \omega_{|I|}\}$ contains more than $l(k)$ elements of some $\Omega^{k,c}$. Then, in view of Step 7 of the design algorithm, we deduce that $v_I^q(x_0) = o(x_0)$. This contradicts our initial assumption according to which Property a) in Lemma 1 is not satisfied. On the other hand, if the set $\{\omega_1, \dots, \omega_{|I|}\}$ contains some set $\Omega^{k,c}$ then, either these two sets are equal and, from (12.16) and (12.19), v_I^q is a linear function (in contradiction with the assumption that Property b) of Lemma 1 is not satisfied), or $\{\omega_1, \dots, \omega_{|I|}\}$ contains $\Omega^{k,c}$ plus extra terms, in which case $v_I^q(x_0) = o(x_0)$ (again in contradiction with our initial assumption).

Having proved Claim 1, we return to the proof of the lemma. In order to simplify the notation, we assume from now on that $T = 2\pi$. For different values of T, the proof follows by a simple change of time variable. Let $\alpha_{N,i}$ denote any sampled sine or cosine function. Away from points of discontinuity,

$$\alpha_{N,i}(t) = \sum_{n=-\infty}^{+\infty} c_n e^{jnt} \,, \tag{12.26}$$

with

$$c_n = \frac{1}{2\pi} \int_0^{2\pi} \alpha_{N,i}(t) e^{-jnt}\, dt \,.$$

First, consider the case when $\alpha_i(t) = \cos \omega_i t$. Then, denoting $\Delta \triangleq T/N = 2\pi/N$,

$$\begin{aligned}
c_n &= \frac{1}{2\pi}\sum_{k=0}^{N-1}\int_{k\Delta}^{(k+1)\Delta}\frac{e^{j\omega_i k\Delta}+e^{-j\omega_i k\Delta}}{2}e^{-jnt}\,dt \\
&= -\frac{1}{4jn\pi}\sum_{k=0}^{N-1}\left(e^{j\omega_i k\Delta}+e^{-j\omega_i k\Delta}\right)\left(e^{-jn(k+1)\Delta}-e^{-jnk\Delta}\right) \\
&= -\frac{1}{4jn\pi}\left(e^{-jn\Delta}-1\right)\sum_{k=0}^{N-1} e^{j(\omega_i-n)k\Delta}+e^{-j(\omega_i+n)k\Delta}.
\end{aligned}$$

If $n-\omega_i \notin N\mathbf{Z}$, then

$$\sum_{k=0}^{N-1} e^{j(\omega_i-n)k\Delta}=\frac{1-e^{j(\omega_i-n)N\Delta}}{1-e^{j(\omega_i-n)\Delta}}=\frac{1-e^{j(\omega_i-n)2\pi}}{1-e^{j(\omega_i-n)\Delta}}=0\,,$$

where the last equality comes from the fact that, from Step 5, $\omega_i \in \mathbf{Z}$. Similarly, if $n+\omega_i \notin N\mathbf{Z}$

$$\sum_{k=0}^{N-1} e^{-j(\omega_i+n)k\Delta}=0\,.$$

Therefore, c_n is possibly different from zero only if $n=\pm\omega_i(mod N)$, so that (12.26) may be rewritten as

$$\begin{aligned}
\cos_N \omega_i t &= \sum_{k=-\infty}^{+\infty}\eta^{1}_{i,k}e^{j(\omega_i+kN)t}+\sum_{k=-\infty}^{+\infty}\eta^{-1}_{i,k}e^{-j(\omega_i+kN)t} \\
&= \sum_{k=-\infty}^{+\infty}\sum_{s\in\{-1,1\}}\eta^{s}_{i,k}e^{sj(\omega_i+kN)t}\,,
\end{aligned} \tag{12.27}$$

where the $\eta^s_{i,k}$ are complex coefficients which depend on ω_i, N, k, and s. Similarly,

$$\sin_N \omega_i t = \sum_{k=-\infty}^{+\infty}\sum_{s\in\{-1,1\}}\eta^{s}_{i,k}e^{sj(\omega_i+kN)t}\,, \tag{12.28}$$

where the $\eta^s_{i,k}$ are other complex coefficients. In view of (12.27) and (12.28), we can rewrite (12.25) as

$$\int_0^T \alpha^q_{N,I}=\sum_{(k_1,\dots,k_{|I|})\in\mathbf{Z}^{|I|}} J_{N,I}(k_1,\dots,k_{|I|})\,,$$

with

$$\begin{aligned}
J_{N,I}(k_1,\dots,k_{|I|}) \triangleq \int_0^{2\pi}&\sum_{s_{|I|}\in\{-1,1\}}\eta^{s_{|I|}}_{i_{|I|},k_{|I|}}e^{s_{|I|}j(\omega_{i_{|I|}}+k_{|I|}N)\tau_{|I|}} \\
&\int_0^{\tau_{|I|}}\cdots\int_0^{\tau_2}\sum_{s_1\in\{-1,1\}}\eta^{s_1}_{i_1,k_1}e^{s_1 j(\omega_{i_1}+k_1 N)\tau_1}\,d\tau_1\dots d\tau_{|I|}\,.
\end{aligned} \tag{12.29}$$

The above expression is to be compared with the following one, derived when the sine and cosine functions are not sampled:

$$\begin{aligned}\int_0^T \alpha_I^q = \int_0^{2\pi} \sum_{s_{|I|}\in\{-1,1\}} \eta_{i_{|I|}}^{s_{|I|}} e^{s_{|I|}j\omega_{i_{|I|}}\tau_{|I|}} \int_0^{\tau_{|I|}} \cdots \\ \int_0^{\tau_2} \sum_{s_1\in\{-1,1\}} \eta_{i_1}^{s_1} e^{s_1 j\omega_{i_1}\tau_1}\, d\tau_1 \ldots d\tau_{|I|}\,,\end{aligned} \tag{12.30}$$

with $\eta_i^1 = \eta_i^{-1} = 1/2$ if α_i is a cosine function, and $\eta_i^1 = -\eta_i^{-1} = -i/2$ if α_i is a sine function. We have proved in [11, Lemma 2] that, when the condition of Claim1 is satisfied, the integral (12.30) is zero. We claim that each iterated integral (12.29) is also equal to zero provided that

$$N > \sum_{i=1}^{|I|} |\omega_i|\,. \tag{12.31}$$

This condition is needed in order to ensure the following property:

$$\left.\begin{array}{r}\sum_{p=1}^{|I|} \lambda_p(\omega_{i_p} + k_p N) = 0 \\ \lambda_p \in \{-1,0,1\}\end{array}\right\} \Longrightarrow \sum_{p=1}^{|I|} \lambda_p \omega_{i_p} = 0\,.$$

We leave to the reader the task of verifying that this property allows a direct transposition of the proof given in [11, Lemma 2] for the integral (12.30). Therefore, both integrals involved in (12.24) are equal to zero when (12.31) holds, and Property c) of Lemma 1 is verified. Note that imposing

$$N > \max_k l(k) \max_{i,k,c} |\omega_i^{k,c}|$$

automatically ensures (12.31) since, from (12.22) and Step 7, $|I| \leq 1/\alpha = \max_k l(k)$. ∎

12.3.2 Control design from a homogeneous approximation

It is often convenient and simpler to work with approximations of control systems. For instance, linear approximations are commonly used for feedback control design when they are controllable (or at least stabilizable). When the linear approximation of the system, evaluated at the equilibrium which feedback control is in charge of stabilizing, is not stabilizable, the extension of the notion of linear approximation yields to homogeneous controllable approximations. Using such an approximation is particularly well adapted to the design of continuous homogeneous feedbacks which render the closed-loop system homogeneous of degree zero. The reason is that asymptotic stabilization of the origin of the homogeneous approximation automatically ensures that the origin of the initial control system is also asymptotically (locally) stabilized by the same feedback control law. It is however important to realize that this property *does not necessarily hold* when using hybrid controllers such as those which we are considering here, and it is not difficult to work out simple examples which illustrate this fact. Nevertheless, it is proved in [11] that a robust controller for the system (S_0) can be derived from the knowledge of a homogeneous approximation of this system, provided that some extra condition is satisfied by the control law. This condition will be stated in a theorem, after recalling a few definitions and properties about homogeneous systems. A complementary proposition will indicate how the control design algorithm previously described can be completed in order to cope with the use of homogeneous approximations.

Given $\lambda > 0$ and a *weight vector* $r = (r_1, \ldots, r_n)$ $(r_i > 0\ \forall i)$, a *dilation* δ^r_λ is a map from $\mathbf{R}^n$ to $\mathbf{R}^n$ defined by

$$\delta^r_\lambda(z_1, \ldots, z_n) = (\lambda^{r_1} z_1, \ldots, \lambda^{r_n} z_n)\,.$$

A function $f \in C^0(\mathbf{R}^n; \mathbf{R})$ is *homogeneous of degree l with respect to the family of dilations* δ^r_λ $(\lambda > 0)$, or, more concisely, *δ^r-homogeneous of degree l*, if

$$\forall \lambda > 0, \quad f(\delta^r_\lambda(z)) = \lambda^l f(z)\,.$$

A *δ^r-homogeneous norm* can be defined as a positive definite function on $\mathbf{R}^n$, δ^r-homogeneous of degree one. Although this is not a "true" norm when the weight coefficients are not all equal, it still provides a means of "measuring" the size of the state.

A continuous vector field X on $\mathbf{R}^n$ is *δ^r-homogeneous of degree d* if, for all $i = 1, \ldots, n$, the function $z \longmapsto X_i(z)$ is δ^r-homogeneous of degree $r_i + d$. According to these definitions, homogeneity is coordinate dependent, however it is possible to define the above concepts in a coordinate independent framework [5, 13].

Finally, we say that the system

$$\dot{z} = \sum_{i=1}^{m} b_i(z) u_i \tag{12.32}$$

is a *δ^r-homogeneous approximation* of (S_0) if:

1. the change of coordinates $\phi :\ x \longmapsto z$ transforms (S_0) into

$$\dot{z} = \sum_{i=1}^{m} (b_i(z) + g_i(z))\, u_i\,, \tag{12.33}$$

where b_i is δ^r-homogeneous of some degree $d_i < 0$, and g_i denotes higher-order terms, i.e. such that $g_{i,j}$ (the j-th component of g_i) satisfies

$$g_{i,j} = o(\rho^{r_j+d_i}), \quad (j = 1, \ldots, n). \tag{12.34}$$

where ρ is a δ^r-homogeneous norm;

2. the system (12.32) is controllable.

Hermes [3] and Stefani [15] have shown that any driftless system (S_0) satisfying the LARC (Lie Algebra Rank Condition) at the origin (12.2) has a homogeneous approximation (which is not unique in general).

Theorem 12.3.2. *Consider a δ^r-homogeneous approximation (12.32) of (S_0), with $d_i \triangleq deg(b_i)$ $(i = 1, \ldots, m)$, and a control function*

$$u \in C^0(U \times [0,T]; \mathbf{R}^m)$$

such that the three assumptions in Theorem 12.2.1 are verified for this approximating system. Assume furthermore that the following assumption, which is a stronger version of the third assumption in Theorem 12.2.1, is also verified for the approximating system:

3-bis. for any multi-index $I = (i_1, \ldots, i_{|I|})$ of length $|I| \leq 1/\alpha$,

$$\int_0^T u_I(z) = \sum_{k:r_k \geq \|I\|} a_{I,k} z_k + o(z), \tag{12.35}$$

where $\|I\| \triangleq -\sum_{j=1}^{|I|} d_{i_j}$, and the $a_{I,k}$'s are some scalars.

Then, the three assumptions of Theorem 12.2.1 are verified for the system (12.33).

When applying the algorithm of Section 12.3.1 to the approximation (12.32), the control law u given by (12.20) may not satisfy the extra condition 3-bis of Theorem 12.3.2. However, it is possible to impose extra requirements on the matrix G defined in Step 2 so as to guarantee the satisfaction of this condition. For instance, the following result is proved in [11].

Proposition 3 *Consider a δ^r-homogeneous approximation (12.32) of (S_0), with every control vector field of this system being δ^r-homogeneous of degree -1. Without loss of generality, we assume that the variables z_i are ordered by increasing weight, i.e.*

$$r_1 \leq r_2 \leq \ldots \leq r_n \,,$$

and decompose z as $z = (z^1, \ldots, z^P)$, where each z^p $(1 \leq p \leq P)$ is the sub-vector of z whose components have same weight r^p $(r_1 \leq r^p \leq r_n)$ with

$$r_1 = r^1 < r^2 < \ldots < r^P = r_n \,.$$

Consider the control design algorithm described in Section 12.3.1 and applied to (12.32). Let $\tilde{b}_j$ $(j \in \{1, \ldots, n\})$ denote the vector fields defined according to Step 1 of the algorithm, and

$$\tilde{B}(z) \triangleq (\tilde{b}_1(z), \ldots, \tilde{b}_n(z)) \,.$$

Due to the ordering of the variables z_i, the matrix $\tilde{B}(z)$ is block lower triangular, and block diagonal at $z = 0$, i.e.

$$\tilde{B}(0) = \begin{pmatrix} \tilde{B}^{11} & 0 & \cdots & 0 \\ 0 & \tilde{B}^{22} & & \vdots \\ \vdots & \vdots & \ddots & 0 \\ 0 & \cdots & \cdots & \tilde{B}^{PP} \end{pmatrix} .$$

Assume that the control gain matrix G involved in Step 2 of the algorithm is chosen as follows

$$G = \tilde{B}(0)^{-1}(H - I_n)$$

with the matrix H being block upper triangular, i.e.

$$A = \begin{pmatrix} H^{11} & \star & \cdots & \star \\ 0 & H^{22} & \ddots & \vdots \\ \vdots & \ddots & \ddots & \star \\ 0 & \cdots & 0 & H^{PP} \end{pmatrix} ,$$

and discrete-stable ($\Leftrightarrow$ H^{ii} is discrete-stable for $i \in \{1, \ldots, P\}$).
Then, the three assumptions of Theorem 12.2.1 are verified for the system (S_0).

12.3.3 Stabilizers for chained systems

In some cases, it is possible to take advantage of specific structural properties associated with the control system under consideration, in order to derive robust control laws that are simpler than those obtained by application of the general algorithm presented in Section 12.3.1. We illustrate this possibility in the case of the following n-dimensional chained system

$$(S_0)\quad\begin{cases}\dot{x}_1 = u_1\\ \dot{x}_2 = u_2\\ \dot{x}_3 = u_1 x_2\\ \quad\vdots\\ \dot{x}_n = u_1 x_{n-1}\,.\end{cases}\tag{12.36}$$

The next result points out a set of robust exponential stabilizers for this system.

Theorem 12.3.3. *With the control function $u \in C^0(\mathbf{R}^n \times [0,T]; \mathbf{R}^2)$ defined by*

$$\begin{cases} u_1(x,t) = \frac{1}{T}[(g_1-1)x_1 + 2\pi\rho_q(x)\sin(\bar{\omega}t)]\\ u_2(x,t) = \frac{1}{T}[(g_2-1)x_2\\ \qquad + \sum_{i=3}^n 2^{i-2}(i-2)!(g_i-1)\frac{x_i}{\rho_q^{i-2}(x)}\cos((i-2)\bar{\omega}t)]\,, \end{cases}\tag{12.37}$$

with

$$\begin{aligned} &T = 2\pi/\bar{\omega}\quad(\bar{\omega}\neq 0)\,,\\ &\rho_q(x) = \sum_{j=3}^n \alpha_j |x_j|^{\frac{1}{q+j-2}}\,,\qquad (q \geq n-2\,,\ \alpha_j > 0)\,,\\ &|g_i| < 1\,,\quad \forall i = 1,\ldots,n\,, \end{aligned}\tag{12.38}$$

the three assumptions in Theorem 12.2.1, and the extra assumption in Theorem 12.3.2, are verified for the system (12.36).

Corollary 1 *(of Theorems 12.3.2 and 12.3.3) With the control function (12.37), the three assumptions in Theorem 12.2.1 are verified for any analytic driftless system for which the chained system (12.36) is a δ^r-homogeneous approximation, with $r = (1, q, \ldots, q+n-2)$ and $q \geq n-2$.*

12.4 Control laws for a dynamic extension

In mechanics, systems with non-holonomic constraints (wheeled mobile-robots, systems with rolling parts,...) give rise to driftless systems like (S_0). In this case, x represents the configuration vector, and the control, u, is a vector of admissible velocities. In practice, it is however more realistic to consider torque control inputs rather than velocity control inputs. Since torques are homogeneous to accelerations, it is then natural to consider the following system (compare with (S_0))

$$(D_0): \begin{cases} \dot{x} = \sum_{i=1}^{m} f_i(x) u_i \\ \dot{u} = w, \end{cases}$$

where $u = (u_1, \dots, u_m)$, (x, u) is the state vector, and $w = (w_1, \dots, w_m)$ is now taken as the control variable. If $\bar{u}$ denotes an exponential robust stabilizer for (S_0) (as derived in the previous section for instance), we would like to deduce an exponential stabilizer w for (D_0), which conserves the robustness properties of $\bar{u}$. More precisely, we look for a feedback $w(y, v, t)$ such that the origin of the controlled system

$$(\bar{D}_\varepsilon): \begin{cases} \dot{x} = \sum_{i=1}^{m} (f_i(x) + h_i(\varepsilon, x)) u_i \\ \dot{u} = w(y, v, t) \\ \dot{y} = (\sum_{k \in \mathbf{N}} \delta_{kT})(x - y_{-\alpha}) \\ \dot{v} = (\sum_{k \in \mathbf{N}} \delta_{kT})(u - v_{-\alpha}) \quad 0 < \alpha < T \end{cases}$$

is exponentially stable when $|\varepsilon|$ is small enough. We say that such a controller is an *exponential robust stabilizer* for (D_0). Let us remark that this is a somewhat simplified problem since we do not consider perturbations on the dynamic part. More precisely, having in mind the dynamic equations of mechanical systems, it would be justified to complement the perturbed system (S_ε) with an equation such as

$$\dot{u} = (I_m + g_1(\varepsilon, x, u)) w + g_0(\varepsilon, x, u),$$

with $g_1(0, ., .) = g_0(0., .,) \equiv 0$, and $g_0(., 0, 0) \equiv 0$ (so that $(x, u) = (0, 0)$ remains an equilibrium point). Beside the possibility that there may not exist controllers which ensure robustness with respect to such general perturbations, the analysis appears much more difficulty in this case. For this reason, the present analysis is limited to perturbations on the kinematic part only. Nonetheless, it is not very difficult to show that the control laws proposed below are also robust with respect to less general perturbations (such as these modeled by a function g_1 which depends on ε only).

The following proposition provides exponential robust stabilizers for (D_0).

Proposition 4 *Let $\bar{u} \in C^0(U \times \mathbf{R}^+; \mathbf{R}^m)$ denote a function Hlder-continuous with respect to x, differentiable and periodic of period T with respect to t. Assume further that $\bar{u}$ is an (hybrid) exponential robust stabilizer for (S_0). Denote α the function*

$$t \longmapsto \alpha(t) = t - \frac{T}{2\pi} \sin \frac{2\pi t}{T}. \tag{12.39}$$

Then,

1. *the function* $\bar{u}_c \in C^0(U \times \mathbf{R}^+; \mathbf{R}^m)$ *defined by*

$$\bar{u}_c(y,t) = \dot{\alpha}(t)\bar{u}(y,\alpha(t)) \tag{12.40}$$

is also an exponential robust stabilizer for (S_0), *with the function* $t \longmapsto \bar{u}_c(y(t),t)$ *being continuous along the trajectories of the closed-loop system* $(\bar{S}_0)$,

2. *the function* $w \in C^0(U \times \mathbf{R}^m \times \mathbf{R}^+; \mathbf{R}^m)$ *defined by*

$$w(y,v,t) = \frac{\partial}{\partial t}\bar{u}_c(y,t) - \frac{v}{T} \tag{12.41}$$

is an exponential robust stabilizer for (D_0) .

Proof: First, we show that Property 1 is satisfied. From (12.39), α defines a time-scaling on $\mathbf{R}^+$ which leaves each $t = kT$ invariant (i.e. $\alpha(kT) = kT$ for all $k \in \mathbf{N}$). One readily verifies that this time-scaling maps the solutions of (S_0) controlled by $\bar{u}$ to the solutions of (S_0) controlled by $\bar{u}_c$, i.e.

$$\begin{aligned}\dot{x}(t) = \sum_{i=1}^{m} f_i(x(t))\bar{u}(x_0,t) \Longrightarrow \frac{d}{dt}x(\alpha(t)) &= \sum_{i=1}^{m} f_i(x(t))\dot{\alpha}(t)\bar{u}(x_0,\alpha(t)) \\ &= \sum_{i=1}^{m} f_i(x(t))\bar{u}_c(x_0,\alpha(t))\,.\end{aligned}$$

Since this time-scaling also "preserves" the solutions of the perturbed systems (S_ε), we conclude that $\bar{u}_c$ is a robust exponential stabilizer for (S_0). Finally, $\bar{u}_c$ is continuous along the trajectories of $(\bar{S}_0)$ because $\dot{\alpha}(kT) = 0$ for all k, so that

$$\bar{u}_c(y(kT),kT) = 0 = \lim_{t \to kT} \bar{u}_c(y(t),t)\,.$$

Now we show that Property 2 is verified. We only prove exponential convergence to the origin of the closed-loop systems' solutions. Existence of these solutions and uniform stability of the origin can be proved via a simple adaptation of the proof of [11, Theorem 1], in the case of driftless systems. Let $(x_\varepsilon, u_\varepsilon, y_\varepsilon, v_\varepsilon)(.,t_0,x_0,u_0,y_0,v_0)$ denote the solution of the controlled system $(\bar{D}_\varepsilon)$ with initial conditions (t_0,x_0,u_0,y_0,v_0), $t_0 \in [k_0T,(k_0+1)T)$, $k_0 \in \mathbf{N}$. Then, for any $k \in \mathbf{N}$ such that $k_0 < k$, and any $t \in [kT,(k+1)T)$, this solution satisfies

$$\left\{\begin{array}{l} \dot{x} = \sum_{i=1}^{m}(f_i(x) + h_i(\varepsilon,x))u_i(t) \\ \dot{u} = w(x(kT),u(kT),t) \\ \dot{y} = 0\,, \quad y(t) = x(kT) \\ \dot{v} = 0\,, \quad v(t) = u(kT) \end{array}\right. \tag{12.42}$$

From (12.42) and (12.41),

$$u(t) = u(kT) + \bar{u}_c(x(kT), t) - \bar{u}_c(x(kT), kT) - \frac{u(kT)}{T}(t - kT). \quad (12.43)$$

Using the fact that $\bar{u}_c(., kT) \equiv 0$ for all k, we deduce that

$$u((k+1)T) = 0. \quad (12.44)$$

As a consequence, for $t \in [kT, (k+1)T)$ and $k \geq k_0 + 2$, we deduce from (12.43) and (12.44) that

$$u(t) = \bar{u}_c(x(kT), t). \quad (12.45)$$

Thus, for $t \geq (k_0 + 2)T$, the x component of the solution of (12.42) coincides with the solution of the system $(\bar{S}_\varepsilon)$ controlled by $\bar{u}_c(y, t)$. Since, from Property 1, $\bar{u}_c$ is an exponential stabilizer for $(\bar{S}_0)$, we deduce that $|x(t)|$ converges exponentially to zero. Then, using the fact that $\bar{u}$ (and therefore $\bar{u}_c$) is Hlder-continuous w.r.t. x, we deduce from (12.45) that $|u(t)|$ also converges exponentially to zero. ∎

References

1. M. K. Bennani and P. Rouchon. Robust stabilization of flat and chained systems. In *European Control Conference (ECC)*, pages 2642–2646, 1995.

2. M. Fliess, J. Lévine, P. Martin, and P. Rouchon. Flatness and defect of non-linear systems: introductory theory and examples. *International Journal of Control*, 61:1327–1361, 1995.

3. H. Hermes. Nilpotent and high-order approximations of vector field systems. *SIAM Review*, 33:238–264, 1991.

4. A. Isidori. *Nonlinear Control Systems*. Springer Verlag, third edition, 1995.

5. M. Kawski. Geometric homogeneity and stabilization. In *IFAC Nonlinear Control Systems Design Symp. (NOLCOS)*, pages 164–169, 1995.

6. W. Liu. An approximation algorithm for nonholonomic systems. *SIAM Journal on Control and Optimization*, 35:1328–1365, 1997.

7. D.A. Lizárraga, P. Morin, and C. Samson. Non-robustness of continuous homogeneous stabilizers for affine systems. Technical Report 3508, INRIA, 1998. Available at http://www.inria.fr/RRRT/RR-3508.html.

8. R.T. M'Closkey and R.M. Murray. Exponential stabilization of driftless nonlinear control systems using homogeneous feedback. *IEEE Trans. on Automatic Control*, 42:614–628, 1997.

9. S. Monaco and D. Normand-Cyrot. An introduction to motion planning using multirate digital control. In *IEEE Conf. on Decision and Control (CDC)*, pages 1780–1785, 1991.

10. P. Morin, J.-B. Pomet, and C. Samson. Design of homogeneous time-varying stabilizing control laws for driftless controllable systems via oscillatory approximation of lie brackets in closed-loop. *SIAM Journal on Control and Optimization*. To appear.

11. P. Morin and C. Samson. Exponential stabilization of nonlinear driftless systems with robustness to unmodeled dynamics. Technical Report 3477, INRIA, 1998. To appear in COCV.

12. R.M. Murray and S.S. Sastry. Nonholonomic motion planning: Steering using sinusoids. *IEEE Trans. on Automatic Control*, 38:700–716, 1993.

13. L. Rosier. *Etude de quelques problèmes de stabilisation*. PhD thesis, Ecole Normale de Cachan, 1993.

14. O. J. Sørdalen and O. Egeland. Exponential stabilization of nonholonomic chained systems. *IEEE Trans. on Automatic Control*, 40:35–49, 1995.

15. G. Stefani. Polynomial approximations to control systems and local controllability. In *IEEE Conf. on Decision and Control (CDC)*, pages 33–38, 1985.

16. H. J. Sussmann and W. Liu. Limits of highly oscillatory controls ans approximation of general paths by admissible trajectories. In *IEEE Conf. on Decision and Control (CDC)*, pages 437–442, 1991.

17. H.J. Sussmann. Lie brackets and local controllability: a sufficient condition for scalar-input systems. *SIAM Journal on Control and Optimization*, 21:686–713, 1983.

18. H.J. Sussmann. A general theorem on local controllability. *SIAM Journal on Control and Optimization*, 25:158–194, 1987.

13. Stabilization of port–controlled Hamiltonian systems via energy balancing

Romeo Ortega, Arjan J. van der Schaft** and Bernhard M. Maschke****

*Laboratoire des Signaux et Systèmes, C.N.R.S.
SUPELEC
91192 Gif-sur-Yvette, France
ortega@lss.supelec.fr

**Faculty of Applied Mathematics
University of Twente, P.O. Box 217
7500 AE Enschede, The Netherlands
A.J.vanderSchaft@math.utwente.nl

***Centre National des Arts et Métiers
Automatisme Industriel
21, rue Pinel
75013, Paris, France
maschke@ensam-paris.fr

Summary.

Passivity–based control (PBC) for regulation of mechanical systems is a well established tehcnique that yields robust controllers that have a clear physical interpretation in terms of interconnection of the system with its environment. In particular, the total energy of the closed–loop is the difference between the energy of the system and the energy supplied by the controller. Furthermore, since the Euler–Lagrange (EL) structure is preserved in closed–loop, PBC is robustly stable *vis á vis* unmodeled dissipative effects and inherits some robust performance measures from its inverse optimality. Unfortunately, these nice properties are lost when PBC is used in other applications, for instance, in electrical and electromechanical systems. Our main objective in this paper is to develop a new PBC theory for port–controlled Hamiltonian (PCH) systems, which result from the network modeling of energy-conserving lumped-parameter physical systems with independent storage elements, and strictly contain the class of EL models. We identify a class of PCH models for which PBC ensures the Hamiltonian structure is preserved, with storage function the energy balance. One final advantage of the method is that it is rather systematic and the controller can be easily derived using symbolic computation.

13.1 Introduction

The term passivity-based control (PBC) was first introduced in [15] to define a controller design methodology which achieves stabilization by rendering *passive* a suitably defined map. This idea has been very successful to control physical systems described by Euler–Lagrange (EL) equations of motion, which as thoroughly detailed in [16], includes mechanical, electrical and electromechanical applications. PBC has its roots in the ground-breaking work of Takegaki and Arimoto [23] on state-feedback regulation of fully actuated robot manipulators. For such (so-called simple) mechanical systems the controller design proceeds along two basic stages. First, an *energy shaping* stage where we modify the potential energy of the system in such a way that the new potential energy function has a strict local minimum in the desired equilibrium.[1] Second, a *damping injection* stage where we now modify the dissipation function to ensure asymptotic stability.

A central feature of this technique is that the closed-loop dynamics remain in Lagrangian form. There are three important advantages of requiring the closed-loop to be an EL system which, to a large extent, explain the practical success of PBC:

1. The control action has a clear physical interpretation as an interconnection of the system with the controller. In particular, stabilization can be understood in terms of energy balance between them. Indeed, we will show in this paper that the total energy of the closed-loop EL system is the difference between the energy of the open-loop and the energy provided to the system from its environment.[2]
2. Since EL systems are passive with respect to physically meaningful outputs, a margin of robustness *vis a vis* uncertain parameters and unmodeled dynamics is ensured. Furthermore, the closed-loop inherits some robust performance measures from its inverse optimality.
3. For mechanical systems the controller is a simple PD-like law with the controller parameters playing the role of dampers and springs. This property can hardly be overestimated in engineering applications where commissioning of the controller for a robust behaviour is an issue of prime importance.

[1] It is clear that, similarly to the choice of a Lyapunov function, no systematic procedure exists to select the desired potential energy. An important advantage of the method proposed here is that this step is considerably simplified, and sometimes even obviated.

[2] With an obvious abuse of notation in the sequel we will refer to this function as "energy-balancing function".

PBC has been extended, within the class of simple mechanical systems, to consider regulation with output feedback [17], [22], underactuation [1] and the presence of input constraints [9]. PBC ideas were also applied to electrical and electromechanical systems described by EL models, as well as to solve tracking problems -for a complete set of references see [16]. While in regulation problems for mechanical systems it suffices to shape the potential energy, to address the other applications (even in regulation tasks) we had to modify also the kinetic energy. Unfortunately, this modification could not be achieved preserving the Lagrangian structure. That is, in these cases, the closed-loop -although still defining a passive operator- is no longer an EL system, and the storage function of the passive map does not have the interpretation of total energy. Consequently these designs will not, in general, enjoy the three nice features mentioned above. As explained in Section 10.3.1 of [16], this situation stems from the fact that, to shape the kinetic energy, we carry out an inversion of the system along the reference trajectories that destroys the EL structure.[3] Another shortcoming of the EL approach is that the "desired" storage function for the closed-loop map is defined in terms of some error quantities whose physical interpretation is far from obvious.

The *main contribution* of this paper (see also [19]) is the development of a controller design methodology that extends, to a broader class of systems the nice features of PBC of simple mechanical systems described above. Towards this end, we develop a new systematic technique to achieve energy-shaping and damping injection in PBC for set-point regulation of systems described by *port-controlled Hamiltonian (PCH) models.* An important advantage of the method is that the basic step of PBC of choosing the "desired" storage function -being now a true energy function- becomes more natural. Actually, as we will see in the paper, we don't even need to know it explicitly, but the method provides the means to verify its existence. We also have that, if the damping satisfies some structural conditions (or if it is zero), the total energy is the "energy-balancing function". Finally, the design is rather systematic and the controller can be easily derived using symbolic computation.

The remaining of the paper is organized as follows. In Section 2 we briefly describe the class of PCH models studied in the paper. The main result of our work, namely a procedure to design a stabilizing PBC for PCH systems which preserves the Hamiltonian structure, is presented in Section 3. In Sections 4 and 5 we give two alternative interpretations of the stabilization mechanism of the proposed PBC in terms of: 1) the overall systems energy-balance, and 2) the method of Energy-Casimir functions [10]. For the former we show that, if the damping satisfies some structural conditions (or if it is zero), then the storage function assigned to the closed-loop is the "energy balancing function". To provide the second interpretation, we follow [14], [22], and view

[3] It also imposes a stable invertibility requirement to the system which is obviated in the approach presented here.

our controller as a PCH system in a power–preserving interconnection with the plant. In this way the plant is embedded in a higher dimensional system for which a series of Casimir functions can be constructed. We wrap up the paper with some open problems and concluding remarks.

13.2 Port controlled Hamiltonian systems

13.2.1 Systems model

Network modeling of energy-conserving lumped-parameter physical systems [12] with independent storage elements leads to models of the form –called port controlled Hamiltonian systems [11], [26]–

$$\Sigma : \begin{cases} \dot{x} = J(x)\frac{\partial H}{\partial x}(x) + g(x)u \\ y = g^T(x)\frac{\partial H}{\partial x}(x) \end{cases} \tag{13.1}$$

where $x \in \mathcal{R}^n$ are the energy variables, the smooth function $H(x) : \mathcal{R}^n \to \mathcal{R}$ represents the total stored energy, which we assume is bounded from below, and $u, y \in \mathcal{R}^m$ are the port power variables. (All vectors defined in the paper are *column* vectors, even the gradient of a scalar function.) u and y are conjugated variables, for instance currents and voltages in electrical circuits or forces and velocities in mechanical systems. The interconnection structure is captured in the $n \times n$ matrix $J(x)$ and the $n \times m$ matrix $g(x)$, both depending smoothly on the state x. Because of the assumption of energy-conservation, the matrix $J(x)$ is skew-symmetric, that is,

$$J(x) = -J^T(x), \quad \forall\, x \in \mathcal{R}^n \tag{13.2}$$

The geometric structure of Hamiltonian systems has been thoroughly studied in the literature, we refer the interested reader to [7], [10]. The matrix $J(x)$ defines a generalized Poisson bracket on the state manifold (generalized because it need not satisfy the Jacobi-identity [26]).

Energy-dissipation is included by terminating some of the ports by resistive elements, see e.g. [26]. Indeed, consider instead of $g(x)u$ in (13.1) a term

$$\begin{bmatrix} g(x) & g_R(x) \end{bmatrix} \begin{bmatrix} u \\ u_R \end{bmatrix} = g(x)u + g_R(x)u_R$$

and extend correspondingly $y = g^T(x)\frac{\partial H}{\partial x}(x)$ to

$$\begin{bmatrix} y \\ y_R \end{bmatrix} = \begin{bmatrix} g^T(x)\frac{\partial H}{\partial x}(x) \\ g_R^T(x)\frac{\partial H}{\partial x}(x) \end{bmatrix}$$

Here u_R, y_R denote the power variables at the ports which are terminated by (linear) resistive elements

$$u_R = -Sy_R$$

for some positive semi–definite symmetric matrix S. Substitution in (13.1) leads to models of the form

$$\Sigma : \begin{cases} \dot{x} = [J(x) - R(x)]\frac{\partial H}{\partial x}(x) + g(x)u \\ y = g^T(x)\frac{\partial H}{\partial x}(x) \end{cases}$$

where

$$R(x) \triangleq g_R(x)Sg_R^T(x)$$

which is a non-negative *symmetric* matrix depending smoothly on x, *i.e.*

$$R(x) = R^T(x) \geq 0, \quad \forall\, x \in \mathcal{R}^n \tag{13.3}$$

The autonomous dynamics (13.2.1) is the addition of a Hamilton vector field defined with respect to the pseudo–Poisson brackett associated with $J(x)$ and a gradient vector field defined with respect to the metric associated to $R(x)$.

We want to study also systems where the control acts through the *interconnection structure*. These are typically systems with switches where the controller commutes between different topologies. Assuming a sufficiently fast sampling and (for instance) a PWM implementation of the control action we can approximate the average behaviour of the switched system by a smooth system, where the control is now the PWM duty ratio. This situation, which is very common in power electronic devices [16], [4], leads us to consider systems of the form

$$\dot{x} = [J(x,u) - R(x)]\frac{\partial H}{\partial x}(x) + g(x,u) \tag{13.4}$$

where

$$J(x,u) = -J^T(x,u), \quad \forall\, x \in \mathcal{R}^n,\ u \in \mathcal{R}^m$$

The vector function $g(x,u)$ is introduced to capture two kind of interconnections, the standard $g(x)u$ and "constant source inputs", where u denotes the switching of the source input. See, for instance, the model of the Ćuk converter in [4].

13.2.2 Energy balance and passivity

PCH systems, as EL systems, define passive operators with storage function the total energy function $H(x)$. This can be easily established evaluating the rate of change of the total energy and using (13.1) and (13.2) to obtain, for the case when $R(x) = 0$, the power-balance

$$\frac{d}{dt}H = u^T y \tag{13.5}$$

with $u^T y$ the power externally supplied to the system. For the system with dissipation (13.2.1) the power-balance (13.5) extends to

$$\frac{d}{dt}H = -\left[\frac{\partial H}{\partial x}(x)\right]^T R(x)\frac{\partial H}{\partial x}(x) + u^T y \tag{13.6}$$

where the first term on the right-hand (which is non-positive by (13.3)) represents the *dissipation* due to the resistive (friction) elements in the system. Integrating (13.6), taking into account (13.3) and the fact that the total energy is bounded from below, we see that the map $\Sigma : u \mapsto y$ is *passive* [24]. That is, for all square integrable inputs $u(t)$, and all $t \geq 0$, we have the energy–balance equation

$$\underbrace{\int_0^t u^T(s)y(s)ds}_{supplied} = \underbrace{H[x(t)] - H[x(0)]}_{stored} + \underbrace{\int_0^t \left[\frac{\partial H}{\partial x}[x(s)]\right]^T R[x(s)]\frac{\partial H}{\partial x}[x(s)]ds}_{dissipated\ energy} \tag{13.7}$$

which expresses the fact that a passive system cannot store more energy than it is supplied to it from the outside, with the difference being the dissipated energy.

Notice that for the more general class of systems (13.4), since u is *not a port variable*, the passivity property is not established with respect to this signal, but between suitable elements of $\frac{\partial H}{\partial x}$ and $g(x,u)$.

13.2.3 Problem formulation

In the light of the discussion of the introduction we formulate our PBC stabilization objective as follows

- Given the PCH system (13.2.1) (or (13.4)) and a desired equilibrium $\bar{x} \in \mathcal{R}^n$. Find a control law u (which may be a static or dynamic state–output feedback) such that the closed–loop is still a PCH system and the equilibrium is asymptotically stable.

Keeping up with the spirit of PBC, one of our main concerns in this paper is to provide an energy–balance interpretation of the stabilization mechanism. Namely, we want to know under which conditions the *"energy-balance function"*[4]

$$H[x(t)] - \int_0^t u^T(s)y(s)ds$$

is a storage function to the closed–loop PCH system. In Section 5 we will show that this is the right candidate for being a storage function, because it is the total energy of the system connected to the controller.

13.3 Controller design procedure

In this section we will present a procedure to design a PBC with the aim of stabilizing a desired equilibrium point $\bar{x} \in \mathcal{R}^n$ for the PCH system (13.2.1) (or (13.4)). We will consider first the case of *static feedback*, see [19] for the dynamic feedback case. To avoid cluttering the notation we will make the control function of the *full state* but, as will become clear below, all results apply as well to the partial state–feedback case. See [18], [21] for two application examples.

13.3.1 Rationale

Motivated by the discussion of the introduction we want to design our stabilizing PBC in such a way that the closed–loop system is also a PCH system with dissipation. In this paper we will further require that the internal interconnection structure of the open–loop system (i.e. the matrix $J(x)$) is also preserved. See, however, point 3 of Subsection 3.3 for the case where we change also the interconnection structure.

Following the energy–shaping plus damping injection principles of PBC [16], [24] we will achieve this objective by:

1. Assigning to the closed–loop an energy function $H_d(x)$, which should have a strict local minimum at the desired equilibrium $\bar{x}$. (That is, there exists an open neighbourhood $\mathcal{B}$ of $\bar{x}$ such that $H_d(x) > H_d(\bar{x})$ for all $x \in \mathcal{B}$.) We will define

$$H_d(x) \triangleq H(x) + H_a(x) \tag{13.8}$$

[4] Notice that this function(al) coincides, up to an additive constant and with opposite sign, with the third right hand dissipation term of (13.7). In other words, we want the dissipation function to be non–decreasing.

where $H_a(x)$, which is a function to be defined, plays the role of energy function for the controller.

2. Injecting some additional damping such that $R_a(x)$ to get

$$R_d(x) \triangleq R(x) + R_a(x) \geq 0, \ \ \forall\, x \in \mathcal{R}^n \tag{13.9}$$

That is, we are looking for a static state–feedback control $u = \beta(x)$ such that the following identity holds

$$[J(x,\beta(x)) - R(x)]\frac{\partial H}{\partial x}(x) + g(x,\beta(x)) = [J(x,\beta(x)) - R_d(x)]\frac{\partial H_d}{\partial x}(x) \tag{13.10}$$

In this way, the closed–loop dynamics will be a PCH system with dissipation of the form

$$\dot{x} = [J(x,\beta(x)) - R_d(x)]\frac{\partial H_d}{\partial x}(x) \tag{13.11}$$

Clearly, along the dynamics (13.11) we will have

$$\frac{d}{dt}H_d = -\left[\frac{\partial H_d}{\partial x}(x)\right]^T R_d(x)\frac{\partial H_d}{\partial x}(x) \leq 0, \ \forall x \in \mathcal{R}^n \tag{13.12}$$

Hence, if the conditions 1 and 2 above are satisfied, $\bar{x}$ will be a *stable* equilibrium.

13.3.2 Main result

Proposition 13.3.1. *Given $J(x,u), R(x), H(x), g(x,u)$ and the desired equilibrium to be stabilized $\bar{x} \in \mathcal{R}^n$. Assume we can find functions $\beta(x), R_a(x)$ such that $R(x) + R_a(x) \geq 0, \ \ \forall\, x \in \mathcal{R}^n$, and a vector function $K(x)$ satisfying*[5]

$$[J(x,\beta(x)) - (R(x) + R_a(x))]K(x) = R_a(x)\frac{\partial H}{\partial x}(x) + g(x,\beta(x)) \tag{13.13}$$

and such that

- (Integrability) *$K(x)$ is the gradient of a scalar function. That is,*

$$\frac{\partial K}{\partial x}(x) = \left[\frac{\partial K}{\partial x}(x)\right]^T \tag{13.14}$$

- (Equilibrium assignment) *$K(x)$, at $\bar{x}$, verifies*

[5] Compare this equation with (4.49) of [24].

$$K(\bar{x}) = -\frac{\partial H}{\partial x}(\bar{x}) \tag{13.15}$$

- (Lyapunov stability) *The Jacobian of $K(x)$, at $\bar{x}$, satisfies the bound*

$$\frac{\partial K}{\partial x}(\bar{x}) > -\frac{\partial^2 H}{\partial x^2}(\bar{x}) \tag{13.16}$$

Under these conditions, the closed-loop system $u = \beta(x)$ will be a PCH system with dissipation of the form (13.11), where $H_d(x)$ is given by (13.8), and

$$\frac{\partial H_a}{\partial x}(x) = K(x) \tag{13.17}$$

Furthermore, $\bar{x}$ will be a (locally) stable *equilibrium of the closed-loop. It will be* asymptotically *stable if, in addition, the largest invariant set under the closed-loop dynamics contained in*

$$\left\{ x \in \mathcal{R}^n \cap \mathcal{B} \mid \left[\frac{\partial H_d}{\partial x}(x)\right]^T R_d(x) \frac{\partial H_d}{\partial x}(x) = 0 \right\} \tag{13.18}$$

equals $\{\bar{x}\}$.

Proof
For every given $\beta(x)$, $R_a(x)$, (and on any contractible neighbourhood of $\mathcal{R}^n$), the solution of equation (13.13) is a gradient of the form (13.17) if and only if the integrability condition (13.14) of the proposition is satisfied. Using (13.8) and (13.9) it is easy to see that, in this case, the closed-loop is a PCH system of the form (13.11) and total energy (13.8).

We will now prove that, under (13.15), (13.16), the stability of the equilibrium is ensured. To this end, notice that the equilibrium assignment condition (13.15) ensures $H_d(x)$ has an extremum at $\bar{x}$, while the Lyapunov stability condition (13.16) shows that it is actually an isolated minimum. On the other hand, from (13.12) we have that, along the trajectories of the closed-loop, $H_d(x(t))$ is non-increasing, hence it qualifies as a Lyapunov function, and we can conclude that $\bar{x}$ is a stable equilibrium. Asymptotic stability follows immediately invoking La Salle's invariance principle [5] and the condition (13.18).

The corollary below establishes the *passivation properties* of the PBC derived above as applied to the PCH system (13.2.1), i.e., when the control action does not affect the interconnection structure of the system.

Corollary 13.3.1. *Consider the PCH system (13.2.1). Let the function $\beta(x)$ be defined as in Proposition 3.1 above, and set $u = \beta(x)+v$. Then, the closed-loop system, given as*

$$\dot{x} = [J(x) - R_d(x)]\frac{\partial H_d}{\partial x}(x) + g(x)v$$

defines a passive map $v \mapsto y_a$*, with storage function* $H_d(x)$*, and the new output defined as*

$$y_a \triangleq y + g^T(x)K(x)$$

with $K(x)$ *solution of (13.13).*

13.3.3 Discussion

1. Notice that the construction of Proposition 3.1 *does not require* the explicit derivation of the Lyapunov (storage) function $H_d(x)$. This can be obtained, though, as a by–product integrating $\frac{\partial H_a}{\partial x}(x)$.

2. The PBC of Proposition 3.1 does not ensure, in general, passivity with respect to the natural output y, but to a new augmented output y_a. It is clear that, if $g^T(x)K(x) \equiv 0$, we have y as passive output and we recover the robustness features mentioned in the introduction. This will be, in particular, the case for (simple) mechanical systems.

3. In Proposition 3.1 we have decided to preserve in closed–loop the same internal interconnection structure as the open–loop. However, there are applications, which will be reported elsewhere, where it is interesting to modify it. That is, we want to obtain in closed–loop a PCH system of the form

$$\dot{x} = [J_d(x, \beta(x)) - R_d(x)]\frac{\partial H_d}{\partial x}(x)$$

where $J_d(x, \beta(x)) \triangleq J(x, \beta(x)) + J_a(x)$. In this case, the vector function $K(x)$ should satisfy

$$\begin{aligned}&[J(x, \beta(x)) + J_a(x) - R(x) - R_a(x)]K(x)\\ &= -[J_a(x) - R_a(x)]\frac{\partial H}{\partial x}(x) - g(x, \beta(x))\end{aligned}$$

The proposition applies *verbatim* to this case. It is possible to show that, with this additional degree of freedom, we can establish some connections between our PBC and the controllers obtained, with the technique of controlled Lagrangians, in the interesting paper [2]. See Section 6 for additional remarks on this point.

4. The choice of "admissible" damping injection matrices $R_a(x)$ is clearly tied with the span of the input interconnection matrix $g(x, u)$. In general, we will take $R_a(x) \geq 0$. However, in some particular examples, e.g., [18], [21], injection of *positive* damping has proven instrumental to solve some output feedback problems.

5. If for given $\beta(x), R_a(x)$ the integrability and equilibrium assignment conditions are satisfied, but the Hessian $\frac{\partial K}{\partial x}(\bar{x}) + \frac{\partial^2 H}{\partial x^2}(\bar{x})$ is a sign indefinite matrix, then the equlibrium will be *unstable*.

13.4 Stabilization via energy–balancing

To gain further insight into the derivations above we view the proposed PBC from two alternative perspectives. First, in this section, we give an interpretation of its stabilization mechanism in terms of the overall systems energy–balance. Then, in the next section, we will view the control action as a PCH system in a power preserving interconnection with the plant and establish the relationship with the method of Energy–Casimir functions [10].

Throughout these two sections we will restrict ourselves to the case when the control does not affect the interconnection structure of the system, that is, we will consider only the PCH system (13.2.1). Also, for the sake of clarity, we will not add damping to the system. See the point 3 of the discussion of Subsection 5.1.

We show below that, if the damping satisfies a structural condition (or if it is zero), then the storage function assigned to the closed–loop (i.e., its total energy) is the energy balance function. That is,

$$H_d(x(t)) = H(x(t)) - \int_0^t u^T(s)y(s)ds + c \tag{13.19}$$

which is the difference between the total energy of the open–loop and the energy provided to the system from the controller.

Proposition 13.4.1. *Consider the PBC of Proposition 3.1 applied to the PCH system (13.2.1) without damping injection, i.e.,* $R_a(x) = 0$*. Assume the natural damping of the system verifies*[6]

$$R(x)K(x) = 0 \tag{13.20}$$

Then, the total energy of the closed–loop $H_d(x)$ *satisfies (13.19).*

Proof
If $R_a(x) = 0$ the closed–loop system is given by

[6] We will show in Subsection 6.1 that the condition, which essentially means that the supplied energy is not affected by the dissipation, is satisfied for simple mechanical systems.

$$\dot{x} = [J(x) - R(x)]\frac{\partial H_d}{\partial x}(x)$$

where $H_d(x)$ is given by (13.8), with $H_a(x)$ defined as (13.17). Hence, along the trajectories of the closed–loop, we have

$$\begin{aligned}
\frac{d}{dt}H_d &= -\left[\frac{\partial H_d}{\partial x}(x)\right]^T R(x)\frac{\partial H_d}{\partial x}(x) \\
&= -\left[\frac{\partial H}{\partial x}(x) + \frac{\partial H_a}{\partial x}(x)\right]^T R(x)\left[\frac{\partial H}{\partial x}(x) + \frac{\partial H_a}{\partial x}(x)\right] \\
&= -\left[\frac{\partial H}{\partial x}(x)\right]^T R(x)\frac{\partial H}{\partial x}(x)
\end{aligned}$$

where we have used (13.20) with (13.17) to get the last identity. Now, replacing above the power balance equation for the open–loop system (13.6) we get

$$\frac{d}{dt}H_d = \frac{d}{dt}H - u^T y$$

The proof is completed integrating the equation above.

13.5 Casimir functions method

In this section we provide an interpretation of our controller in terms of Casimir functions, which are first integrals of the system dynamics. Towards this end, we view our controller as a PCH system in a power–preserving interconnection with the plant, in this way, the plant is embedded in a higher dimensional system. To study the stability of the closed–loop we restrict the behaviour of this augmented system to an invariant subspace defined by a series of Casimir functions. This approach has been explored in [14] within the context of Lyapunov function generation for PCH systems with dissipation and constant forcing inputs. For the case of simple mechanical systems, it is discussed in [3] for static state feedback control, and in [22] with a dynamic extension for output feedback damping injection. The latter work provides a nice power–preserving interpretation, hence a more natural derivation, of the results in [17], which are presented in an EL framework. As pointed out in [14] the construction has some close connections with the Energy–Casimir method of mechanics [10], see also [8], [2].

In all the works cited above the simplest "constant" power–preserving interconnection (13.21) –with u_C, y_C the controllers input and output, respectively- is considered. As will be shown below to treat the controllers proposed in this paper we have to extend this interconnection to include a "state modulation", albeit retaining the power–preservation feature. To put in perspective this new class of interconnections and, at the same time, underscore the limitations of the constant one, we explain first the rationale of the Casimir functions for the constant interconnection case. Throughout the section we give only a brief, and rather informal, presentation of the material. We refer the interested reader to the literature cited above for further technical details.

13.5.1 Port–controlled Hamiltonian controllers: constant interconnection

We consider the system described by (13.2.1) in interconnection with a *PCH controller*

$$\Sigma_C \ : \ \begin{cases} \dot{\zeta} = u_C, \\ y_C = \frac{\partial H_C}{\partial \zeta}(\zeta) \end{cases}$$

with state $\zeta \in \mathcal{R}^m$, input u_C, output y_C, and $H_C(\zeta)$ the energy of the controller –which we assume bounded from below.[7] (See point 4 in the discussion below for an explanation for the choice of this structure of the PCH controller.)

The interconnection constraints are power–preserving of the form

$$\begin{aligned} u_C &= y \\ u &= -y_C \end{aligned} \tag{13.21}$$

The composed system is clearly still Hamiltonian and can be written as

$$\begin{bmatrix} \dot{x} \\ \dot{\zeta} \end{bmatrix} = \begin{bmatrix} J(x) - R(x) & -g(x) \\ g^T(x) & 0 \end{bmatrix} \begin{bmatrix} \frac{\partial H_d}{\partial x} \\ \frac{\partial H_d}{\partial \zeta} \end{bmatrix} \tag{13.22}$$

with $H_d(x, \zeta)$ the *desired* energy function (defined in an extended state space (x, ζ))

$$H_d(x, \zeta) \triangleq H(x) + H_C(\zeta) \tag{13.23}$$

We can easily see that this energy function is non–increasing, since

$$\frac{d}{dt} H_d = - \left[\frac{\partial H}{\partial x}(x) \right]^T R(x) \frac{\partial H}{\partial x}(x) \leq 0$$

[7] We have introduced the notation H_C here to highlight its interpretation as controller energy. We will see later, that this function play essentially the same role as H_a in Proposition 3.1.

Once we have embedded the system in a higher dimensional state the main idea of the procedure is to study its stability looking at its behaviour in some *invariant subspaces* of the extended state space. More specifically, we look for functions (called Casimir functions) $F_i(x,\zeta)$, $i = 1, \cdots, m$, whose derivatives along the closed-loop dynamics are zero. Without loss of generality[8] we will consider functions of the form

$$F_i(x,\zeta_i) = c_i(x) - \zeta_i, \quad i = 1, \cdots, m \tag{13.24}$$

In this case, a sufficient condition for the Casimirs to exist, that is, to verify

$$\dot{F}_i(x(t),\zeta(t)) = 0, \quad i = 1, \cdots, m$$

is the solvability of the PDEs

$$\begin{bmatrix} [\frac{\partial C}{\partial x}(x)]^T \vdots -I_m \end{bmatrix} \begin{bmatrix} J(x) - R(x) & -g(x) \\ g^T(x) & 0 \end{bmatrix} = 0 \tag{13.25}$$

with $C(x) \triangleq [c_1(x), \cdots, c_m(x)]^T$. If we can solve these PDEs, then we can express the behaviour of the "controller state" $\zeta(t)$ as functions of the system state $x(t)$. That is, we can then write

$$\zeta(t) = C(x(t)) + c$$

with c a vector of constants that can be set to zero. See Fig. 2. Under these conditions, we can restrict the desired energy function (13.23) to the systems state space as

$$H_d(x, C(x)) = H(x) + H_C(C(x)) \tag{13.26}$$

and look for some suitable function H_C to assign a strict local minimum to $H_d(x, C(x))$ at the desired equilibrium. This is the essence of the controller design method studied in [3] and [22].

Discussion 1. It is easy to show that for a controller designed with the procedure above the energy balance equation (13.19) holds. This follows from the fact that, by construction, H_C is the energy supplied by the controller to the system. Hence,

$$\frac{d}{dt}H_C = \left[\frac{\partial H_C}{\partial \zeta}(\zeta)\right]^T \dot{\zeta} = -u^T y$$

which, upon integration and utilization of the Casimir functions, yields

[8] This is because $g(x)$ is full rank, hence (13.24)generates (locally) all Casimir functions.

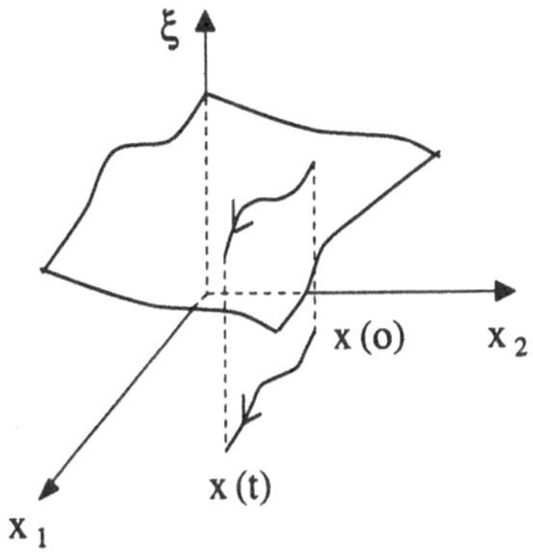

Fig. 13.1. Augmented state space (ζ, x), invariant subspace $\zeta = C(x) + c$, and a state trajectory $t \mapsto x(t)$.

$$H_C(C(x(t))) = -\int_0^t u(s)^T y(s) ds + c$$

with c some constant, that again can be set equal to zero.

2. We will now prove that a necessary condition for the existence of Casimir functions is that the damping satisfies the energy–balancing constraint (13.20). Towards this end, we spell out the PDEs (13.25) to get

$$\left[\frac{\partial C}{\partial x}(x)\right]^T [J(x) - R(x)] = g^T(x) \tag{13.27}$$

$$-\left[\frac{\partial C}{\partial x}(x)\right]^T g(x) = 0 \tag{13.28}$$

These equations can be alternatively expressed as

$$J(x)\frac{\partial C}{\partial x}(x) = -g(x) \tag{13.29}$$

$$R(x)\frac{\partial C}{\partial x}(x) = 0 \tag{13.30}$$

Now, if (13.30) holds then $R(x)\frac{\partial H_C(C)}{\partial x}(x) = 0$ for any function H_C. The prove of our claim is completed with (13.26). We will elaborate further on this point on the next subsection.

3. It has been argued in [14] that the class of PCH systems that admit Casimir functions (as presented above) is quite restrictive. Indeed, replacing (13.27) in (13.28), and assuming that $J(x) - R(x)$ is invertible, we get

$$g^T(x)[J(x) - R(x)]^{-1} g(x) = 0 \tag{13.31}$$

Now, replacing (13.31) in the system dynamics (13.2.1) and rearranging some terms we have that the output function y must satisfy

$$g^T(x)\frac{\partial H}{\partial x}(x) = g^T(x)[J(x) - R(x)]^{-1}\dot{x}$$

Consequently, y must be equal to zero when evaluated at an equilibrium. (Recall that in mechanical systems y are the generalized velocities, which are indeed zero at an equilibrium.) The linear RLC examples of [14] illustrate that even in simple linear systems this might not be the case. As is it argued in that paper the generation of Lyapunov functions with this method is hampered by the fact that the supplied energy must be bounded (otherwise H_d is not bounded from below). Hence, we can (roughly speaking) say that the method applies only to systems that drain a finite amount of energy from the source. A generalisation of the method –with a more complicated embedding system– has been proposed in [14] to generate Lyapunov functions when the inputs are constant.

4. For the PCH controller structure (13.5.1) we have taken the simplest Hamiltonian dynamics, with a pure integrator. This is done without loss of generality, since it is clear that our analysis applies as well to the more general case:

$$\Sigma_C : \begin{cases} \dot{\zeta} = [J_C(\zeta) - R_C(\zeta)]\frac{\partial H_C}{\partial \zeta}(\zeta) + g_C(\zeta)u_C \\ y_C = g_C^T(\zeta)\frac{\partial H_C}{\partial \zeta}(\zeta) + D_C(\zeta)u_C \end{cases}$$

for any skew–symmetric matrix $J_C(\zeta)$, any positive-semidefinite matrices $R_C(\zeta)$, $D_C(\zeta)$, and any function $g_C(\zeta)$. The effect of $D_C(\zeta)$ is simply to add *damping* to the system as can be seen from the closed–loop equations

$$\begin{bmatrix} \dot{x} \\ \dot{\zeta} \end{bmatrix} = \begin{bmatrix} J(x) - R(x) - g(x)D_C(\zeta)g^T(x) & -g(x)g_C^T(\zeta) \\ g_C(\zeta)g^T(x) & J_C(\zeta) - R_C(\zeta) \end{bmatrix} \begin{bmatrix} \frac{\partial H_d}{\partial x} \\ \frac{\partial H_d}{\partial \zeta} \end{bmatrix}$$

As shown in [22], see also [17], damping injection can also be achieved with a "true" dynamic extension, this case has been studied in [19].

5. It is important to stress that the interconnection point of view exposed in this subsection yields a clear energy picture, and yields several passivity properties. For instance, we have closed-loop passivity from u (in fact, an external signal) to y, as well as closed-loop passivity to other pairs of port-variables –like additional ports modelling the interaction with the environment.

6. When the PCH system is linear with quadratic Hamiltonian function the condition (13.31) holds *if and only if* the systems transfer matrix $\hat{u}(s) \mapsto \hat{y}(s)$ has a blocking zero (and hence a transmission zero) at $s = 0$.[9]

[9] The first author is grateful to A. Astolfi for this interesting remark.

13.5.2 State–modulated interconnection

In this subsection we will give a Casimir function interpretation to the PBC of Proposition 3.1 without damping injection and with natural damping satisfying the energy–balancing condition (13.20). The latter is a natural requirement since, as remarked in point 1 of the discussion of Subsection 5.1, the closed–loop total energy of a PCH system under PCH control is -by construction- the "energy–balance function".

Towards this end we keep the same PCH controller (13.5.1), choose its total energy as

$$H_C(\zeta) = -\zeta \tag{13.32}$$

and modify the interconnection structure to include a "state modulation"[10]

$$\begin{aligned} u_C &= \beta^T(x)y \\ u &= -\beta(x)y_C \end{aligned}$$

with $\beta(x)$ as defined in Proposition 3.1. See Fig. 13.2.

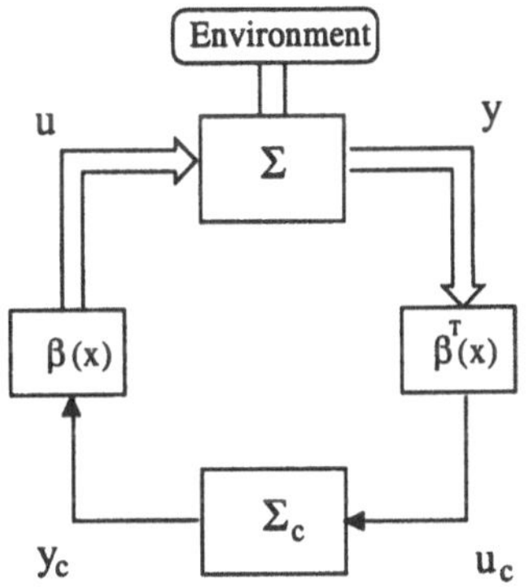

Fig. 13.2. State–modulated interconnection.

The composed system is clearly still Hamiltonian and can be written as

$$\begin{bmatrix} \dot{x} \\ \dot{\zeta} \end{bmatrix} = \begin{bmatrix} J(x) - R(x) & -g(x)\beta(x) \\ \beta^T(x)g^T(x) & 0 \end{bmatrix} \begin{bmatrix} \frac{\partial H_d}{\partial x} \\ \frac{\partial H_d}{\partial \zeta} \end{bmatrix} \tag{13.33}$$

[10] This yields, of course, the control $u = \beta(x)$.

with total energy $H_d(x, \zeta)$ defined by (13.23) and (13.32). The proposition below gives necessary and sufficient conditions for the Casimir functions method to apply.

Proposition 13.5.1. *Consider the PCH system (13.2.1) with a state–feedback control $u = \beta(x)$. The closed-loop, which can be represented as the augmented system (13.33), (13.23), (13.32), admits a Casimir function of the form*

$$F(x, \zeta) = -H_a(x) - \zeta \tag{13.34}$$

if and only if

- *the integrability condition (13.13), (13.14) with $R_a(x) = 0$ of Proposition 3.1 holds, that is*

$$[J(x) - R(x)]\frac{\partial H_a}{\partial x}(x) = g(x)\beta(x) \tag{13.35}$$

- *the damping matrix $R(x)$ verifies the energy–balancing constraint (13.20).*

Proof
Differentiating (13.34) along the trajectories of the augmented system (13.33) and setting it equal to zero we get

$$[J(x) + R(x)]\frac{\partial H_a}{\partial x}(x) = g(x)\beta(x)$$
$$\beta^T(x)g^T(x)\frac{\partial H_a}{\partial x}(x) = 0$$

Using the skew–symmetry of $J(x)$ and the non–negativity of $R(x)$ it is easy to show that these two equations are equivalent to

$$R(x)\frac{\partial H_a}{\partial x}(x) = 0$$
$$J(x)\frac{\partial H_a}{\partial x}(x) = g(x)\beta(x)$$

which –in view of (13.17)– are precisely the energy–balancing constraint (13.20) and the integrability condition (13.13), (13.14), respectively.

13.6 Concluding remarks and future research

We have presented in this paper a procedure to design stabilizing controllers for PCH models which effectively exploits the structural properties of the system, in particular its passivity. The main features of the proposed scheme may be summarized as follows:

1. The Hamiltonian structure is preserved in closed–loop, which allows for an energy interpretation of the control action.
2. Conditions on the damping are given to ensure that the closed–loop storage function is precisely the "energy–balance function".
3. Given the clear–cut definition of the interconnection structure and the damping (captured in the matrices $J(x)$ and $R(x)$, respectively) the incorporation of the physical intuition is effectively enhanced. This aspect is very important, not only for the definition of the "desired dynamics", but also for the commissioning of the controller.
4. In many applications there is no need to explicitly derive the Lyapunov function, only its existence need be ascertained.
5. The procedure is amenable for symbolic calculations.

The results reported here are restricted to the case of stabilization of fixed points. In some cases it is possible to adapt the procedure to treat the stabilization of periodic orbits, see e.g. [21]. Other applications, for instance the induction machine where the periodic orbit is not known *a priori*, have proved more elusive. Current research is under way to extend our approach to stabilization of general periodic orbits, and eventually to handle the more challenging *tracking* problem.

In all our developments here we have kept invariant the interconnection structure. However, as pointed out in Subsection 3.3, our procedure allows for the possibility of modifying it. With this additional degree of freedom we can recover the controller for the inverted pendulum reported in [8]. In particular, it is possible to show that modifying the kinetic energy of a simple mechanical system without affecting the potential energy nor the damping (as done in [8]) is tantamount –in our formulation– to selecting the closed–loop interconnection matrix as

$$J_d(q,p) = \begin{bmatrix} 0 & M_d^{-1}(q)M(q) \\ -M(q)M_d^{-1}(q) & Z(q,p) \end{bmatrix}$$

where $M_d(q)$, $M(q)$ are the closed–loop and open–loop inertia matrices, respectively, and the elements of $Z(q,p)$ are computed as

$$Z(q,p)_{i,j} = -p^T M^{-1}(q)M_d(q)\left[(M_d^{-1}M)_{\cdot i}, (M_d^{-1}M)_{\cdot j}\right](q)$$

with $(M_d^{-1}M)_{\cdot i}$ the i–th column of $M_d^{-1}M$ and $[\cdot,\cdot]$ the standard Lie brackett, see [13]. With this simple structural modification it is possible, for instance, to stabilize the upward position of the pendulum. The stabilization mechanism is not very clear because –to be consistent with Lagrange's principle– this results in a closed–loop EL system with sign–indefinite inertia matrix. In any case, this principle seems very promising for systems with mixed dynamics, (e.g., electromechanical systems), where the interconnection between the (electrical and the mechanical) dynamics has to be modified to be able to "transfer" the action of the control from one subsystem to the other.

References

1. A. Ailon and R. Ortega, An observer-based set-point controller for robot manipulators with flexible joints, *System & Control Letters*, Vol. 21, No. 4, pp. 329-335, 1993.

2. A. Bloch, P. Krishnaprasad, J. Marsden and G. Sanchez, Stabilization of rigid body dynamics by internal nad external torques, *Automatica*, Vol. 28, No. 4, pp. 745–756, 1992.

3. M. Dalsmo and A.J. van der Schaft, On representations and integrability of mathematical structures in energy-conserving physical systems, *SIAM J. on Optimization and Control*), Vol.37, No. 1, 1999.

4. G. Escobar, A. van der Schaft and R. Ortega, A hamiltonian viewpoint in the modeling of switching power converters, *Automatica*, Vol. 35, 1999.

5. H. Khalil, **Nonlinear systems**, Second Ed. Prentice–Hall, New Jersey, 1996.

6. P. C. Krause, *Analysis of Electric Machinery*, McGraw-Hill, 1986.

7. P. Libermann and C.M. Marle, **Symplectic Geometry and Analytical Mechanics**. Reidel, Dordrecht, 1987.

8. A. Bloch, N. Leonhard and J. Marsden, Controlled Lagrangians and the stabilization of mechanical systems, *Proc. IEEE Conf. Decision and Control*, Tampa, FL, USA, Dec. 1998.

9. A. Loria, R. Kelly, R. Ortega and V. Santibanez, On output feedback control of Euler–Lagrange systems with bounded inputs, *IEEE Trans. Automat. Contr.*, Vol. 42, No. 8, 1997, pp. 1138-1143.

10. J. Marsden and T. Ratiu, **Introduction to mechanics and symmetry**, Springer, NY, 1994.

11. B. M. Maschke and A.J. van der Schaft, Port controlled Hamiltonian systems: modeling origins and system theoretic properties, *Proc. 2nd IFAC Symp. on Nonlinear Control Systems design*, NOLCOS'92, pp.282-288, Bordeaux, June 1992.

12. B. M. Maschke, A. J.van der Schaft and P. Breedveld, An intrinsic Hamiltonian formulation of network dynamics: Non–standard Poisson structures and gyrators, *J. Franklin Inst.*, 329 (1992), pp. 923–926.

13. B. M. Maschke, Interconnection and structure in physical systems' dynamics, *4th IFAC Symp. on Nonlinear Control Systems Design*, NOLCOS'98, pp.291-296, Enschede, the Netherlands, July 1-3, 1998

14. B. M. Maschke, R. Ortega and A. J. van der Schaft, Energy–based Lyapunov functions for forced Hamiltonian systems with dissipation, *IEEE Conf. Dec. and Control*, Tampa, FL, USA, Dec. 1998.

15. R. Ortega and M. Spong, Adaptive motion control of rigid robots: A tutorial, *Automatica*, Vol. 25, No.6, pp. 877-888, 1989.

16. R. Ortega, A. Loria, P. J. Nicklasson and H. Sira–Ramirez, **Passivity–based control of Euler–Lagrange systems**, Springer-Verlag, Berlin, Communications and Control Engineering, Sept. 1998.

17. R. Ortega, A. Loria, R. Kelly and L. Praly, On output feedback global stabilization of Euler-Lagrange systems, *Int. J. of Robust and Nonlinear Cont.*, Special Issue on Mechanical Systems, Eds. H. Nijmeijer and A. van der Schaft, Vol. 5, No. 4, pp. 313-324, July 1995.

18. R. Ortega, A. Astolfi, G. Bastin and H. Rodriguez, Output feedback stabilization of mass–balance systems, in **Output–feedback stabilization of nonlinear systems**, Eds. H. Nijmeijer and T. Fossen, Springer–Verlag, 1999.

19. R. Ortega, A.J. van der Schaft, B. Maschke and G. Escobar, Stabilization of port–controlled Hamiltonian systems: Energy–balancing and passivation, *CDC'99*, Phoenix, AZ, USA, Dec. 7–10, 1999.

20. V. Petrovic, R. Ortega and A. Stankovic, An energy–based globally stable controller for PM synchronous motors, *Northeastern University Int. Report*, Boston, USA, 1999.

21. H. Rodriguez, R. Ortega and G. Escobar, A Robustly Stable Output Feedback Saturated Controller for the Boost DC–to–DC Converter, *Rap. Int. LSS-Supelec*, May 1999.

22. S. Stramigioli, B. M. Maschke and A. J. van der Schaft, Passive output feedback and port interconnection, *Proc. 4th IFAC Symp. on Nonlinear Control Systems design*, NOLCOS'98, pp. 613–618, Enschede, NL, July 1–3, 1998.

23. M. Takegaki and S. Arimoto, A new feedback method for dynamic control of manipulators, *ASME J. Dyn. Syst. Meas. Cont.*, Vol. 102, pp. 119-125, 1981.

24. van der Schaft, A. J., **L_2–Gain and Passivity Techniques in Nonlinear Control**, Lect. Notes in Contr. and Inf. Sc., Vol. 218, Springer–Verlag, Berlin, 1996.

25. van der Schaft, A. J., System theory and mechanics, in **Three Decades of Mathematical System Theory**, H. Nijmeijer and J. M. Schumacher eds., Lect. Notes Contr. Inf. Sci., Vol. 135, pp. 426-452, Springer, Berlin, 1989.

26. A. van der Schaft and B. Maschke, The Hamiltonian formulation of energy–

conserving physical systems with external ports, *Archiv für Elektronik und Übertragungstechnik*, 49 (1995), pp. 362–371.

14. Invariant tracking and stabilization: problem formulation and examples

Pierre Rouchon and Joachim Rudolph***

*Centre Automatique et Systèmes
École des Mines de Paris
60, bd. Saint-Michel
75272 Paris Cedex 06, France
`rouchon@cas.ensmp.fr`

**Institut für Regelungs- und Steuerungstheorie
TU Dresden, Mommsenstr. 13
01062 Dresden, Germany
`rudolph@erss11.et.tu-dresden.de`

Summary.

The problems of invariant tracking and invariant stabilization are considered: Design a state feedback for tracking or stabilization, respectively, such that the closed-loop dynamics is invariant under the action of a given group. Errors are then defined as invariants of the group under consideration. The approach is illustrated on two classical examples: the nonholonomic car and a continuous stirred chemical reactor. In both cases the differential flatness of the models allows for a systematic design of feedback laws for invariant tracking. The feedback synthesis is simplified by using implicit system descriptions.

14.1 Introduction

Symmetries play a major role in physics. Nonetheless, they seem not yet having attracted the same attention in control theory – as instances of investigations of the concept see, however, [2, 6, 5].

When designing a feedback control loop for a system possessing a symmetry group, the following most natural question arises: *How could we design the feedback control loop in such a way that the closed-loop system admits the same symmetries as the open loop?* In other words: *How to design an invariant feedback, i.e. a symmetry preserving one?* This question may be generalized as follows. Given any group G of transformations on a system, considered as being relevant for some (arbitrary) reason: *How to design a* *feedback in such a way that the closed loop is invariant under G, i.e. that G is a group of symmetries of the closed-loop system?*

The formulation of these problems and partial answers are the subject of the present paper. We consider stabilization and tracking; as for the latter we concentrate on differentially flat systems [3, 4] which admit a systematic approach.

As a particularly interesting consequence of the design of invariant feedback it results that the commonly employed definition of the error as the difference between the actual value of the controlled variables and their reference (or setpoint) is probably not the best-suited one for nonlinear systems. While this somewhat "negative observation" is not surprising in principle, we can also give a "positive answer" by proposing an alternative definition of the tracking error: a set of invariants of the group under consideration.

Two classical examples are considered in some detail: the nonholonomic car and a continuous stirred chemical reactor. In both examples the invariant tracking is possible by the fact that the system models are differentially flat. For the nonholonomic car the symmetry group is the group $SE(2)$ of planar translations and rotations – see [6] and [5], where also the observer design is considered. The two components of the tracking error are defined as the projections of the vector from the desired to the actual position of the center of the rear axle (the flat output) on the tangent and on the normal of the desired trajectory, respectively – cf. [8]. This allows us to design a time-varying static state feedback for the stabilization of the trajectory tracking via quasi-static state feedback design methods [1]. The time scale used for this is the arc length – cf. [3].

For the chemical reactor with two chemical species the group defined by the transformation of molar fractions into mass fractions is considered. The tracking error is defined as the difference between the natural logarithms of the ratio of the molar fractions and the one of the ratio of the desired molar fractions.

The paper is structured as follows. In the next section we formulate the problem of invariant tracking and stabilization, related notions of invariant errors, and introduce a concept of invariant feedback. In section 3, we consider the invariant tracking for the nonholonomic car, in section 4 the same problem for the chemical reactor. We conclude by indicating some of the open questions.

14.2 Invariant tracking and stabilization

Consider a system of the form

$$\dot{x} = f(x, u), \qquad x \in M \ (\dim M = n), u \in \mathbb{R}^m \tag{14.1}$$

where u is the input. Let G be a (local) symmetry group of finite order on $M \times \mathbb{R}^m$, which transforms (x, u) into (X, U) according to

$$X = \varphi_g(x), \quad U = \psi_g(x, u), \quad g \in G.$$

For the tracking problem, assume that desired (or reference) trajectories $[0, t_*] \ni t \mapsto (x_d, u_d) \in M \times \mathbb{R}^m$ can be computed; then

$$\dot{x}_d = f(x_d, u_d)$$

– this is always the case for differentially flat systems (see below). Collecting an arbitrary number of successive derivatives in vectors denoted with bars, like $\overline{u}_d = (u_d, \dot{u}_d, \ldots, u_d^{(\rho)})$, the differential prolongation of ψ_g is written as $\overline{U} = \overline{\psi}_g(x, \overline{u})$.

Definition 14.2.1. *An* invariant static state feedback *is a mapping defined by*

$$u = k(x, x_d, \overline{u}_d),$$

such that for all $g \in G$ and for arbitrary trajectories $[0, t_*] \ni t \mapsto (x_d, u_d) \in M \times \mathbb{R}^m$ the following commutativity condition holds true:

$$k(\varphi_g(x), \varphi_g(x_d), \overline{\psi}_g(x_d, \overline{u}_d)) = \psi_g(x, k(x, x_d, \overline{u}_d)).$$

Thus, an invariant feedback commutes with the elements of the group G whatever reference trajectory is considered.

Choosing a setpoint as the desired trajectory and supposing that the feedback achieves stabilization of the setpoint we obtain a corresponding definition.

Definition 14.2.2. *For a system* (14.1), *an* invariant static state feedback that is locally (resp. globally) asymptotically stabilizing at $(x_s, u_s) \in M \times \mathbb{R}^m$ *is a mapping*

$$u = k(x, x_s, u_s),$$

such that for all $g \in G$:

$$k(\varphi_g(x), \varphi_g(x_s), \psi_g(x_s, u_s)) = \psi(x, k(x, x_s, u_s))$$

and such that $(x, u) \to (x_s, u_s) \in M \times \mathbb{R}^m$ with $t \to \infty$ locally (resp. globally).

The design of the closed-loop dynamics relying on an error notion, we introduce:

Definition 14.2.3. *A* local (resp. global) invariant state error *with respect to a point $(x_s, u_s) \in M \times \mathbb{R}^m$ is a set of n invariants of G*

$$i(x, x_s, u_s) = (i_1(x, x_s, u_s), \ldots, i_n(x, x_s, u_s))$$

such that $(i_1, \ldots, i_n)$ is a local (resp. global) diffeomorphism on M and $i(x_s, x_s, u_s) = 0$.

The following *problem of invariant stabilization* then occurs: *Given a system (14.1) with a group G acting on $M \times \mathbb{R}^m$, design an asymptotically stabilizing feedback law in such a way that the (closed-loop) error dynamics is invariant under G.* If G is a symmetry group of the system, it is also of interest to define the feedback as an invariant asymptotically stabilizing feedback.

Obviously, with $\imath$ an invariant state error w.r.t. to (x_s, u_s), the problem of invariant stabilization at (x_s, u_s) is solved if an asymptotically stable invariant error dynamics $d/dt\ \imath = e(\imath), \imath \in M$ is obtained. Suppose the system was affine in u and $m = 1$, for the sake of simplicity. Use the error coordinates $\imath = \imath_1(x, x_s, u_s)$ with $x = \tau(\imath, x_s, u_s)$ in order to write (14.1) as

$$\frac{d}{dt}\imath = \left(\frac{\partial \tau}{\partial \imath}\right)^{-1} (\imath, x_s, u_s) \left(g_0(\tau(\imath, x_s, u_s)) + g_1(\tau(\imath, x_s, u_s))u\right).$$

In a neighborhood of $\imath = 0$, let $V(\imath)$ be a control Lyapunov function. Then a static feedback can be defined in such a way that the closed loop is locally asymptotically stable around $\imath = 0$; let its vector field be $e(\imath)$. One has $L_{g_0}V + u\, L_{g_1}V = L_e V$ and the feedback is of the form

$$u = \frac{L_e V(\imath) - L_{g_0}V(\imath)}{L_{g_1}V(\imath)}.$$

Obviously, this is an invariant static state feedback asymptotically stabilizing the system at (x_s, u_s).

On the academic example

$$\begin{aligned}\dot{x} &= \frac{(x - x_s)^2}{x_s} + \frac{x_s}{z_s}(z - z_s)\\ \dot{z} &= (z - z_s) + z_s u\end{aligned}$$

consider the stabilization at $(x_s, z_s, 0) \neq (0, 0, 0)$. The system admits the scaling group defined by $X = px$, $Z = qz$ as a symmetry group. In particular, this permits to normalize the representation with $p_1 = 1/x_s$ and $q_1 = 1/z_s$ which yields $X_1 = x/x_s$, $Z_1 = z/z_s$ and

$$\begin{aligned}\dot{X}_1 &= (X_1 - 1)^2 + (Z_1 - 1)\\ \dot{Z}_1 &= (Z_1 - 1) + u.\end{aligned}$$

Invariant errors are given by

$$\imath_1 = \frac{x - x_s}{x_s} \qquad \text{and} \qquad \imath_2 = \frac{z - z_s}{z_s}.$$

With these coordinates the system reads

$$\frac{d}{dt}i_1 = i_1^2 + i_2$$
$$\frac{d}{dt}i_2 = i_2 + u$$

and an invariant static state feedback asymptotically stabilizing at $i = 0$ is readily obtained by feedback linearization:

$$u = -(k_1 + 2i_1)(i_1^2 + i_2) + k_0 i_1 - i_2$$

with $0 < k_0, k_1 \in \mathbb{R}$. The invariance of the error dynamics is obvious.

14.2.1 Invariant tracking and differential flatness

If the system (14.1) under consideration is differentially flat, reference trajectories can be calculated as

$$x_d = \phi_x(y_d, \dot{y}_d, \ldots, y_d^{(\alpha)})$$
$$u_d = \phi_u(y_d, \dot{y}_d, \ldots, y_d^{(\beta)})$$

with $[0, t_*] \ni t \mapsto y_d \in \mathbb{R}^m$ arbitrary trajectories of the so-called flat output $y = h(x, \overline{u})$ – see e.g. [3, 4] for a thorough discussion of differential flatness.

This can be directly generalized to implicit systems, as for instance those of the form

$$\dot{x} = f(x, z, u)$$
$$0 = g(x, z, u),$$

where $x \in M_x \times M_z$ $\dim M_x = n, \dim M_z = p, u \in \mathbb{R}^m$, $f = (f_1, \ldots, f_n)$, and $g = (g_1, \ldots, g_p)$. This system is called differentially flat if there exist m scalar functions h_i defining $y = (y_1, \ldots, y_m)$ – called a flat output – via

$$y_i = h_i(x, z, u, \dot{x}, \dot{z}, \dot{u}, \ldots, x^{(\tilde{\alpha})}, z^{(\tilde{\beta})}, u^{(\tilde{\gamma})}), \quad i = 1, \ldots, m$$

such that

$$x = \phi_x(y, \dot{y}, \ldots, y^{(\alpha)})$$
$$z = \phi_z(y, \dot{y}, \ldots, y^{(\beta)})$$
$$u = \phi_u(y, \dot{y}, \ldots, y^{(\gamma)}).$$

Thus, trajectories of y can again be freely assigned and the trajectories of all other variables can be calculated without integration.

For the tracking problem of flat systems we may now define invariant errors as follows.

Definition 14.2.4. *Consider a differentially flat system with a flat output* $y \in \mathbb{R}^m$ *and a group* G *acting on* $\mathbb{R}^m$ *with* $Y = \Psi_g(y)$. *A local (resp. global)* invariant tracking error *is a set of* m *invariants of* G*:*

$$I(y, \overline{y}_d) = (I_1(y, \overline{y}_d), \ldots, I_m(y, \overline{y}_d)),$$

in the sense that

$$I_j(y, \overline{y}_d) = I_j(\Psi_g(y), \overline{\Psi}_g(\overline{y}_d)), \quad j = 1, \ldots, m,$$

such that I *is a local (resp. global) diffeomorphism on* $\mathbb{R}^m$ *for any reference trajectory* $[0, t_*] \ni t \mapsto y_d(t) \in \mathbb{R}^m$, *and such that* $I(y_d, \overline{y_d}) = 0$, *i.e. this error is zero if the system evolves on the reference trajectory.*

One may now consider the following problem of invariant tracking for flat systems: *Given a differentially flat system with flat output* $y \in \mathbb{R}^m$ *and a group* G *acting on* $y \in \mathbb{R}^m$, *design a feedback law in such a way that the closed-loop dynamics is invariant under* G. Obviously, this may be achieved with a feedback law designed by assigning an autonomous system in an invariant tracking error I as the closed-loop dynamics. Asymptotically stable tracking is then achieved by choosing this closed-loop dynamics asymptotically stable at $I = 0$. Finally, stabilization at a setpoint is achieved by choosing $t \mapsto y_s$, with $y_s \in \mathbb{R}^m$ a point corresponding to the setpoint, as the reference trajectory of the flat output.

14.3 The nonholonomic car

Consider the classical nonholonomic car, rolling without slipping on a horizontal plane, as depicted in Fig. 14.1.

The system is by now well-known to be flat, with the position of the center of the rear axle H as a flat output. The absolute value v of the velocity of this point together with the steering angle φ can be used as the control input. Alternatively, one may use the curvature κ of the path C followed by H instead of φ: they are related via $\kappa = \tan\varphi / l$, with l the distance between the front and the rear axle. The classical explicit model in a Cartesian coordinate frame fixed in the plane then reads:

$$\begin{aligned} \dot{x}_1 &= v \cos\theta \\ \dot{x}_2 &= v \sin\theta \\ \dot{\theta} &= v \frac{\tan\varphi}{l}. \end{aligned}$$

The coordinate representation of H, which forms a flat output, is $H = (x_1, x_2)$.

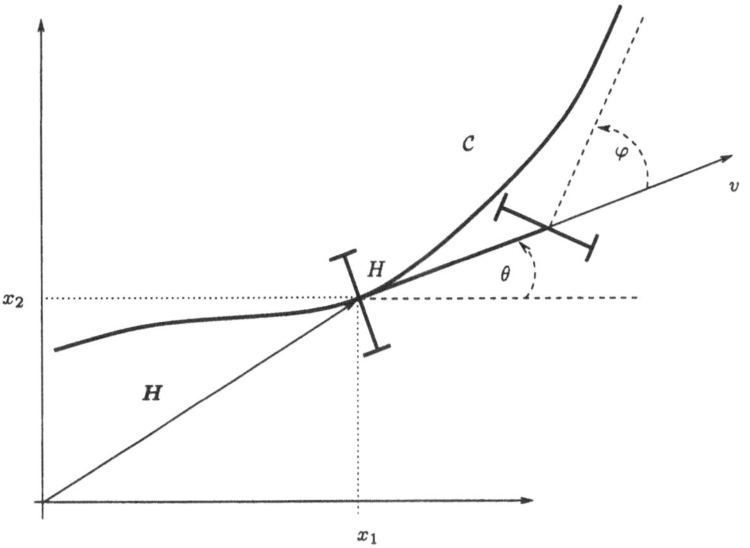

Fig. 14.1. The car in an inertial Cartesian frame.

To avoid the well-known difficulties at $v = 0$, it is convenient to use the arc length s_d of the reference curve $\mathcal{C}_d$ with $[0, L] \ni s_d \mapsto H_d(s_d) \in \mathbb{R}^2$ instead of the physical time t as the parameter, see e.g. [3]. Denoting differentiation w.r.t. s_d by primes, with

$$\dot{z} = \frac{dz}{dt} = \frac{dz}{ds_d}\frac{ds_d}{dt} = z'\dot{s}_d,$$

the model becomes:

$$\begin{aligned} x_1' &= u\cos\theta \\ x_2' &= u\sin\theta \\ \theta' &= u\kappa. \end{aligned}$$

with $u = v/\dot{s}_d$ as the new control variable[1]. The time t can be reintroduced via the parametrization $[0, t_*] \ni t \mapsto s_d(t) \in [0, L]$.

It is obvious that the path followed by the car when a (open-loop) control $[0, L] \ni s_d \mapsto (v(s_d), \kappa(s_d)) \in \mathbb{R}^2$ is applied is the same whatever coordinate frame is used to represent the model. Analogously, when the same control is applied starting at two different initial conditions, which are related by

[1] Obviously, there is a "time scaling symmetry" defined by the time scaling introduced together with the input transformation from v to u. This interesting property is not to be confounded with the Euclidian symmetry considered in the problem of invariant tracking.

a translation of the starting point and a rotation of the initial orientation, the corresponding curve followed by H is obtained by applying the same translation and rotation. This means that the system is invariant under planar translations and rotations, i.e. under the action of the elements of the group $SE(2)$ – cf. [6, 5]. (This can be easily verified by calculation.)

It is, therefore, useful to use a coordinate free representation of the model. For this, denote as $\boldsymbol{H}$ the vector pointing to H, as $\boldsymbol{\tau}$ the normalized tangent of the curve $\mathcal{C}$ at H (oriented in the direction of increasing arc length s), and as $\boldsymbol{\nu}$ the normalized normal to $\mathcal{C}$ at H (oriented outwards) – cf. Fig. 14.2. With these vectors the model of the car can be written in an implicit form,

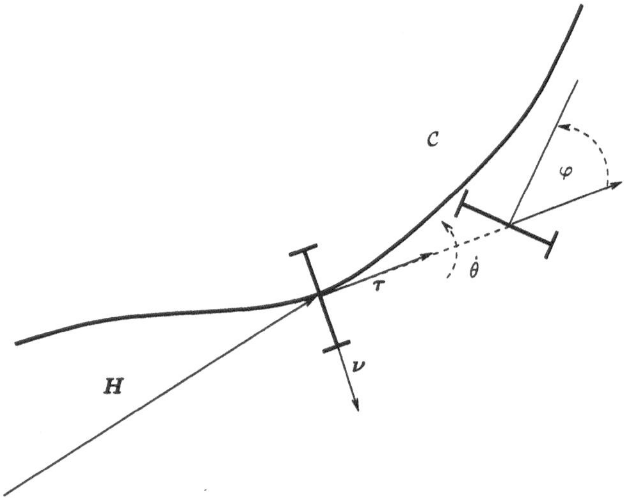

Fig. 14.2. The car with the tangent vector $\boldsymbol{\tau}$ and the normal vector $\boldsymbol{\nu}$ at $\mathcal{C}$ in H.

which is independent of the choice of the plane-fixed coordinate frame, as:

$$\begin{aligned} \boldsymbol{H}' &= u\boldsymbol{\tau} \\ \boldsymbol{\tau}' &= u\kappa\boldsymbol{\nu} \\ \boldsymbol{\nu}' &= -u\kappa\boldsymbol{\tau} \\ \langle\boldsymbol{\tau},\boldsymbol{\nu}\rangle = 0, \quad &\langle\boldsymbol{\tau},\boldsymbol{\tau}\rangle = 1, \quad \langle\boldsymbol{\nu},\boldsymbol{\nu}\rangle = 1. \end{aligned}$$

Now, for any reference trajectory $[0, L] \ni s_d \mapsto \boldsymbol{H}_d \in \mathbb{R}^2$ of the flat output the corresponding curvature can be computed as[2]: $\kappa_d = \langle\boldsymbol{\tau}'_d, \boldsymbol{\nu}_d\rangle$.

[2] As a consequence of the time scaling symmetry observed above, the equations of the controls w.r.t. the time t are of the same form, with v_d instead of u_d and d/dt instead of d/ds_d.

A first solution to the invariant tracking problem can now be easily obtained by choosing the error $\boldsymbol{e} = \boldsymbol{H} - \boldsymbol{H}_d$ and by using dynamic state feedback to assign the error dynamics

$$\boldsymbol{e}'' + k_1\boldsymbol{e}' + k_0\boldsymbol{e} = 0, \quad 0 < k_0, k_1 \in \mathbb{R}.$$

Clearly, this simple solution is independent of the choice of the coordinate frame used to represent the vectors. The thus-defined feedback is obtained by solving the error equation for the control inputs. One has

$$\boldsymbol{H}'' = u'\boldsymbol{\tau} + u^2\kappa\boldsymbol{\nu} = \boldsymbol{H}_d'' - k_1\boldsymbol{e}' - k_0\boldsymbol{e}.$$

Thus, the dynamic feedback law (with the state u) is

$$\begin{aligned} u' &= \langle \boldsymbol{H}_d'' - k_1\boldsymbol{e}' - k_0\boldsymbol{e}, \boldsymbol{\tau}\rangle \\ \kappa &= \langle \boldsymbol{H}_d'' - k_1\boldsymbol{e}' - k_0\boldsymbol{e}, \boldsymbol{\nu}\rangle / u^2. \end{aligned}$$

(Notice that $u \approx 1$ in the neighborhood of the reference path.)

One might, however, consider two properties of this solution as being inconvenient: Firstly, the feedback is a dynamic one, which means that an integration is required in the controller. Secondly, there are only two design parameters for the error dynamics, while the original system is of dimension three.

These drawbacks can be avoided by designing a tracking control law with the techniques of quasi-static state feedback design. In order to obtain an invariant solution, the coordinates of the tracking error are here defined as invariants. This is achieved by using a moving frame defined by the reference curve C_d – see also [8]. For this, define vectors $\boldsymbol{H}_d$, $\boldsymbol{\tau}_d$, and $\boldsymbol{\nu}_d$ on C_d analogous to $\boldsymbol{H}$, $\boldsymbol{\tau}$, and $\boldsymbol{\nu}$. The invariant tracking error is now defined as

$$I = (e_{\|}, e_{\perp}) = (\langle \boldsymbol{e}, \boldsymbol{\tau}_d\rangle, \langle \boldsymbol{e}, \boldsymbol{\nu}_d\rangle).$$

Its components are the projections of $\boldsymbol{e}$ on the tangent and on the normal of C_d at H_d, respectively.

The feedback law can now be defined using the quasi-static state feedback design method from [1], as follows. Differentiating I twice w.r.t. s_d yields:

$$\begin{aligned} e_{\|}' &= u\langle \boldsymbol{\tau}, \boldsymbol{\tau}_d\rangle + \kappa_d e_{\perp} = u\cos\delta + \kappa_d e_{\perp} \\ e_{\perp}' &= u\langle \boldsymbol{\tau}, \boldsymbol{\nu}_d\rangle - \kappa_d e_{\|} = u\sin\delta - \kappa_d e_{\|} \\ e_{\|}'' &= u'\cos\delta - u(u\kappa - \kappa_d)\sin\delta + \kappa_d' e_{\perp} + \kappa_d e_{\perp}' \\ e_{\perp}'' &= u'\sin\delta + u(u\kappa - \kappa_d)\cos\delta - \kappa_d' e_{\|} - \kappa_d e_{\|}' \end{aligned}$$

where δ is the angle between $\boldsymbol{\tau}$ and $\boldsymbol{\tau}_d$, whence

$$-\delta'\sin\delta = (\cos\delta)' = (\langle \boldsymbol{\tau}, \boldsymbol{\tau}_d\rangle)' = -\sin\delta(-\kappa_d + u\kappa),$$

and $-\delta' = -\kappa_d + u\kappa$. Introducing

$$w_\parallel = e'_\parallel \qquad \text{and} \qquad w_\perp = e''_\perp$$

linearizes the error dynamics. This defines the quasi-static state feedback with

$$u = (-\kappa_d e_\perp + e'_\parallel)\cos\delta$$
$$u(u\kappa - \kappa_d) = \begin{pmatrix} -\sin\delta & \cos\delta \end{pmatrix} \left[\begin{pmatrix} -e_\perp & -e'_\perp \\ e_\parallel & e'_\parallel \end{pmatrix} \begin{pmatrix} \kappa'_d \\ \kappa_d \end{pmatrix} + \begin{pmatrix} w'_\parallel \\ w_\perp \end{pmatrix} \right],$$

where κ may be easily computed by solving the second equation. Exponential stabilization is then achieved with

$$w_\parallel = -\lambda_\parallel e_\parallel$$
$$w_\perp = -\lambda^1_\perp e'_\perp - \lambda^0_\perp e_\perp,$$

where $0 < \lambda_\parallel, \lambda^0_\perp, \lambda^1_\perp \in \mathbb{R}$. Obviously, the derivative $w'_\parallel = -\lambda_\parallel w_\parallel$ required in the feedback law is easily computed without integration – cf. [1].

Clearly, an invariant tracking error dynamics has been obtained. Both drawbacks observed on the dynamic solution proposed first have been eliminated: no integration is required in the feedback and three independent design parameters are available. However, the calculations are slightly less simple now and, furthermore, the derivative of the reference curvature κ'_d is required with this solution. It is also interesting to observe that singularities which would be encountered at $\theta = 0$ or $\theta = \pm\pi/2$ when using the error $\boldsymbol{e}$ (which is not invariant) and quasi-static state feedback are avoided. Corresponding singularities exist now at $\delta = \pm\pi/2$, which corresponds to large tracking errors: we have obtained a local solution only. Notice that for large errors there is a singularity, at $u = 0$, in the dynamic feedback solution, too.

14.4 A chemical reactor

We consider a continuous stirred chemical reactor with a single liquid phase where both pression and reaction volume are hold constant. Two species, A and B, are present in the reactor; they are involved in a reaction of the type A → B. Both the density and the molar enthalpy are considered as strongly depending on the temperature and the concentrations.

Energy and molar balances then lead to the following implicit model:

$$\dot{N}_a = F x_F - L x - N r(x,T)$$
$$\dot{N}_b = F(1 - x_F) - L(1 - x) + N r(x,T)$$
$$\dot{H} = F h_F - L h(x,T) + \Delta h(x,T)\, N r(x,T) + Q$$
$$N_a = N x, \quad N_b = N(1 - x), \quad \mathcal{V} = N \rho(x,T) = \text{const.}$$

where F, x_F, h_F, and $\mathcal{V}$ are constant parameters. Here N_a and N_b are the respective molar hold-ups of the species A and B; L is the molar flow leaving the reactor, F the entering one, x_F the molar fraction of A at the inflow, $H = Nh$ the enthalpy hold-up, $h(x,T)$ the molar enthalpy of the fluid in the reactor, h_F the one in the feed flow, $\Delta h(x,T)$ the reaction enthalpy, and $r(x,T)$ its velocity. The reactor is controlled via the heat exchanged through a cooling device: Q is the input.

The state dimension of this implicit system (with index 2) is equal to 2, and (x,T) can be considered as a state. However, it is convenient to use the implicit model for the control design by exploiting its flatness – compare with the gantry crane example in [3].

Indeed, the reactor model is flat, and x can be used as a flat output, as can be seen by the following reasoning. Denote as x_d a desired trajectory of x, continuously differentiable at least twice. With $N_a = xN$ and $\dot{N} = F - L$, one has

$$Fx_F - Lx - Nr(x,T) = \dot{N}_a = x\dot{N} + \dot{x}N = x(F-L) + \dot{x}N,$$

thus $N\dot{x} = F(x_F - x) - Nr(x,T)$. Now, with $N = \mathcal{V}/\rho(x,T)$ it is clear that a reference trajectory T_d of T can be obtained by solving the implicit equation

$$\dot{x}_d = (F/\mathcal{V})\rho(x_d,T_d)(x_F - x_d) - r(x_d,T_d).$$

The reference of N satisfies $N_d = \mathcal{V}/\rho(x_d,T_d)$, whence $L_d = F - \dot{N}_d$ depends on $x_d, \dot{x}_d$, and $\ddot{x}_d$. Therefore, $H_d = N_d h(x_d,T_d)$ can be expressed in terms of $x_d, \dot{x}_d$, and $\ddot{x}_d$, too. Finally, the same holds for the control corresponding to x_d which satisfies

$$Q_d = \dot{H}_d - Fh_F + L_d h(x_d,T_d) - \Delta h(x_d,T_d)\, r(x_d,T_d) =: q(x_d,\dot{x}_d,\ddot{x}_d).$$

The above calculations allow us to define a feedback that locally exponentially stabilizes the system around the trajectory x_d. One has $Q = q(x,\dot{x},\ddot{x})$ and, thus, with

$$Q = q(x,\dot{x},\ddot{x}_d - k_1(\dot{x} - \dot{x}_d) - k_0(x - x_d)) \tag{14.2}$$

it (locally) follows

$$(\ddot{x} - \ddot{x}_d) + k_1(\dot{x} - \dot{x}_d) + k_0(x - x_d) = 0$$

which is exponentially stable if $0 < k_0, k_1 \in \mathbb{R}$. (Notice that using the fact that $\dot{x} = (F/\mathcal{V})\rho(x,T)(x_F - x) - r(x,T)$ the feedback can be realized if x and T are measured – or observed.)

The above model has been written by using molar quantities. However, an analogous model can be obtained by using mass quantities. Both are related via the ratio $\mu = M_b/M_a$ of the molar masses M_a and M_b of the species A and B. As the model is flat with x as a flat output, it is sufficient to define how the mass fraction of A, which we denote as w, is related to x; one has:

$$w = \frac{x}{x + (1-x)\mu}, \qquad x = \frac{w}{w + (1-w)\mu^{-1}}, \quad \text{with } \mu^{-1} = \frac{1}{\mu}.$$

Obviously, this defines a one parameter local group of transformations on the flat output space, which by the definition of x is $(0,1) \subset \mathbb{R}$.

The problem of invariant tracking defined above appears natural here: Define the stabilizing feedback in such a way that the system behavior does not depend on whether molar quantities or masses are used. In fact one may observe that the feedback (14.2) proposed above does not provide a solution. However, an invariant tracking error can be defined as

$$I(x, x_d) = \ln\left(\frac{x}{1-x}\frac{1-x_d}{x_d}\right).$$

The invariance can be easily seen using $w/(1-w) = \mu\, x/(1-x)$. This invariant satisfies $I(x_d, x_d) = 0$. The problem of invariant tracking can now be solved with

$$Q = q(x, \dot{x}, \gamma(x, \dot{x}, x_d, \dot{x}_d, \ddot{x}_d))$$

where γ is defined by solving the invariant error dynamics

$$\ddot{I} + k_1\dot{I} + k_0 I = 0$$

for $\ddot{x}$.

14.5 Conclusion

Aiming at invariant closed-loop dynamics both in stabilization and in tracking problems may be considered as one way "to put physics into control". Therefore, we expect the loops thus-obtained to be "more natural" then those which do not respect such a property. Several aspects of the examples considered confirm this interpretation, so the fact that the singularity is easily avoided in the local quasi-static state feedback design of the car and also that the physical domain of the flat output is reflected in the invariant tracking error of the chemical reactor, for instance. These properties ought be further investigated.

Of course, many other questions remain open. Let us mention just four of them. Firstly, systematic methods for the determination of symmetry groups in underdetermined systems, i.e. in controlled ones, should be developed.

Secondly, when calculating the invariant feedback for the chemical reactor the formulae are "of the same structure" when using mass quantities and when using the molar ones. This observation, of striking practical interest, has to be formalized. Thirdly, given a group acting on the flat output space: How can we systematically compute an invariant tracking error? Finally, invariant tracking errors of different order may be obtained, in the sense that time derivatives of different orders of the reference trajectory are involved in the definition – compare the quasi-static feedback solution for the car and the solution for the chemical reactor. What are the consequences of this fact?

References

1. E. Delaleau and J. Rudolph. Control of flat systems by quasi-static feedback of generalized states. *Internat. J. Control*, 71:745–765, 1998.

2. F. Fagnani and J. Willems. Representations of symmetric linear dynamical systems. *SIAM J. Control and Optim.*, 31:1267–1293, 1993.

3. M. Fliess, J. Lévine, P. Martin, and P. Rouchon. Flatness and defect of non-linear systems: Introductory theory and examples. *Internat. J. Control*, 61:1327–1361, 1995.

4. M. Fliess, J. Lévine, P. Martin, and P. Rouchon. A Lie-Bäcklund approach to equivalence and flatness of nonlinear systems. *IEEE Trans. Automat. Control*, 1999. To appear.

5. D. Guillaume and P. Rouchon. Observation and control of a simplified car. In *Proceedings IFAC Motion Control '98, Grenoble*, 1998.

6. P. Martin, R. M. Murray, and P. Rouchon. *Flat systems*, pages 211–264. 1997.

7. P. J. Olver. *Equivalence, Invariants and Symmetry*. Cambridge University Press, 1995.

8. C. Woernle. Flatness-based control of a nonholonomic mobile platform. *Z. angew. Math. Mech.*, 78 Suppl. 1:43–46, 1998.

15. Control of mechanical structures by piezoelectric actuators and sensors

Kurt Schlacher and Andreas Kugi

Institute of Automatic Control and Electrical Drives
Johannes Kepler University of LinzAltenbergerstrasse 69
A-4040, Linz Auhof, Austria
Tel: +43 732 2468 9730;Fax: +43 732 2468 9734
`kurt@regpro.mechatronik.uni-linz.ac.at`
`andi@regpro.mechatronik.uni.linz.ac.at`

Summary.

Smart structures based on piezoelectricity represent an important new group of actuators and sensors for active vibration control systems. This technology allows to construct spatially distributed devices. Since the design of the spatially distributed sensors and actuators becomes part of the controller design, special design methods for the controllers are required. Several well established approaches like PD-, H_2- and H_∞-design are adapted to solve this problem. They are based on infinite dimensional Lagrangian systems in conjunction with collocated actuator and sensor pairing. Finally, applications to beams and plates demonstrate the power and effectiveness of the proposed methods.

15.1 Introduction

Piezoelectric devices represent an important new group of actuators and sensors for active vibration control of mechanical systems. Indeed, this technology allows to construct spatially distributed devices, [7], [8]. This fact requires special control techniques to improve the dynamical behavior of this kind of smart structures, e.g., [1], [6], because one has not only to design the control law, but in addition the spatial distribution of the sensors and actuators. Therefore, the design of the controller has to be considered together with the design of the actuators and sensors.

The second part of this contribution is concerned with a short introduction to infinite dimensional systems, which can be derived from a Lagrangian functional. Since we restrict ourselves to the time invariant case, there exists a conservation law, which simply says, that the change of the total energy of the system with respect to the time equals the flow of power into the system.

The third part is devoted to the control of Lagrangian systems. Based on the conservation law, we present a dramatic simplification of design methods, which are well established in the finite dimensional case, like e.g., the PD-, H_2- and H_∞-design. The main point is that one is able to convert the partial differential equations of the Hamilton-Jacobi type into simple algebraic ones. The collocation of sensors and actuators is the price, which one has to pay. The fourth part presents some basics of piezoelectricity and mechanical structures, and the fifth and the sixths part are devoted to the application of the presented methods to beams and plates.

15.2 Some remarks on Lagrangian systems

Let us consider a mechanical system $\mathcal{M}$, which is a Lagrangian system with $p+1$ independent coordinates t, x^i, $i = 1, \ldots, p$ and q dependent coordinates u^j, $j = 1, \ldots q$. To shorten the notation, we use the symmetric multi-indices notation with $J = (j_1, \ldots, j_k)$, $0 \leq j_i \leq p$ and $k = \#J$ to describe the kth order partial derivatives

$$f_J = f_{j_1,\ldots,j_k} = \frac{\partial^k}{\partial x^{j_1} \ldots \partial x^{j_k}} f \,, \quad f_0 = \frac{\partial}{\partial t} f$$

of a smooth function $f : R^{p+1} \to R$ [5]. The case $\#J = 0$, or the zero order partial derivative of f, is defined by $\partial_J f = f$. Let L,

$$L = \int_{\mathcal{D}} l\left(t, x, u^{(n)}\right) \omega \,, \quad \omega = \mathrm{d}x^1 \wedge \ldots \wedge \mathrm{d}x^p \tag{15.1}$$

denote the Lagrangian of $\mathcal{M}$, where $u^{(n)}$ denotes all possible partial derivatives of the dependent variables u^j with respect to the independent variables t, x^i up to order n. Furthermore, we assume that the smooth Lagrangian density l is well defined on the domain $\mathcal{D} \subset R^p$ for $t \geq 0$. To complete the problem, we have to add suitable conditions for u^j on the boundary $\partial\mathcal{D}$. For the sake of simplicity, we will specify them in the applications only because they are not relevant for the rest. But, we assume that no energy transport across the boundary is possible.

Hamilton's principles states that the action A,

$$A = \int_{t_1}^{t_2} L\mathrm{d}t$$

is extremized [11]. The well known solution of this problem is given by the equations of motion [5]

$$\phi_\alpha\left(l\right) = 0 \;, \quad \alpha = 1, \ldots, q \tag{15.2}$$

with the Euler-Lagrange operators

$$\phi_\alpha = \sum_J (-1)^{\# J} D_J \frac{\partial}{\partial u_J^\alpha} \; . \tag{15.3}$$

D_J is a shortcut for

$$D_J = \frac{\mathrm{d}}{\mathrm{d}x^{j_1}} \cdots \frac{\mathrm{d}}{\mathrm{d}x^{j_k}}$$

with $J = (j_1, \ldots, j_k)$ and the total derivative

$$\begin{aligned} \tfrac{\mathrm{d}}{\mathrm{d}x^i} &= \tfrac{\partial}{\partial x^i} + \textstyle\sum_{\alpha=1}^q \sum_J u_{J,i}^\alpha \tfrac{\partial}{\partial u_J^\alpha} \;, \quad i = 1, \ldots, p \\ \tfrac{\mathrm{d}}{\mathrm{d}t} &= \tfrac{\partial}{\partial t} + \textstyle\sum_{\alpha=1}^q \sum_J u_{J,0}^\alpha \tfrac{\partial}{\partial u_J^\alpha} \end{aligned} \tag{15.4}$$

with $J, i = j_1, \ldots, j_k, i$. The sum in (15.3) and (15.4) is over all symmetric multi-indices J up to order n.

We are interested in Lagrangian systems only, which describe a time invariant mechanical system $\mathcal{M}$ with inputs or generalized external forces f^i, $i = 1, \ldots, m$, such that the Lagrangian density l of (15.1) is given by

$$l\left(t, x, u^{(n)}\right) = l^0\left(x, u^{(n)}\right) - \sum_{i=1}^m l^i\left(x, u^{(n)}\right) f^i\left(t\right) \; . \tag{15.5}$$

Furthermore, we have

$$\frac{\partial}{\partial t} l^i = 0 \;, \quad i = 0, \ldots, m \;,$$

and Noether's theorem [5] implies that there exists a conservation law of the free system, $f^i = 0$, such that

$$\frac{\mathrm{d}}{\mathrm{d}t} E = 0 \;, \quad E = \int_{\mathcal{D}} e\omega \;, \quad e = a - l^0 \tag{15.6}$$

is met for a suitable function $a\left(x, u^{(n)}\right)$. Since E is time independent, we get

$$\frac{\mathrm{d}}{\mathrm{d}t} E = \int_{\mathcal{D}} \sum_{\alpha=1}^q \sum_J u_{J,0}^\alpha \frac{\partial}{\partial u_J^\alpha} \left(a - l^0 + \sum_{i=1}^m l^i f^i \right) \omega$$

or

$$\frac{\mathrm{d}}{\mathrm{d}t} E = \sum_{i=1}^m f^i \frac{\mathrm{d}}{\mathrm{d}t} L^i \;, \quad L^i = \int_{\mathcal{D}} l^i \omega \tag{15.7}$$

in addition. Of course, E is nothing else than the sum of kinetic and potential energy of $\mathcal{M}$. It is worth to mention that in general the derivation of E in accordance with (15.6) can be a laborious task [5].

Let E be a positive definite function of the state (u, u_0) of $\mathcal{M}$. From now on, we assume that the requirement

$$\frac{\mathrm{d}}{\mathrm{d}t} E \leq 0$$

implies the stability of the system $\mathcal{M}$. In contrast to the finite dimensional case, more investigations are necessary for infinite dimensional systems. But this hypothesis applies for the presented examples [2].

15.3 Control of Lagrangian systems

Let us consider a mechanical system $\mathcal{M}$ with the Lagrangian density l of (15.5). In contrast to the previous section, we denote the external forces by u^i, which are the control input acting on the structure. We complete the system by the choice for the output y^i,

$$y^i = \int_{\mathcal{D}} l^i \omega \,, \quad i = 1, \ldots, m \tag{15.8}$$

to obtain the Lagrangian control system $\mathcal{L}$ with collocated input and output. The special choice (15.8) for the output is called natural output, too [3]. Obviously, (15.7) is given by

$$\frac{\mathrm{d}}{\mathrm{d}t} E = \sum_{i=1}^{m} u^i \dot{y}^i \,, \quad \frac{\mathrm{d}}{\mathrm{d}t} y^i = \dot{y}^i \,, \quad i = 1, \ldots m \,. \tag{15.9}$$

Under the conditions of section 15.2, one can easily demonstrate some nice properties of the system $\mathcal{L}$. E.g., the control law (see [3])

$$u^i = -\sum_{j=1}^{m} \left(K_{ij} y^j + D_{ij} \dot{y}^j \right) \tag{15.10}$$

with positive (semi)definite matrices K and D preserves stability, because (15.9) can be rewritten as

$$\frac{\mathrm{d}}{\mathrm{d}t} \left(E + \frac{1}{2} \sum_{i,j=1}^{m} y^i K_{ij} y^j \right) = - \sum_{i,j=1}^{m} \dot{y}^i D_{ij} \dot{y}^j \,.$$

Next, let us try to find a solution of $\mathcal{L}$, which minimizes the objective function[1]

$$J_2 = \sup_{T\in[0,\infty)} \inf_{u\in L_2^m[0,T]} \frac{1}{2}\int_0^T \left(\|\dot{y}\|^2 + \|u\|^2\right) \mathrm{d}t \tag{15.11}$$

with euclidean norm $\|\ \|$. The Hamilton-Jacobi-Belman inequality of this H_2-problem is given by

$$\inf_u \left(\frac{\mathrm{d}}{\mathrm{d}t}V + \frac{1}{2}\left(\|\dot{y}\|^2 + \|u\|^2\right)\right) \leq 0\,. \tag{15.12}$$

A short calculation shows that the ansatz $V = \rho E$, $\rho > 0$ leads to the simple control law

$$u^i = -\rho \dot{y}^i\ ,\ i = 1, \ldots, m$$

and converts (15.12) to

$$\frac{1-\rho^2}{2}\|\dot{y}\|^2 \leq 0\,.$$

The choice $\rho = 1$ solves the problem exactly and the objective function J_2,

$$J_2 = -\int_0^\infty \sum_{j=1}^m \dot{y}^j u^j \mathrm{d}t$$

is equal to the energy dissipated by the controller.

Let us assume that we can split the external forces f^i of (15.5) into two parts such that u^i, $i = 1, \ldots, m$ acts as the control input and d^i, $i = 1, \ldots, m$ acts as the disturbance input on the structure. Analogously to (15.9), we rewrite (15.7) and get

$$\frac{\mathrm{d}}{\mathrm{d}t}E = \sum_{i=1}^m \left(u^i \dot{y}^i + d^i \dot{z}^i\right)\ ,\quad \frac{\mathrm{d}}{\mathrm{d}t}y^i = \dot{y}^i\ ,\quad \frac{\mathrm{d}}{\mathrm{d}t}z^i = \dot{z}^i\ ,\quad i = 1, \ldots, m\,. \tag{15.13}$$

Now, we consider the H_∞-design problem (see [9])

$$J_\infty = \sup_{T\in[0,\infty)} \inf_{u\in L_2^m[0,T]} \sup_{d\in L_2^m[0,T]} \frac{1}{2}\int_0^T \left(\|\dot{y}\|^2 + \|u\|^2 - \gamma\|d\|^2\right) \mathrm{d}t\ , \tag{15.14}$$

[1] The objective function can be changed by any transform $y = A\bar{y}$, $u = B\bar{u}$ with regular matrices A, B and $B^T A = \lambda I$, because (15.9) is preseved by the new energy $\bar{E} = E\lambda$.

$\gamma > 0$, with the Hamilton-Jacobi-Belman-Isaacs inequality

$$\inf_u \sup_d \left(\frac{\mathrm{d}}{\mathrm{d}t} V + \frac{1}{2} \left(\|\dot{y}\|^2 + \|u\|^2 - \gamma \|d\|^2 \right) \right) \leq 0 . \tag{15.15}$$

Let us assume that the disturbances d^i act in the same way on the structure as the control inputs u^i or the relation $y^i = z^i$, $i = 1, \ldots, m$ is met. Then the ansatz $V = \rho E$, $\rho > 0$ leads to the simple equations

$$u^i = -\rho \dot{y}^i \ , \quad \gamma d^i = \rho \dot{y}^i \ , \quad i = 1, \ldots, m$$

and simplifies (15.15) to

$$\sum_{i=1}^{m} \frac{1}{2} \left(1 - \rho^2 \frac{\gamma - 1}{\gamma}) \|\dot{y}\|^2 \right) \leq 0 .$$

Of course, one can reach even equality, iff the relation $\gamma > 1$ is met. Again, the objective function J_∞,

$$J_\infty = -\sqrt{\frac{\gamma - 1}{\gamma}} \int_0^\infty \sum_{j=1}^{m} \dot{y}^i u^j \mathrm{d}t$$

is proportional to the energy dissipated in the controller.

¿From the properties above follows that the Lagrangian control system $\mathcal{L}$ fits also the requirements of the problem of rendering the closed loop input-output L_2-stable [10]. E.g., let us assume that the plant with Lagrangian L, input u and output $\dot{y}$ is in the steady state for $t = 0$. Furthermore, let $\mathcal{L}$ be passive or equivalently let the inequality

$$\int_0^T \sum_{i=1}^{m} \dot{y}^i u^i \mathrm{d}t \geq 0$$

be met for all $T \geq 0$. One can show that any strictly passive controller $\mathcal{P}$ with finite gain renders the closed loop input-output L_2-stable with bias [10]. Let u_C denote the input and y_C the output of the controller, which is in the equilibrium at $t = 0$. Strictly passivity means that the inequality

$$\int_0^T \sum_{i=1}^{m} u_C^i y_C^i \mathrm{d}t \geq \varepsilon \int_0^T \|u_C\|^2 \mathrm{d}t$$

is met with $\varepsilon > 0$ for all $T \geq 0$ and finite gain implies that the inequality

$$\left(\int_0^T \|y_C\|^2 \mathrm{d}t \right)^{1/2} \leq \gamma \left(\int_0^T \|u_C\|^2 \mathrm{d}t \right)^{1/2} + b$$

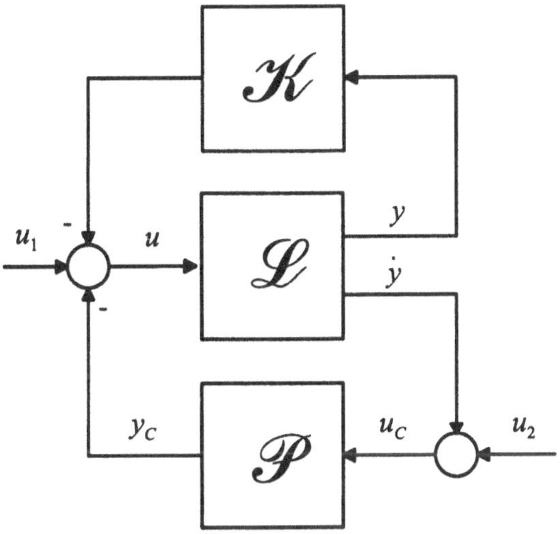

Fig. 15.1. Control system

is fulfilled for $\gamma, b > 0$. It is worth to mention that strict passivity of linear, time invariant controllers can easily be checked in the frequency domain [10].

One can combine the controller $\mathcal{P}$ with a controller $\mathcal{K}$ such that $\mathcal{K}$ can be derived from a potential $l_{\mathcal{K}}$ or the relation

$$u^i_{\mathcal{K}} = \phi_i\left(l_{\mathcal{K}}\right) \ , \quad i = 1, \ldots, q$$

with ϕ_i from (15.3) is met. In this case we must augment the Lagrangian of $\mathcal{L}$ by the additional term $-l_{\mathcal{K}}$ only. If $l_{\mathcal{K}}$ meets in addition the relation $l_{\mathcal{K}} \geq 0$, then stability is preserved. Figure 15.1 summarizes these approaches. Of course, the PD-law (15.10) is nothing else than a special case of this approach.

The optimization problems (15.11) and (15.14) offer another interesting extension. We solve the problems $J_2\left(\varepsilon\right)$, $J_\infty\left(\varepsilon\right)$, $\varepsilon \geq 0$,

$$J_2\left(\varepsilon\right) = \tfrac{1}{2}\int_0^\infty \left(\varepsilon\left\|y\right\|^2 + \left\|\dot{y}\right\|^2 + \left\|u\right\|^2\right) \mathrm{d}t$$

$$J_\infty\left(\varepsilon\right) = \tfrac{1}{2}\int_0^\infty \left(\varepsilon\left\|y\right\|^2 + \left\|\dot{y}\right\|^2 + \left\|u\right\|^2 - \gamma\left\|d\right\|^2\right) \mathrm{d}t$$

for a linearized finite state approximation $\mathcal{L}_a$ of $\mathcal{L}$. Since the solutions of the problems $J_2\left(0\right)$, $J_\infty\left(0\right)$ with output feedback are simple D-laws, which are L_2-stable and strictly passive and J_2, J_∞ depend continuously on ε, we can expect that these properties are preserved for the solutions of the problems $J_2\left(\varepsilon\right)$, $J_\infty\left(\varepsilon\right)$ for ε sufficiently small. Now, the derivation of the optimal control laws of the linearized and approximated problem with output feedback is a straightforward procedure. According to the considerations above, one can use these control laws in combination with the original plant $\mathcal{L}$, because their L_2-stability and strict passivity guarantee the L_2-stability of the closed loop. Since we derive an output feedback law, we avoid the common problem to measure the state of the finite state approximation $\mathcal{L}_a$, which has no physical meaning in general. It is worth to mention that derivation of the equations (15.2) (15.1) it is not necessary, because all the presented design methods uses the Lagrangian (15.1) only.

15.4 The mechanical model

The mechanical structures under investigation are beams and plates. Since a beam can be regarded as a limit case of a plate, we restrict the considerations to plates only. Subsequently, we use a 3-dimensional Euclidean space with orthonormal basis $\mathcal{B} = \{e_1, e_2, e_3\}$, $(e_i, e_j) = \delta^{ij}$ and coordinates x^i, $i = 1,2,3$ with $x = \sum_{i=1}^{3} x^i e_i$. The independent coordinates are $t = x^0$, x^i, $i = 1,2,3$ and the dependent mechanical coordinates are the displacements u^j, $j = 1,2,3$.

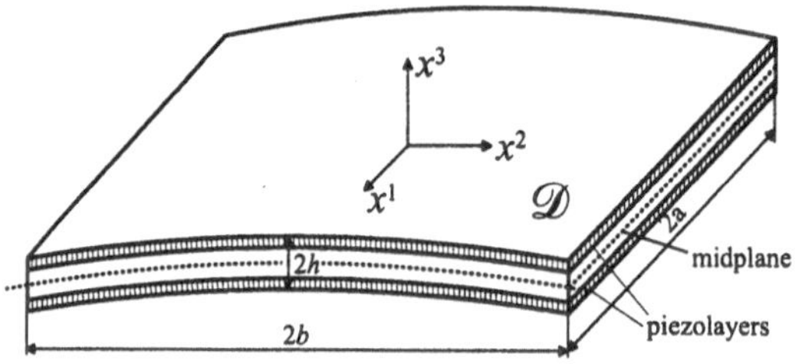

Fig. 15.2. Simply supported straight plate

Let u denote the displacement of the midplane and $\mathcal{D}$ the surface of the midplane with the corresponding surface element $\omega = \mathrm{d}x^1 \wedge \mathrm{d}x^2$ at the equilibrium position $u = u_0 = 0$ (see figure 15.2). If rotational inertias are neglected, then the kinetic energy of the plate is given by

$$W_k = \int_{\mathcal{D}} \tilde{w}_k \omega \ , \quad \tilde{w}_k = \frac{\mu}{2}\left(u_0, u_0\right) \ , \quad \mu = \int_{-h}^{h} \rho \mathrm{d}x^3 \tag{15.16}$$

with the mass density ρ.

The smart plate consists of several piezoelectric layers, which are covered by metallic electrodes, where a voltage is applied. Inside the piezoelectric substrate the electrical flux density D fulfills the equation

$$\sum_{i=1}^{3} \frac{\partial}{\partial x^i} D^i = 0 \,. \tag{15.17}$$

The quasi-static electrical field density E is connected to the electrical potential P by

$$E^i = -\frac{\partial}{\partial x^i} P \,, \quad i = 1,2,3 \,. \tag{15.18}$$

P is constant along the metallic electrodes of the capacitor and the tangential components of E vanish there.

For the calculation of the potential energy, we use the constitutive equations of the form

$$\sigma^{ij} = \sum_{k,l=1}^{3} c_{ijkl}\varepsilon^{kl} - \sum_{k} a_{kij} D^k \,, \quad i,j = 1,2,3 \tag{15.19}$$

and

$$E^i = -\sum_{kl} a_{ikl}\varepsilon^{kl} + \sum_{k} d_{ik} D^k \tag{15.20}$$

with the integrability conditions $c_{ijkl} = c_{jikl} = c_{ijlk} = c_{klij}$, $a_{kij} = a_{kji}$ and $d_{ik} = d_{ki}$ to describe the relation between stress σ, strain ε and electrical flux density D in a piezoelectric lamina [4]. The latter conditions allow to derive the volume energy density w_p

$$w_p = \frac{1}{2}\left(\sum_{ijkl} c_{ijkl}\varepsilon^{kl}\varepsilon^{ij} - 2\sum_{ikl} a_{ikl}\varepsilon^{kl} D^i + \sum_{ik} d_{ik} D^i D^k \right) \,. \tag{15.21}$$

By means of Kirchhoff's assumptions the strains ε^{11}, ε^{12}, ε^{22} are related to the strains $\bar{\varepsilon}^{11}$, $\bar{\varepsilon}^{12}$, $\bar{\varepsilon}^{22}$ of the midplane by

$$\begin{aligned}
\varepsilon^{11} &= \bar{\varepsilon}^{11} - x^3 u^3_{11} \,, \; \bar{\varepsilon}^{11} = u^1_1 + \tfrac{1}{2}\left(u^3_1\right)^2 \\
\varepsilon^{22} &= \bar{\varepsilon}^{22} - x^3 u^3_{22} \,, \; \bar{\varepsilon}^{22} = u^2_2 + \tfrac{1}{2}\left(u^3_2\right)^2 \\
\varepsilon^{12} &= \bar{\varepsilon}^{12} - x^3 u^3_{12} \,, \; \bar{\varepsilon}^{12} = \tfrac{1}{2}\left(u^2_1 + u^1_2\right) + \tfrac{1}{2} u^3_1 u^3_2 \,,
\end{aligned} \tag{15.22}$$

(see [11]). Furthermore, a nonlinear formulation in the sense of v. Karman is used for $\bar{\varepsilon}^{11}$, $\bar{\varepsilon}^{12}$, $\bar{\varepsilon}^{22}$. The solution of the electrical field equations (15.17), (15.18) for the plate are too expensive. Therefore, some simplifications in accordance with the Kirchhoff's assumptions and the special properties of the considered material are assumed. The strains ε^{ij} vanish for $i = 3$ and $D^i = 0$ holds for $i = 1,2$. Now, from (15.17) follows $\frac{\partial}{\partial x^3} D^3 = 0$.

The piezoelectric material, which is the basis of the subsequent considerations, allows the simplification

$$\begin{aligned}\sigma^{11} &= \frac{Y}{1-\nu^2}\varepsilon^{11} + \frac{Y\nu}{1-\nu^2}\varepsilon^{22} - a_{311}D^3 \\ \sigma^{22} &= \frac{Y\nu}{1-\nu^2}\varepsilon^{11} + \frac{Y}{1-\nu^2}\varepsilon^{22} - a_{322}D^3 \\ \sigma^{12} &= \frac{Y}{1+\nu}\varepsilon^{12}\end{aligned} \tag{15.23}$$

of (15.19) as well as the simplification

$$E^3 = -a_{311}\varepsilon^{11} - a_{322}\varepsilon^{22} + d_{33}D^3 \tag{15.24}$$

of (15.20). Here, Y denotes Young's modulus and ν Poisson's ratio, respectively. The energy density w_p (15.21) for such a material is given by

$$\begin{aligned}w_p &= w_a + w_m + w_e \\ w_a &= -D^3\left(a_{311}\varepsilon^{11} + a_{322}\varepsilon^{22}\right) \\ w_m &= \tfrac{1}{2}\tfrac{Y}{1-\nu^2}\left(\left(\varepsilon^{11}\right)^2 + 2\nu\varepsilon^{11}\varepsilon^{22} + \left(\varepsilon^{22}\right)^2\right) + \tfrac{Y}{2(1+\nu)}\left(\varepsilon^{12}\right)^2 \\ w_e &= \tfrac{1}{2}d_{33}\left(D^3\right)^2 \ .\end{aligned} \tag{15.25}$$

The material parameters Y, ν, a_{311}, a_{322} and d_{33} may vary from layer to layer, however they are assumed to be constant within each layer. For the sake of simplicity, we assume that the plate is built up symmetrically with respect to the midplane $x^3 = 0$ and that the different piezoelectric and structural layers are perfectly bonded to the substrate. Despite this symmetry, we can apply the voltage U_i symmetric or antisymmetric with respect to the midplane. Let $\mathcal{H}$ denotes the lines corresponding to the heights of a layer couple. If we take the integral of w_p,

$$\tilde{w}_p = \int_{\mathcal{H}} w_p \mathrm{d}x^3$$

with w_p from (15.25), then we get the area densities

$$\tilde{w}_a^s = D^3\left(a_{311}\bar{\varepsilon}^{11} + a_{322}\bar{\varepsilon}^{22}\right)\Lambda^s\left(x^1, x^2\right) \tag{15.26}$$

in the symmetric and

$$\tilde{w}_a^a = D^3\left(a_{311}u_{11}^3 + a_{322}u_{22}^3\right)\Lambda^a\left(x^1, x^2\right) \tag{15.27}$$

in the antisymmetric case because of (15.21), (15.22) and (15.23). Hereby, the functions Λ^a, Λ^s depend on the special design of the layers like the pattern of the metallic surface, etc.. Now, let us consider a plate with $2m$ layers, where m_a layer couples are supplied antisymmetrically by a voltage U_i^a, $i = 1, \ldots, m_a$ and m_s layer couples are supplied symmetrically by a voltage U_i^s, $i = 1, \ldots, m_s$. Hereafter, the symbol s stands for symmetric and

a for antisymmetric. According to these considerations, we obtain for $\tilde{w}_p$ the expression

$$\begin{aligned}\tilde{w}_p = \tilde{w}_m + \tilde{w}_e - \sum_{i=1}^{m_s} \Lambda_i^s \left(x^1, x^2\right) \left(a_{311}\bar{\varepsilon}^{11} + a_{322}\bar{\varepsilon}^{22}\right) f_i^s \left(U_i^s\right) \\ + \sum_{i=1}^{m_a} \Lambda_i^a (x^1, x^2) \left(a_{311} u_{11}^3 + a_{322} u_{22}^3\right) f_i^a \left(U_i^a\right) .\end{aligned} \tag{15.28}$$

Again, the area densities $\tilde{w}_m$, $\tilde{w}_e$ contain the pure elastic and pure electric part. The index i refers to the ith layer couple. Furthermore, we assume that the electrical flux density and the voltage are connected by a linear relation, which implies

$$f_i^s \left(U_i^s\right) = \lambda_i U_i^s \ , \quad f_i^a \left(U_i^a\right) = \lambda_i U_i^a \ . \tag{15.29}$$

The results above indicate that by means of this piezoelectric material only bending and stretching motion and no shear strain can be actuated. This restriction can be changed by introducing a skew angle between the principal axis of the piezoelectric lamina and the reference axis of the laminate.

The sensor is based on the relation (15.24). Since we short-circuit the corresponding electrodes of a piezoelectric sensor layer, the relation $E^3 = 0$ is met. By integration over the effective metallic surface of the electrodes of the sensor layer, we get the electrical charge Q

$$Q = \int_{\mathcal{D}} \left(a_{311}\varepsilon^{11} + a_{322}\varepsilon^{22}\right) \Gamma \left(x^1, x^2\right) \omega \ .$$

Also here, the function $\Gamma\left(x^1, x^2\right)$ depends on the design of the metallic surface. Since the layers of the piezoelectric plate are arranged symmetrically with respect to the midplane, we have again two possibilities for measuring the charge. If we take the sum of the charges of the two corresponding layers of a sensor layer couple j, we get

$$Q_j^s = \int_{\mathcal{D}} \Gamma_j^s \left(x^1, x^2\right) \left(a_{311}\bar{\varepsilon}^{11} + a_{322}\bar{\varepsilon}^{22}\right) \omega \ , \tag{15.30}$$

and we get

$$Q_j^a = \int_{\mathcal{D}} \Gamma_j^a \left(x^1, x^2\right) \left(a_{311} u_{11}^3 + a_{322} u_{22}^3\right) \omega \ , \tag{15.31}$$

if we take their difference. The index j indicates the jth layer couple in both cases. The functions Γ_j^a and Γ_j^s depend on the position $\left(x^1, x^2\right)$ only. Comparing (15.30), (15.31) with (15.28), we see that collocation of the piezoelectric actuator with the sensor can be achieved in a straightforward manner.

15.5 Control of beams

As mentioned above, we consider beams as a limit case of a plate. The independent coordinates are t, x^1 and the dependent mechanical coordinates are the displacements u^1 and u^3. Now, the kinetic energy (15.16) simplifies to

$$W_k = \int_{\mathcal{D}} \tilde{w}_k \mathrm{d}x^1 \;, \quad \tilde{w}_k = \frac{\mu}{2} b \left(\left(u_0^1\right)^2 + \left(u_0^3\right)^2 \right) \;,$$

relation (15.28) is given by

$$\begin{aligned} \tilde{w}_p = \tilde{w}_m + \tilde{w}_e - \sum_{i=1}^{m_s} \Lambda_i^s \left(x^1\right) a_{311} \bar{\varepsilon}^{11} f_i^s \left(U_i^s\right) \\ + \sum_{i=1}^{m_a} \Lambda_i^a (x^1) a_{311} u_{11}^3 f_i^a \left(U_i^a\right) \;. \end{aligned} \tag{15.32}$$

and (15.30), (15.31) yield to

$$Q_j^s = \int_{\mathcal{D}} \Gamma_j^s \left(x^1\right) a_{311} \bar{\varepsilon}^{11} \mathrm{d}x^1 \;, \quad Q_j^a = \int_{\mathcal{D}} \Gamma_j^a \left(x^1\right) a_{311} u_{11}^3 \mathrm{d}x^1 \;, \tag{15.33}$$

respectively. Here, $\mathcal{D}$ denotes the line of length $2a$ and b the width of the beam.

15.5.1 H_∞-control of cantilever beam

Let us consider the cantilever beam of figure 15.3 of length $2a$ with a tip force f^d acting on the end $x^1 = a$. The boundary conditions of this problem are

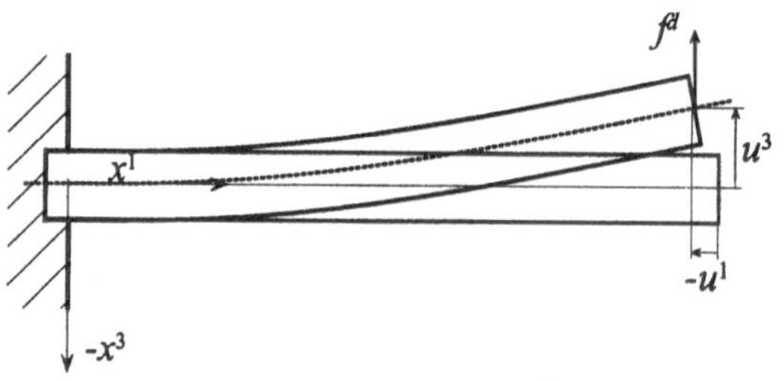

Fig. 15.3. Cantilever beam with a vertical tip force

$$\begin{gathered} u^3 (t, -a) = u_1^3 (t, -a) = u_2^3 (t, a) = u_3^3 (t, a) = 0 \\ u^1 (t, -a) = u_1^1 (t, a) = 0 \end{gathered}$$

To counteract the influence of the tip force f^d, we design a controller following the H_∞-approach of section 15.3. Since the design goal is the stabilization of the vertical displacement $u^3(t,a)$ of the beam, we take the objective function (15.14),

$$J_\infty = \frac{1}{2}\int_0^T \left(\left\|u_0^3(t,a)\right\|^2 + \left\|U^a\right\|^2 - \gamma\left\|f^d\right\|^2\right) \mathrm{d}t \ ,$$

where U denotes the voltage of an actuator layer which has to be designed now. The tip force is taken by the term

$$f^d(t)\, L^d = f^d(t)\int_{-a}^{a} u_1^3 \mathrm{d}x^1 = f^d(t)\, u^3(t,a)$$

into account. If the actuator is built up by one antisymmetric layer couple (see (15.29), (15.32)) with the supply voltage $U = U^a$, we have to enlarge the Lagrangian by

$$U^a L^a = U^a \int_{-a}^{a} \lambda_a a_{311} u_{11}^3 \Lambda^a\left(x^1\right) \mathrm{d}x^1 \ .$$

Following the considerations of section 15.3, we design the actuator such that the control input U^a acts in the same way as the disturbance input f^d. This implies that the relation

$$\phi_3\left(l^a - l^d\right) = 0$$

with ϕ_3 from (15.2) is met. Now, simple integration by parts

$$L^d = \int_{-a}^{a} u_1^3 \mathrm{d}x^1 = \int_{-a}^{a} (a - x)\, u_{11}^3 \mathrm{d}x^1$$

shows that the choice

$$\lambda^a a_{311} \Lambda^a\left(x^1\right) = (a - x)$$

guarantees that the actuator voltage U^a acts as the disturbance f^d on the beam. The collocated sensor follows immediately from (15.33), if one chooses

$$\Gamma^a\left(x^1\right) = \lambda^a \Lambda^a\left(x^1\right)$$

for an antisymmetric sensor couple. Now, the control law

$$U^a = -\sqrt{\frac{\gamma}{\gamma - 1}}\frac{\mathrm{d}}{\mathrm{d}t} Q^a$$

solves the H_∞-problem. Since $\frac{\mathrm{d}}{\mathrm{d}t}Q^a$ is nothing else than the current of the sensor, this law can be implemented in a straightforward manner. It is worth to mention that spatially distributed actuators and sensors are used to generate tip forces or to measure displacements of a dicrete point of the structure.

15.5.2 PD-control of simply supported beam

Here, we consider the straight composite beam of figure 15.4 under the action of an arbitrary space-wise distributed lateral loading $p(t)$. The flexural

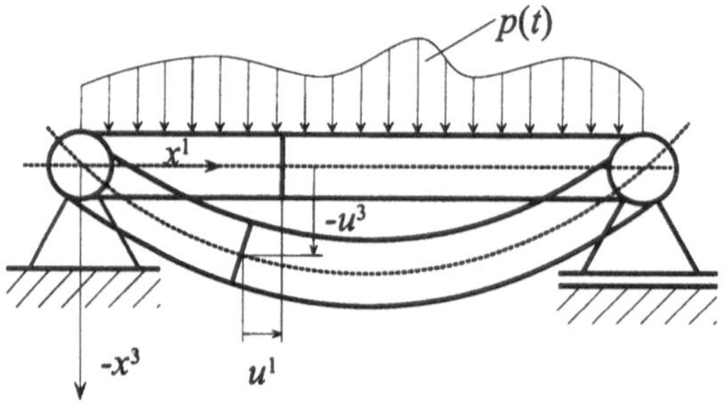

Fig. 15.4. Simply supported beam with arbitrary vertical loadings

boundary conditions for the simply supported beam read as

$$u^3(t,-a) = u^3(t,a) = u^3_2(t,-a) = u^3_2(t,a) = 0$$

and the longitudinal boundary conditions are given by

$$u^1(t,-a) = u^1(t,a) = 0 \ .$$

In order to meet the requirements that the flexural vibrations caused by any lateral loadings have to be suppressed, we use two antisymmetrically supplied actuator layer couples with the shaping functions $\Lambda^a_1(x^1)$ and $\Lambda^a_2(x^1)$, respectively. These shaping functions are designed in such a way that the corresponding actuator voltages U^a_1 and U^a_2 act in the same manner on the structure as a fictitious space-wise constant lateral loading $p_1(t)$ and a space-wise linear lateral loading $p_2(t)$. This is motivated by the fact that by U^a_1 all even and by U^a_2 all odd deflection modes can be influenced using a H_∞- or PD-controller. In the Lagrangian the two fictitious loadings $p_1(t)$ and $p_2(t)$ are taken into account in the form

$$p_1 L^{p_1} = p_1 \int_{-a}^{a} u^3 \mathrm{d}x^1 \quad \text{and} \quad p_2 L^{p_2} = p_2 \int_{-a}^{a} \left(\frac{x^1+a}{2a}\right) u^3 \mathrm{d}x^1 \ .$$

If the actuator is built up by an antisymmetric layer couple (see (15.29), (15.32)), we have to enlarge the Lagrangian by

$$U^a L^a = U^a \int_{-a}^{a} \lambda_a a_{311} u^3_{11} \Lambda^a(x^1)\, \mathrm{d}x^1 \ .$$

The requirements

$$\phi_3\left(l^{a_1} - l^{p_1}\right) = 0 \quad \text{and} \quad \phi_3\left(l^{a_2} - l^{p_2}\right) = 0$$

lead to the shaping functions

$$\lambda_a^1 a_{311} \Lambda_1^a\left(x^1\right) = \frac{\left(\left(x^1\right)^2 - a^2\right)}{2}$$

$$\lambda_a^2 a_{311} \Lambda_2^a\left(x^1\right) = \frac{\left(\left(x^1\right)^2 - a^2\right)\left(3a + x^1\right)}{12a} .$$

The corresponding collocated sensor layer couples follow from (15.33) in the form $\Gamma_j^a\left(x^1\right) = \lambda_a^j \Lambda_j^a\left(x^1\right)$ with the measured charge

$$Q_j^a = \int_{-a}^{a} \Gamma_j^a\left(x^1\right) a_{311} u_{11}^3 \mathrm{d}x^1$$

for $j = 1, 2$. For the simply supported beam and the designed shaping functions the PD-controller (15.10) reads as

$$\begin{bmatrix} U_1^a \\ U_2^a \end{bmatrix} = -K \begin{bmatrix} Q_1^a \\ Q_2^a \end{bmatrix} - D \begin{bmatrix} \frac{\mathrm{d}}{\mathrm{d}t} Q_1^a \\ \frac{\mathrm{d}}{\mathrm{d}t} Q_2^a \end{bmatrix}$$

with positive (semi)definite matrices K and D.

15.6 Plates

We consider the straight, composite rectangular plate of figure 15.2. The geometric dimensions are $(2a, 2b, 2h)$ with $h \ll a$, $h \ll b$. Furthermore, it is assumed that the plate is simply supported, which leads to the boundary conditions

$$\left.\begin{aligned} u^1 = 0 &: \left(\left(x^1\right)^2 - a^2\right) = 0, \\ u^2 = 0 &: \left(\left(x^2\right)^2 - b^2\right) = 0, \\ u^3 = 0 &: \left(\left(x^1\right)^2 - a^2\right)\left(\left(x^2\right)^2 - b^2\right) = 0 \\ \frac{\partial}{\partial u_{11}^3} l = 0 &: \left(\left(x^1\right)^2 - a^2\right) = 0 \\ \frac{\partial}{\partial u_{22}^3} l = 0 &: \left(\left(x^2\right)^2 - b^2\right) = 0 . \end{aligned}\right\} \begin{aligned} -a \le x^1 \le a \\ -b \le x^2 \le b \end{aligned} \tag{15.34}$$

The design goal is to stabilize the vertical displacement u^3, which is caused by a vertical spatially constant time dependent pressure loading $p(t)$. To annihilate the effect of the disturbance p at least in the steady state case, the control input must act precisely in the same way as p on the structure. Here, p enters the Lagrangian by the expression

$$pL^d = p \int_{\mathcal{D}} \varkappa \left(x^1, x^2\right) u^3 \omega$$

with $\varkappa = 1$. The function $\varkappa \left(x^1, x^2\right)$ allows to consider other spatial distributions. To clarify the influence of p, we use the integration by part technique twice and get

$$\int_{\mathcal{D}} \varkappa \left(x^1, x^2\right) u^3 \omega = \int_{\mathcal{D}} \left(g^1 u^3_{11} + g^2 u^3_{22}\right) \omega$$

with

$$g^1_{11} + g^2_{22} = \varkappa$$

and $g^1 = g^2 = 0$ at $\partial\mathcal{D}$. A short calculation shows that

$$g^1 = \upsilon^1 \hat{g} \,, \quad g^2 = \upsilon^2 \hat{g} \,, \quad 2\hat{g} = \left(x^1\right)^2 - a^2 + \left(x^2\right)^2 - b^2 \,, \quad \upsilon^1 + \upsilon^2 = 1 \,.$$

solves the equation above for $\varkappa = 1$. Now, we take one antisymmetric layer couple and choose the function $\Lambda^a(x^1, x^2)$ such that the relations (see (15.28), (15.29))

$$\Lambda^a(x^1, x^2) a_{311} \lambda_a = \upsilon^1 \hat{g} \,, \quad \Lambda^a(x^1, x^2) a_{322} \lambda_a = \upsilon^2 \hat{g}$$

are met. The collocated sensor follows immediately from (15.31), if one chooses

$$\Gamma^a \left(x^1, x^2\right) = \Lambda^a(x^1, x^2) \lambda_a$$

for an antisymmetric sensor couple. Following the considerations above, any PD-law

$$U^a = -K Q^a - D I^a$$

with $K, D > 0$ and the current

$$I^a = \frac{\mathrm{d}}{\mathrm{d}t} Q^a$$

renders the closed loop input-output L_2-stable with bias. Furthermore, any control law of the form

$$\mathcal{L}\left\{\hat{U}^a\right\} = K \mathcal{L}\left\{Q^a\right\} + R(s)\, \mathcal{L}\left\{I^a\right\}$$

with a strictly passive transfer function R of finite gain renders the closed loop input-output L_2-stable with bias. Here, $\mathcal{L}$ denotes the Laplace transform. Of course, $R(s)$ has finite gain, iff $R(s)$ is BIBO-stable and $R(s)$ is strictly passive, iff the inequality

$$0 < \varepsilon = \inf_{\omega} \operatorname{Re}\left(R(j\omega)\right)$$

is met.

15.7 Conclusion

This contribution is concerned with control techniques for mechanical structures with piezoelectric sensor and actuator layers. This technology allows to improve the dynamical behavior of mechanical structures significantly, because spatially distributed sensors and actuators can be constructed. It turns out that collocated control using a suitable actuator and sensor pairing simplifies the controller design a lot. This method is well established for finite dimensional systems, but seems to be even more import for infinite dimensional ones. Spatially distributed piezoelectric devices allow to adjust the actuator such that the influence of the disturbance is totally rejected in the steady state case. Spatially distributed piezoelectric sensors are needed in order to measure the corresponding natural output, which is the input of the control law.

The presented mathematical models of piezoelectric beams and plates take into account a nonlinear formulation of the strains in the sense of v. Karman, but rely on the laws of linear piezoelasticity. These models belong to a class of nonlinear, infinite dimensional Lagrangian systems. It turns out that passivity of the control law guarantees stability of the closed loop, if the Lagrangian system itself is stable. The presented approach is still limited to small displacements. Future work is necessary to extend these models for large displacements. This requires a nonlinear geometric formulation of the problem and the linear piezoelectric laws must be replaced by nonlinear ones. The latter point is important also in the case of small displacements, if low voltage materials are used because of their nonlinear effects due to hysteresis and polarization.

References

1. A. Kugi , K. Schlacher and H. Irschik, *Infinite Dimensional Control of Nonlinear Beam Vibrations by Piezoelectric Actuator and Sensor Layers*, Nonlinear Dynamics, in press, 1999.

2. J.E. Marsden and T.J.R. Hughes, *Mathematical Foundations of Elasticity*, Dover Publications, INC., New-York, 1994.

3. H. Nijmeijer and A.J. van der Schaft, *Nonlinear Dynamical Control Systems*, Springer-Verlag, Berlin, 1991.

4. W. Nowacki, *Dynamic Problems of Thermoelasticity*, Noordhoff International Publishing, Leyden, 1975..

5. P.J. Olver, Applications of Lie Groups to Differential Equations, Springer-Verlag, Wien, 1993.

6. K. Schlacher, H. Irschik and A. Kugi, *Control of Nonlinear Beam Vibrations by Multiple Piezoelectric Layers*, In: IUTAM Symposium on Interaction between Dynamics and Control in Advanced Mechanical Systems, Editor: D.H.van Campen, Kluwer Academic Publishers, pp.355-362, 1996.

7. H.S. Tzou, *Active Piezoelectric Shell Continua*, Intelligent Structural Systems, Editors: Tzou H.S. and Anderson G.L., Kluwer Academic Publishers, pp.9-74, 1992.

8. H.S. Tzou, J.P. Zhong and J.J. Hollkamp, *Spatially Distributed Orthogonal Piezoelectric Shell Actuators: Theory and Applications*, J. of Sound and Vibration, Vol.177, No.3, pp.363-378, 1994.

9. A.J. van der Schaft, *Nonlinear State Space H_∞ Control Theory, Essays on Control: Perspectives in the Theory and its Applications*, Birkhäuser Verlag, 1993.

10. M. Vidyasagar, *Nonlinear System Analysis*, Prentice Hall, 1993.

11. F. Ziegler, *Mechanics of Solids and Fluids*, Springer-Verlag, Wien, 1995.

16. A novel impedance grasping strategy as a generalized hamiltonian system

Stefano Stramigioli

Faculty of Information Technology and Systems
Department of Electrical Engineering
Systems- and Control Laboratory, P.O. Box 5031
NL-2600 GA Delft, The Netherlands
Tel: +31 (15) 278 5768 Fax. +31 (15) 278 6679
`S.Stramigioli@et.tudelft.nl`
Url: `lcewww.et.tudelft.nl/ stramigi`

Summary.

This chapter presents an intrinsically passive control strategy for robotic grasping tasks. This can be seen as a nontrivial application of the techniques presented in [14]. It is shown that robot control design synthesized as a spatial interconnection between the robot and the controller can be easily handled. The idea is based on what is called the *Virtual Object Concept* [13] and can be used both for tips grasp and full grasp. One of the major advantages of the presented strategy is the passive nature of the algorithm and the physical intuition it supplies due to the description of the controller as a spatial interconnection of physical elements.

16.1 Introduction

Most of the grasping strategies known in literature deal with the control of the tip contact forces [8]. Since grasping is concerned with the interaction of a robotic hand with the environment, strategies which consider the control of interaction explicitly seem more appropriate. One of the most problematic phenomena of some force control strategies is that stability cannot be ensured if very restrictive features of the object to be grasped are not assumed like its stiffness and friction. Furthermore, a force control strategy is not suitable to control the change between no-contact and contact. This is due to the fact that force control is only meaningful in contact since it is not possible to apply a force different than zero in free space.

For these reasons, a grasping technique based on more physical reasoning and passivity seems worth to be pursued. The presented technique is based on a strategy which shapes the potential energy of the system in order to achieve a desired compliance and injects some damping to ensure asymptotic stability and a proper behavior. These techniques have been already used in the past [9, 11], but here a nontrivial geometry plays a role.

Such an impedance strategy does not have the shortcomings of other grasping techniques: it is strictly passive in steady-state situations for any passive environment and the supplied energy in moving tasks is directly controllable.

The compliance control of each finger allows for rolling, slipping, and whole-hand grasping in a natural way.

The analysis will be shown without considering the kinematic structure of the hand, which would constrain the mobility. These constraints in the kinematics are not crucial since the controller intrinsically brings the system to the minimum potential configuration the structure of the hand allows. Often, this is not the minimum of the desired energy but the minimum of the constrained system.

16.2 Background

The notation used in this chapter is the one introduced in [12] and will be here briefly reviewed.

To a set of rigid bodies moveing with respect to each other is associated a set of Eucledian spaces $\mathcal{E}_i$ of equal dimensions n. An object B_i is a set of points in the corresponding Eucledian space $\mathcal{E}_i$.

We indicate with the symbol Ψ_k a right handed cartesian coordinate frame fixed in space $\mathcal{E}_k$ and with $\psi_k : \mathcal{E}_k \to \mathbb{R}^{n+1}; q \mapsto (p^T \;\; 1)^T$ the coordinate function associated to it which associates to a point $q \in \mathcal{E}_k$ its homogeneous coordinates. The relative position of object B_i with respect to B_j, or equivalently of their associated Eucledian spaces is a positive isometry $h_i^j : \mathcal{E}_i \to \mathcal{E}_j$ which maps points in one space to the corresponding points in the other space[1]. Such positive isometries are elements of a set which is indicated with $SE_i^j(n)$ and is NOT a group unless $i = j$. In the latter case $SE_i^i(n)$ is indicated with $SE_i(n)$ and corresponds to the *special Eucledian* group of $\mathcal{E}_i$. Associated to each Lie group $SE_i(n)$ there is a Lie algebra $se_i(n)$. In [12] it is shown that there is NOT an intrinsic bijection between the elements of the Lie groups and algebras of different Eucledian spaces.

The local velocity of $\mathcal{E}_i$ with respect to $\mathcal{E}_j$ in a certain configuration h_i^j is an element $\dot{h}_i^j$ belonging to the tangent space $T_{h_i^j} SE_i^j(n)$.

---

[1] In this setting, there are no implicit reference relative positions as it is usually done defining coordinate systems.

We can map the tangent velocity $\dot{h}_i^j$ in an intrinsic way either to $t_i^{i,j} \in se_i(n)$ or to $t_i^j \in se_j(n)$ using respectively the *left intrinsic map*

$$\pi^i_{h_i^j} : T_{h_i^j} SE_i^j(n) \to se_i(n); \dot{h}_i^j \mapsto t_i^{i,j}$$

or the *right intrinsic map*

$$\pi^j_{h_i^j} : T_{h_i^j} SE_i^j(n) \to se_j(n); \dot{h}_i^j \mapsto t_i^j.$$

Elements of $se_i(n)$ are called *twists* and represent the generalised relative velocity between bodies (or associate Eucledian spaces). An element $t_i^{k,j} \in se_k(n)$ represents the twist of $\mathcal{E}_i$ with respect to $\mathcal{E}_j$ as a geometric entity in the space $\mathcal{E}_k$. In the case $k = j$ we will write t_i^j. Based on the right intrinsic map, we can define its adjoint $\pi^{j\,*}_{h_i^j} : se_j^*(n) \to T^*_{h_i^j} SE_i^j(n)$ and their inverses which will be indicated with $\chi_{h_i^j} := \left(\pi^j_{h_i^j}\right)^{-1} : se_j(n) \to T_{h_i^j} SE_i^j(n)$ and $\chi^*_{h_i^j} := \left(\pi^j_{h_i^j}\right)^{-1}$.

Elements of $se_i^*(n)$ are called wrenches and represent the geometric generalisation of a force.

Eventually, the *hybrid Adjoint map* introduced in [12], maps twists from a space to another which are in a relative configuration h_l^k:

$$t_i^{k,j} = Ad_{h_l^k} t_i^{l,j}.$$

16.3 Controllable springs

For the control strategy which will be presented, it is necessary to consider springs with two hinge points for which the minimum potential energy relative position can be varied. This is analog to a spring the natural length of which can be varied. Such an action changes the spring-stored energy and this implies that, in order to properly describe this process in an energetical consistent way, we need an additional power port through which we can control this action.

If we consider as a simple example a linear, one-dimensional spring with finite length x_l, and stiffness k, its energy function can be expressed as:

$$E(x) = \frac{1}{2} k (x - x_l)^2.$$

The energetic port of the spring is then characterized by the effort-flow pair $(k(x - x_l), \dot{x})$. If we consider the possibility of varying the final length x_l, we need to consider an additional port. The new energy function should be considered, then, as a function of $(x - x_l)$ and not only as a function of x and the new energetic port, which is used to control the finite length, should be described by the effort-flow pair $(k(x_l - x), \dot{x}_l)$; the spring's state change is $(\dot{x} - \dot{x}_l)$.

For a geometric spring, as treated in [12], a "finite length" corresponds to the relative configuration $r_i^j \in SE_i^j(n)$ for which the energy function has its minimum.

If we want to describe a variable spring connecting body $B_b \in \mathcal{E}_b$ to body $B_i \in \mathcal{E}_i$, we consider an additional space $\mathcal{E}_{v(i)}$ which will be called the *supporting space* for the variable spring. We then describe an energy function on the relative position of $\mathcal{E}_{v(i)}$ with respect to $\mathcal{E}_i$, which we indicate as $h^i_{v(i)} \in SE^i_{v(i)}(3)$. From a composition of isometries, we obtain:

$$h^i_{v(i)} = h^i_b \; o \; h^b_{v(i)} \tag{16.1}$$

We now control the effective minimal potential energy-relative position of b with respect to i by varying $h^b_{v(i)}$. We can then analyse how the state $h^i_{v(i)}$ of the variable spring changes as a function of time. The following result gives the searched relation.

Theorem 16.3.1 (Variable length spring state). *Given a spring connecting a body of $\mathcal{E}_i$ to a body of $\mathcal{E}_{v(i)}$ and a positive definite energy function $V^i_{v(i)}(\cdot)$ of $h^i_{v(i)} \in SE^i_{v(i)}(3)$ such that $h^i_{v(i)} = h^i_b \; o \; h^b_{v(i)}$, the following identity holds:*

$$t^i_{v(i)} = t^i_b + Ad_{h^i_b} t^b_{v(i)}. \tag{16.2}$$

Proof. The proof can be found in [12].

A representation in coordinates-free bond graphs of such a variable spring is reported in Fig. 16.1, where t^k_b and t^k_i represent the hinge points where the two bodies attached to the spring are connected and $t^b_{v(i)}$ is the twist which is used to change the equilibrium point of the spring. The mapping $\chi_{h^i_{v(i)}}$ represents the inverse of the right intrinsic map as presented in Sect. 16.2.

16.4 Physical controller structure

In this section we describe the structure of the proposed controller from a conceptual point of view. Following the philosophy of Hogan's Physical equivalence principle [6], we are going to create a controller which has a

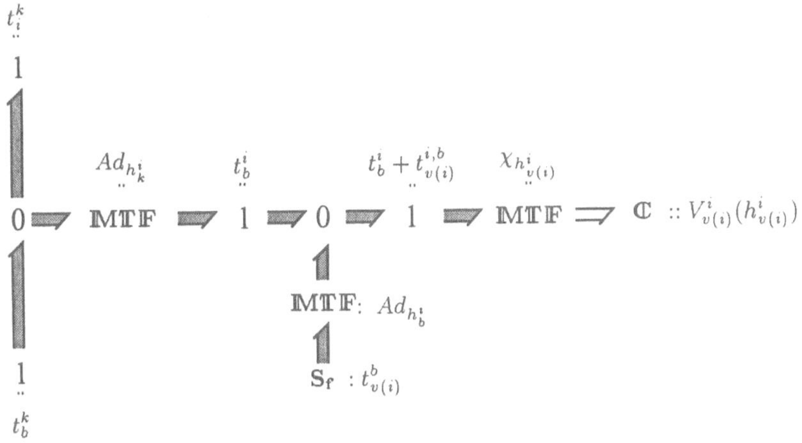

Fig. 16.1. The coordinates free bond graph of a variable length spatial spring

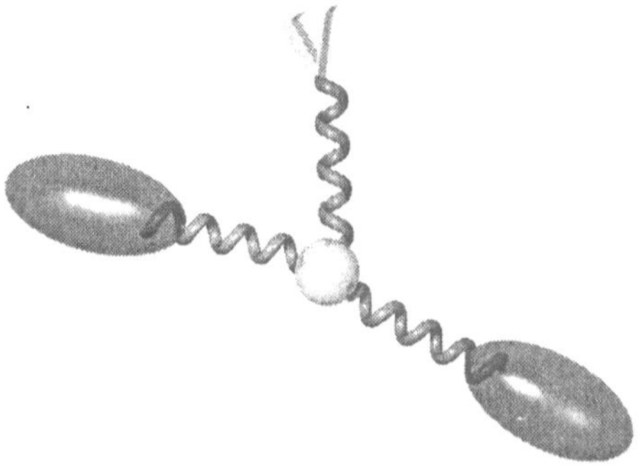

Fig. 16.2. The intuitive idea of the proposed grasping strategy as the spatial interconnection of physical elements

directly interpretable physical equivalent system: it is described as a spatial interconnection of physical elements.

In Fig. 16.2, two fingertips of a robotic hand are represented as ellipsoids. These two tips are the extremities of robotic chains corresponding to the fingers of a robotic hand to be controlled. We always talk about a robotic hand, but the techniques here presented can be equivalently used for the control of coordinated robots.

The goal of the proposed controller is to create the equivalent effect as the drawn springs and the mass corresponding to the sphere in Fig. 16.2. This sphere is called the *virtual object*. This means that the controlled system should dynamically behave as the spatial interconnected system presented in Fig. 16.2 in which for clarity only part of the robot mechanism is shown.

One of the springs, called the *hand configuration spring*, is connected at one side to a V indicating what is known in impedance control [6] as the *virtual position* of the hand. The position of the V can be changed by the supervision system as it will be seen later, and its result is a global motion of the hand.

In Fig. 16.2, only two fingertips are drawn to explain the concepts, but we can have n fingers. The minimum of the potential energy which is function of the relative position of these springs is controllable by the supervisory system using the techniques presented in Sect. 16.3. When the hand is free to move, the robotic system configuration tends to the configuration of minimum potential energy of the system allowed by the kinematic constraints.

If it is ensured that each hand motion also implies motion of the virtual object, we can ensure asymptotically stable behavior by the creation of a damping force on the virtual object which dissipates free energy[2].

In free space, we can therefore control the global position of the hand leaving the minimum energy-relative positions of the fingers' springs unchanged and changing the hand's virtual position, and we can change the configuration of the fingers with respect to each other by changing the equilibrium position of their springs.

In case we are grasping an object, the springs of the fingers will not longer have the minimum potential energy configuration anymore and their stored energy can be used to quantify physically a *grasp energy* representative of the hardness of the grasp. A maximum grasp energy can then be related to the material of the object to be grasped.

It is important to realize that the springs we will define in our controller, are spatial springs [7, 3, 4] and not trivial translational springs. This implies that we can cleverly choose the center of stiffnesses [7, 3, 4] of the springs in such a way that we can specify the way the grasp reacts to disturbing external forces. By means of proper choices, we can easily control rolling contacts around a desired nominal grasp.

16.4.1 The virtual object dynamics

Within the controller, the dynamics of the virtual object is simulated. As shown in [12], we can write the equation of the virtual body as follows:

[2] Energy as such cannot be dissipated because of the first principle of thermodynamics. When it is said that energy is dissipated, we actually mean dissipation of free energy, which can be defined for an isothermal system as the Legendre transformation of the energy with respect to the entropy [1].

$$\begin{pmatrix} \dot{h}_b^0 \\ \dot{m}^b \end{pmatrix} = \begin{pmatrix} 0 & \chi_{h_b^0} \circ Ad_{h_b^0} \\ -Ad_{h_b^0} \circ \chi^*_{h_b^0} & m^b \wedge \end{pmatrix} \begin{pmatrix} \frac{\partial H_B}{\partial h_b^0} \\ \frac{\partial H_B}{\partial m^b} \end{pmatrix} + \begin{pmatrix} 0 \\ Ad^*_{h_b^0} \end{pmatrix} w^0_{\text{tot}} \tag{16.3}$$

$$t_b^0 = (0 \; Ad_{h_b^0}) \begin{pmatrix} \frac{\partial H_B}{\partial h_b^0} \\ \frac{\partial H_B}{\partial m^b} \end{pmatrix}$$

where b indices the virtual body, w^0_{tot} the total wrench applied to the virtual body expressed in the inertial space, t_b^0 the twist of the virtual body with respect to the inertial space, m^b the momentum of body b in its own space, $H_B(h_b^0, m^b)$ the total energy of the virtual body and $m^b\wedge$ represents the Lie-Poisson bracket.

We can then rewrite the previous equations in a more compact form:

$$\begin{aligned} \dot{x}_B &= J_B(x_B)\frac{\partial H_B(x_B)}{\partial x_B} + \phi_B(x_B) w^0_{\text{tot}} \\ t_b^0 &= \phi^*_B(x_B)\frac{\partial H_B(x_B)}{\partial x_B} \end{aligned} \tag{16.4}$$

where $x_B := (h_b^0, m^b)$ and the other assignments should be obvious from Eq. (16.3).

16.4.2 The springs

We can now consider the equations of the $n+1$ springs represented in Fig. 16.2, where n is the number of fingers of the considered robotics hand and the $(n+1)$-th spring is the spring connecting the virtual object to the hand's virtual position (V in Fig. 16.2). This last spring does not have variable length and is a function of $h_b^{v(b)}$ where $v(b)$ indicates the space corresponding to the V of Fig. 16.2.

If we index with b the virtual body space and with $v(i)$ the additional supporting space needed to create the variable spring i, the effective twist of the i–th spring $t^i_{v(i)} = t^i_b + Ad_{h^i_b} t^b_{v(i)}$ (see Sect. 16.3) can be expressed as:

$$t^i_{v(i)} = (Id \; Ad_{h^i_b}) \begin{pmatrix} t^i_b \\ t^b_{v(i)} \end{pmatrix}. \tag{16.5}$$

This implies that the general Hamiltonian equations for the used springs with variable length are (see [12]):

$$\begin{aligned} \dot{h}^i_{v(i)} &= \begin{pmatrix} \chi_{h^i_{v(i)}} & \chi_{h^i_{v(i)}} \circ Ad_{h^i_b} \end{pmatrix} \begin{pmatrix} t^i_b \\ t^b_{v(i)} \end{pmatrix} \\ \begin{pmatrix} w^i_b \\ w^b_{v(i)} \end{pmatrix} &= \begin{pmatrix} \chi^*_{h^i_{v(i)}} \\ Ad^*_{h^i_b} \circ \chi^*_{h^i_{v(i)}} \end{pmatrix} \frac{\partial V^i_{v(i)}(h^i_{v(i)})}{\partial h^i_{v(i)}} \end{aligned} \tag{16.6}$$

where $\chi_{h_i^j}$ and $\chi^*_{h_i^j}$ are defined in Sect. 16.2. and the pair (t_b^i, w_b^i) corresponds to the power port where are attached the bodies connected to the two end points b and i of the spring, whereas the pair $(t^b_{v(i)}, w^b_{v(i)})$ represents the port which is used to change the effective minimal potential relative position (see Fig. 16.1). In the previous equation, w_b^i is the wrench that the body attached to the extreme b of the elastic element applies to the spring, expressed in the space $\mathcal{E}_i$ of the other body.

It is convenient to express the motion of each of the fingertips and of the virtual object in a common space. We use for it the inertial space, which is indexed with 0.

Since it can be seen that

$$t_j^i = Ad_{h_k^j}(t_i^k - t_j^k)$$

and $Ad_{h_k^i}$ is linear, we obtain the following identities:

$$t_b^i = \begin{pmatrix} Ad_{h_0^i} & -Ad_{h_0^i} \end{pmatrix} \begin{pmatrix} t_b^0 \\ t_i^0 \end{pmatrix} \Rightarrow \begin{pmatrix} w_b^0 \\ w_i^0 \end{pmatrix} = \begin{pmatrix} Ad^*_{h_0^i} \\ -Ad^*_{h_0^i} \end{pmatrix} w_b^i \tag{16.7}$$

where w_b^0 is the wrench that the virtual object applies to the i–th spring and w_i^0 is the wrench that tip i applies to the i–th spring, both expressed in space 0.

If furthermore we consider $h_b^{v(b)}$ the state of the $(n+1)$-th spring, it is then possible to collect all the springs equations and give a complete expression which results in:

$$\dot{x}_S = \phi_S(x_S) \begin{pmatrix} t_b^0 \\ t^0_{v(b)} \\ t^0_{\text{tips}} \\ t^b_{\text{var}} \end{pmatrix}$$

$$\begin{pmatrix} w_b^0 \\ w^0_{v(b)} \\ w^0_{\text{tips}} \\ w^b_{\text{var}} \end{pmatrix} = \phi_S^*(x_S) \frac{\partial H_S(x_S)}{\partial x_S} \tag{16.8}$$

where:

$$x_S = (h^1_{v(1)}, \ldots, h^n_{v(n)}, h_b^{v(b)}), \qquad H_S(x_S) = V_b^{v(b)}(h_b^{v(b)}) + \sum_{i=1}^{n} V^i_{v(i)}(h^i_{v(i)}),$$

$$\phi_S = \begin{pmatrix} \phi_b & \phi_{v(b)} & \phi_{\text{tips}} & \phi_{\text{var}} \end{pmatrix},$$

$t^0_{v(b)}$ is the twist of the virtual hand position represented by a V in Fig. 16.2 with respect to the inertial frame and

$$t^0_{\text{tips}} = \begin{pmatrix} t^0_1 \\ \vdots \\ t^0_n \end{pmatrix} \qquad t^b_{\text{var}} = \begin{pmatrix} t^b_{v(1)} \\ \vdots \\ t^b_{v(n)} \end{pmatrix}$$

$$\phi_b = \begin{pmatrix} \chi_{h^1_{v(1)}} \; o \; Ad_{h^1_0} \\ \vdots \\ \chi_{h^n_{v(n)}} \; o \; Ad_{h^n_0} \\ \chi_{h_b^{v(b)}} \; o \; Ad_{h_0^{v(b)}} \end{pmatrix} \quad \phi_{v(b)} = \begin{pmatrix} 0 \\ \vdots \\ 0 \\ -\chi_{h_b^{v(b)}} \; o \; Ad_{h_0^{v(b)}} \end{pmatrix}$$

$$\phi_{\text{tips}} = \begin{pmatrix} -\chi_{h^1_{v(1)}} \; o \; Ad_{h^1_0} & \dots & 0 \\ \vdots & \ddots & \vdots \\ 0 & \dots & -\chi_{h^n_{v(n)}} \; o \; Ad_{h^n_0} \\ 0 & \dots & 0 \end{pmatrix}$$

$$\phi_{\text{var}} = \begin{pmatrix} \chi_{h^1_{v(1)}} \; o \; Ad_{h^1_b} & \dots & 0 \\ \vdots & \ddots & \vdots \\ 0 & \dots & \chi_{h^n_{v(n)}} \; o \; Ad_{h^n_b} \\ 0 & \dots & 0 \end{pmatrix}.$$

In the previous notation, some dependencies have been omitted for the sake of notational clarity.

16.4.3 Interconnection object-springs

By combining Eq. (16.4) and Eq. (16.8), and realizing that $w^0_{\text{tot}} = -w^0_b$ we get:

$$\begin{pmatrix} \dot{x}_B \\ \dot{x}_S \end{pmatrix} = \begin{pmatrix} J_B & -\phi_B \phi_v^* \\ \phi_v \phi_B^* & 0 \end{pmatrix} \begin{pmatrix} \frac{\partial H_C}{\partial x_B} \\ \frac{\partial H_C}{\partial x_S} \end{pmatrix} + \begin{pmatrix} 0 & 0 & 0 \\ \phi_{\text{tips}} & \phi_{v(b)} & \phi_{\text{var}} \end{pmatrix} \begin{pmatrix} t^0_{\text{tips}} \\ t^0_{v(b)} \\ t^b_{\text{var}} \end{pmatrix} \tag{16.9}$$

$$\begin{pmatrix} w^0_{\text{tips}} \\ w^0_{v(b)} \\ w^b_{\text{var}} \end{pmatrix} = \begin{pmatrix} 0 & \phi^*_{\text{tips}} \\ 0 & \phi^*_{v(b)} \\ 0 & \phi^*_{\text{var}} \end{pmatrix} \begin{pmatrix} \frac{\partial H_C}{\partial x_B} \\ \frac{\partial H_C}{\partial x_S} \end{pmatrix}.$$

Where $H_C(x_B, x_S) = H_B(x_B) + H_S(x_S)$. The previous system is a generalized port-controlled Hamiltonian system and it is therefore lossless. In order to achieve an asymptotically stable behavior, we must add damping. Consider the inputs of Eq. (16.9) equal to zero; then we get a holonomic Hamiltonian system with six degrees of freedom corresponding to the configuration of the virtual object in space. This implies that we could find canonical coordinates with six generalized positions, six generalized dual momenta and the other coordinates would be Casimir functions for the Hamiltonian system.

16.4.4 Creating damping

We can inject damping in the physical equivalent system presented in Fig. 16.2: by means of control we can create a viscous friction force applied to the virtual object. Intuitively, we could think of the virtual object as moving in a fluid with high viscosity which would extract energy from its motion irreversibly.

We can add such a viscous effect by considering an antisymmetric two contravariant tensor R_t. We can therefore subtract a symmetric, semi-positive definite tensor R_B from the antisymmetric tensor J_B of Eq. (16.9). R_B has the following form:

$$R_B = \begin{pmatrix} R_h & 0 \\ 0 & R_m \end{pmatrix} \tag{16.10}$$

where the off-diagonal terms have been chosen equal to zero because not physically interpretable. The term R_m is the element representing the usual viscous force since its effect on the virtual object dynamics would be to apply an additional wrench to it, equal to:

$$w_d^b = R_m \frac{\partial H_C}{\partial m^v} = R_m t_b^{b,0} \tag{16.11}$$

and since R_m is positive definite, it represents a dissipative wrench. Implementing such a term, the derivative of the energy of the controller is:

$$\begin{aligned}\dot{H} = -\langle \partial H_C/\partial h_b^0, R_h\, \partial H_C/\partial h_b^0\rangle - \langle t_b^{b,0}, R_m\, t_b^{b,0}\rangle + \\ \langle w_{\text{tips}}^0, t_{\text{tips}}^0\rangle + \langle w_{v(b)}^0, t_{v(b)}^0\rangle + \langle w_{\text{var}}^b, t_{\text{var}}^b\rangle\end{aligned} \tag{16.12}$$

where $\langle , \rangle$ is the natural dual product of a co-vector on a vector. Due to the positive semidefinite hypothesis of R_B, the elements of the first line of Eq. (16.12) will never increase the energy of the controller and if either $t_b^{0,b}$ or $\partial H_C/\partial h_b^0$ are different than zero, these elements will decrease the energy irreversibly if we suppose R_h, R_m to be positive definite.

For practical reasons, we consider only $R_m \neq 0$ and $R_h = 0$, as will be seen in Sect. 16.6.4.

16.5 The controlled hand

Consider now the robotic hand to be controlled as a fully actuated holonomic robotic mechanism with configuration manifold $\mathcal{Q}$. In this specific case, we mean with a fully actuated robotics system a system for which any generalized force $\tau \in T_q^*\mathcal{Q}$ can be applied at any configuration $q \in \mathcal{Q}$.

The *tips Jacobian* for this system is a linear, configuration-dependent mapping of the following form:

$$J_{\text{tips}}(q) : T_q\mathcal{Q} \to \underbrace{se_0(3) \times \ldots \times se_0(3)}_{n-\text{times}} \; ; \; \dot{q} \mapsto J_{\text{tips}}(q)\dot{q} \tag{16.13}$$

which maps a configuration velocity $\dot{q}$ to the twists of the tips of the hand:

$$t^0_{\text{tips}} = J_{\text{tips}}(q)\dot{q} \tag{16.14}$$

We can consider a second Jacobian called the 'interaction Jacobian'. It maps motions of the robot to motions of the links with which the environment can interact. In the general case, for whole-hand manipulation, these could be all the links of the hand. We indicate the interaction Jacobian with $J_I(q)$.

For notational convenience, we define $\hat{\phi}_I(q) := J_I^*(q)$ and $\hat{\phi}_{\text{tips}} := J^*_{\text{tips}}(q)$, which are respectively the adjoints of the interaction Jacobian and the tips Jacobian.

Due to the hypothesis of holonomicity and full actuation of the robotic system, we can write its dynamic equation in the following form:

$$\begin{aligned}
\begin{pmatrix} \dot{q} \\ \dot{p} \end{pmatrix} &= \begin{pmatrix} 0 & I \\ -I & 0 \end{pmatrix} \begin{pmatrix} \frac{\partial H_R(q,p)}{\partial q} \\ \frac{\partial H_R(q,p)}{\partial p} \end{pmatrix} + \begin{pmatrix} 0 & 0 \\ \hat{\phi}_I(q) & I \end{pmatrix} \begin{pmatrix} W \\ \tau \end{pmatrix} \\
\begin{pmatrix} T \\ \dot{q} \end{pmatrix} &= \begin{pmatrix} 0 & \hat{\phi}_I^*(q) \\ 0 & I \end{pmatrix} \begin{pmatrix} \frac{\partial H_R(q,p)}{\partial q} \\ \frac{\partial H_R(q,p)}{\partial p} \end{pmatrix}
\end{aligned} \tag{16.15}$$

where $H_R(q,p) = E_k(q,p) + E_p(q)$ is the total energy of the robotic hand.

The dual pair (T, W) corresponds to the power port through which the robot can exchange energy with the environment. The pair $((q,\dot{q}),(q,\tau))$ corresponds to the energy port of the actuators. It is a dual pair, but it is configuration dependent and not yet suitable to be interconnected with the controller.

In order to solve the problem and to create the desired interconnection as shown in Fig. 16.2, we can consider the dual relation of Eq. (16.14), represented by the adjoint of $J_{\text{tips}}(q)$:

$$\tau = J^*_{\text{tips}}(q)\begin{pmatrix} -w_1^0 \\ \vdots \\ -w_n^0 \end{pmatrix} = -J^*_{\text{tips}}(q)w^0_{\text{tips}} = -\hat{\phi}_{\text{tips}}(q)w^0_{\text{tips}} \tag{16.16}$$

where $-w_i^0$ is the wrench applied by the finger springs i to tip i, which is equal and opposite to the wrench w_i^0 that tip i applies to the spring i once it is connected to it.

After few calculations, it is possible to get the final equations of the interconnected system:

$$\begin{aligned} \dot{x}_T &= (J_T(x_T) - R_T(x_T))\frac{\partial H_T(x_T)}{\partial x_T} + \phi_T(x_T)\begin{pmatrix} W \\ t^0_{v(b)} \\ t^b_{\text{var}} \end{pmatrix} \\ \begin{pmatrix} T \\ w^0_{v(b)} \\ w^b_{\text{var}} \end{pmatrix} &= \phi_T^*(x_T)\frac{\partial H_T(x_T)}{\partial x_T} \end{aligned} \tag{16.17}$$

where $x_T := (q, p, x_B, x_S)$, $H_T(x_T) = H_R(q,p) + H_C(x_B, x_S)$,

$$J_T(x_T) := \begin{pmatrix} 0 & I & 0 & 0 \\ -I & 0 & 0 & -\hat{\phi}_{\text{tips}}\phi^*_{\text{tips}} \\ 0 & 0 & J_B & -\phi_B\phi_v^* \\ 0 & \phi_{\text{tips}}\hat{\phi}^*_{\text{tips}} & \phi_v\phi_B^* & 0 \end{pmatrix}, \quad R_T(x_T) := \begin{pmatrix} 0 & 0 & 0 & 0 \\ 0 & 0 & 0 & 0 \\ 0 & 0 & R_B & 0 \\ 0 & 0 & 0 & 0 \end{pmatrix},$$

$$\phi_T(x_T) := \begin{pmatrix} 0 & 0 & 0 \\ \phi_I(q) & 0 & 0 \\ 0 & 0 & 0 \\ 0 & \phi_{v(b)} & \phi_{\text{var}} \end{pmatrix}.$$

In Fig. 16.3 a bond graph representation of the power-continuous interconnection is shown. In the figure, I.P.C. stands for Intrinsically Passive Controller.

The supervisor can change the virtual position of the hand by supplying a $t^0_{v(b)}$ different than zero and it can change the position of minimal potential energy of the springs by suppling a t^b_{var} different than zero. At the same time it can monitor the energy supplied to the controlled robot. When the twists of the supervisor are zero, only the environment can supply energy to the controlled system.

The dissipation implemented in the I.P.C. and the friction present in the real robot ensures an asymptotically stable behavior.

Remark 16.5.1. Note that the designed controller only uses kinematic information of the robot to be controlled, namely the Jacobian of Eq. (16.14) which

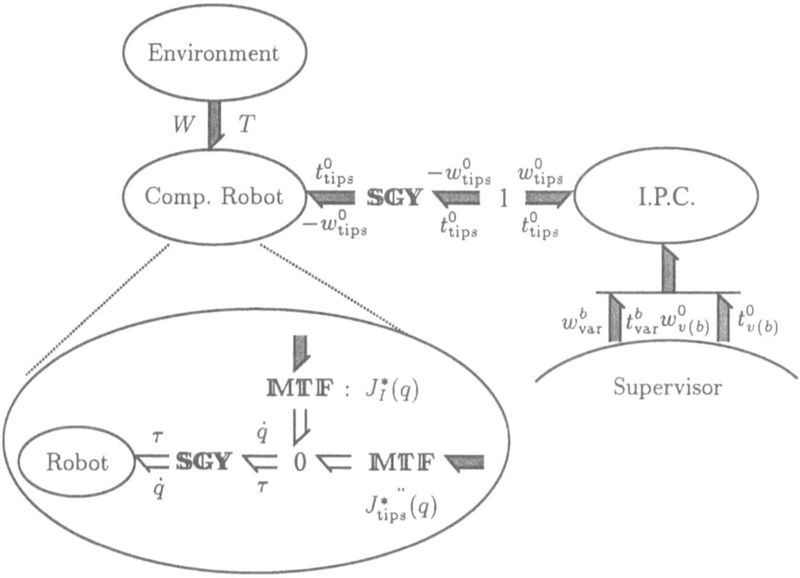

Fig. 16.3. The interconnection between the robot and the intrinsically passive controller.

in Fig. 16.3 has actually been considered as part of the Compensated Robot. This Jacobian is used as a power-continuous transformation and therefore kinematic model mismatching will never create instability.

16.6 Implementation of the control scheme

In the previous section, we analyzed the interconnection from a conceptual point of view. In this section we will choose references and study the control system from an implementational point of view in order to build a procedure which can be implemented directly. In order to do so, we need to choose some references for the various bodies of the system:

- **Tips' frames**
  With reference to Fig. 16.4, assign to each tip i a Cartesian coordinate system Ψ_i, which can be visualized as a frame rigidly connected to the tip. It is useful to introduce a second coordinate system for the i-th tip whose origin is called *center of stiffness*[3]; this second coordinate system is indicated with $\Psi_{c(i)}$.

[3] In a work of [2] this frame is called center of compliance. As shown by Lončarić [7], the center of stiffness and the center of compliance coincide only in certain cases.

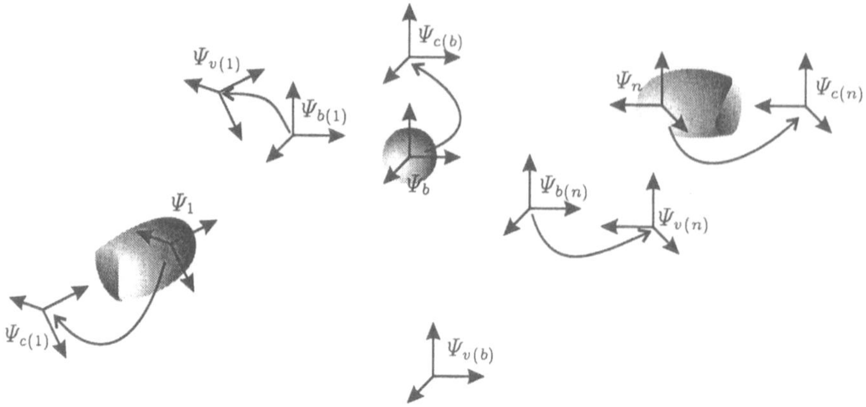

Fig. 16.4. The basic frames which are necessary for the describtion of the algorithm

- **The Virtual Object frame**
  With reference to Fig. 16.4, consider the virtual object to be of spheric shape and uniform density. Consider for it a coordinate frame Ψ_b. Consider its inertial properties described in this frame as a mass m and inertia tensor jI, where I indicates an identity matrix. It is useful to introduce a second coordinate system for the virtual object. This is indicated with $\Psi_{c(b)}$.
- **The Support Spaces frames**
  As shown in Sect. 16.3, to describe variable springs, we need supporting spaces for each spring i which we indicated with $v(i)$. We will have as many of these spaces as the tips we are considering. For these spaces, we consider two possible bases, namely those which in Fig. 16.4 are indicated with $\Psi_{v(i)}$ and $\Psi_{b(i)}$.
- **The Hand Virtual frame**
  In order to control the interaction between the grasped object and the environment, an extra space, which we indicated with $v(b)$, is needed. This space is connected with the V of Fig. 16.2. We choose one coordinate frame for this space, which we indicate with $\Psi_{v(b)}$.

16.6.1 Suitable springs energy functions

We will now give suitable energy functions which can be used to implement the equations of the 3D springs. These energy functions have been introduced in the excellent work of [2] and are here reported in a different way and with many more additional details. As already said, a potential energy function is a function of the relative position of two bodies. These two bodies are for the n tips springs the tips i and the corresponding supporting spaces $v(i)$ and for the hand spring, the body space b and the space $v(b)$.

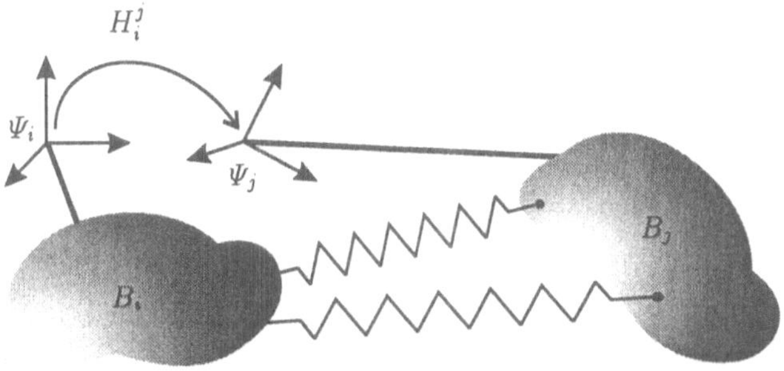

Fig. 16.5. Spatial spring between two bodies: the thick lines indicate that the two frames are rigidly connected to the respective bodies

To express these functions analytically, we have to choose coordinates for the various spaces. To describe the energies of the fingers springs, we choose for the tip the coordinates $\Psi_{c(i)}$ and for the supporting space $v(i)$ the Cartesian coordinates $\Psi_{v(i)}$. The relative position $h_i^{v(i)}$ in these coordinates will therefore be:

$$H_{c(i)}^{v(i)} := \psi_{c(i)} \; o \; h_{v(i)}^{i} \; o \; \psi_{v(i)}^{-1}$$

The energy function we describe is such that it has a minimum in $H_{c(i)}^{v(i)} = I$, which corresponds to a relative position of $\mathcal{E}_{v(i)}$ and $\mathcal{E}_i$ such that the two frames $\Psi_{c(i)}$ and $\Psi_{v(i)}$ coincide.

The common origin of these two frames at equilibrium is called the *center of stiffness* because we choose the stiffness expressed in this frame as having a special form.

For generality of exposition, with reference to Fig. 16.5, we index as i and j the two bodies between which we want to consider a spring, and two coordinates frames Ψ_i and Ψ_j attached respectively to the body i and j such that in a minimum potential energy corresponding to an equilibrium position we can conclude that the relative position H_i^j is equal to I. We can then consider a mapping of the following form as the generalized elastic force:

$$dV : SE(3) \to T^*SE(3); H_i^j \mapsto (H_i^j, F_i^j)$$

The stiffness is the linearization of the previous map at the identity ($H_i^j = I$). In this point, by definition $\Psi_i = \Psi_j$.

A differential of the previous map is a linear map of the following form:

$$K : se(3) \to se^*(3) \; ; \; \delta\bar{T} \mapsto K\delta\bar{T} \tag{16.18}$$

where K is a matrix that can be represented in the coordinates in which $\delta\bar{T} \in se(3)$ is expressed. In our case this is expressed using $\Psi_i = \Psi_j$.

The element $\delta\bar{T}$ of Eq. (16.18) is an infinitesimal twist represented in vector form. To be more specific, consider the representation H_i^j of a relative position. At a certain instant, the corresponding numerical representation of the twist t_i^j will be:

$$T_i^j = \begin{bmatrix} \Omega_i^j & v_i^j \\ 0 & 0 \end{bmatrix} = \dot{H}_i^j H_j^i = \begin{bmatrix} \dot{R}_i^j & \dot{p}_i^j \\ 0 & 0 \end{bmatrix} \begin{bmatrix} R_j^i & p_j^i \\ 0 & 1 \end{bmatrix} \tag{16.19}$$

where Ω_i^j is an antisymmetric matrix. We can associate to Ω_i^j a unique vector $\omega_i^j \in \mathbb{R}^3$ such that for all $x \in \mathbb{R}^3$ we have $\Omega_i^j x = \omega_i^j \wedge x$ where $\wedge$ is the usual vector product of two vectors. In general we indicate such an operation with the operator tilde so that in the case just explained we would indicate:

$$\Omega_i^j = \tilde{\omega}_i^j = \begin{bmatrix} 0 & -\omega_3 & \omega_2 \\ \omega_3 & 0 & -\omega_1 \\ -\omega_2 & \omega_1 & 0 \end{bmatrix} \text{ with } \omega_i^j = \begin{bmatrix} \omega_1 \\ \omega_2 \\ \omega_3 \end{bmatrix}. \tag{16.20}$$

This implies that we can consider a vector representation of the matrix T_i^j, which we indicate with $\bar{T}_i^j$ and define as:

$$\bar{T}_i^j = \begin{bmatrix} \omega_i^j \\ v_i^j \end{bmatrix} \Leftrightarrow T_i^j = \begin{bmatrix} \tilde{\omega}_i^j & v_i^j \\ 0 & 0 \end{bmatrix}.$$

An analogous expression can be given for wrenches, for which we have:

$$\bar{W}_i^j = \begin{bmatrix} m_i^j \\ f_i^j \end{bmatrix} \Leftrightarrow W_i^j = \begin{bmatrix} \tilde{f}_i^j & m_i^j \\ 0 & 0 \end{bmatrix}$$

with m_i^j corresponding to the angular torque. The wrench $w_j^i \in se_i^*(n)$ and the corresponding numeric form W_j^i indicates the wrench that a spring connecting body B_i and body B_j applies to body B_i. We can then partition Eq. (16.18) in order to show the rotational and translational components in the following way:

$$\begin{bmatrix} m_i^j \\ f_i^j \end{bmatrix} = \begin{bmatrix} K_o & K_c \\ K_c^T & K_t \end{bmatrix} \begin{bmatrix} \delta\theta_i^j \\ \delta p_i^j \end{bmatrix} \tag{16.21}$$

where we indicate with $\delta\bar{T} = \left[(\delta\theta_i^j)^T \; (\delta p_i^j)^T\right]^T$. The matrix K as defined here is always symmetric and therefore K_o and K_t, which are respectively called *rotational stiffness* and *translational stiffness* are also symmetric. In [7] it has been shown that $K_c = K_c^T$ corresponds to a maximum decoupling between rotation and translation. In this case, the point corresponding to the coinciding origins of the coordinate systems Ψ_i and Ψ_j at equilibrium is called *center of stiffness*. In our search for proper energy functions, we consider a symmetric K_c, which implies that the origin of our reference frames are implicitly chosen at the center of stiffness at equilibrium. By means of the matrix identity:

$$\tilde{v} = A\tilde{w} + \tilde{w}A^T \Leftrightarrow v = (\mathrm{tr}(A)I - A^T)w \tag{16.22}$$

and the hypothesis that K_o, K_t and K_c are all symmetric, we can express Eq. (16.21) in an equivalent form, namely:

$$\begin{aligned} \tilde{m}_i^j &= 2\,\mathrm{as}(G_o\delta\tilde{\theta}_i^j) + 2\,\mathrm{as}(G_c\delta\tilde{p}_i^j) \\ \tilde{f}_i^j &= 2\,\mathrm{as}(G_c\delta\tilde{\theta}_i^j) + 2\,\mathrm{as}(G_t\delta\tilde{p}_i^j) \end{aligned} \tag{16.23}$$

where $K_x = (\mathrm{tr}(G_x)I - G_x)$ for $x = t, o, c$ and the G_x are called *co-stiffnesses*. The operator as() returns the antisymmetric part of the matrix given as an argument. Since K_x is symmetric, there exist a conformal transformation corresponding to a rotation of the coordinate system, such that we have:

$$K_x = R_x \Gamma_x R_x^T \quad \text{where} \quad R_x^{-1} = R_x^T \tag{16.24}$$

and Γ_x is a diagonal matrix of principal stiffnesses in the directions corresponding to the columns of the orthonormal matrix R_x which are expressed in the coordinates $\Psi_i = \Psi_j$ at equilibrium. It is easy to see that to any diagonal matrix Γ_x, we can associate a unique diagonal matrix Λ_x for which:

$$\Gamma_x = \mathrm{tr}(\Lambda_x)I - \Lambda_x \quad \text{and} \quad \Lambda_x = \frac{1}{2}\mathrm{tr}(\Gamma_x)I - \Gamma_x. \tag{16.25}$$

The Λ_x corresponding to the Γ_x of Eq. (16.24) are called *principal co-stiffnesses.* It is then possible to see that for each K_x, we have a corresponding G_x, given by:

$$G_x = \frac{1}{2}\mathrm{tr}(K_x)I - K_x \quad \text{and} \quad K_x = \mathrm{tr}(G_x)I - G_x. \tag{16.26}$$

Remark 16.6.1. It is therefore possible, from a practical point of view, to choose a center of stiffness and K_o, K_t, K_c by means of choosing principal directions and corresponding stiffness values. From these K_x we can then compute the corresponding co-stiffnesses G_x. We should then find an energy function V parameterized by the G_x such that the linearization of dV around the origin would result in the relations of Eq. (16.23).

We can decompose the total energy in three energies: translational, rotational and couple energies:

$$V(R_i^j, p_i^j) = V_t(R_i^j, p_i^j) + V_o(R_i^j) + V_c(R_i^j, p_i^j). \tag{16.27}$$

For the previous energies, $V_t(\cdot)$ should just depend on K_t and G_t, $V_o(\cdot)$ only on K_o and G_o and $V_c(\cdot)$ only on K_c and G_c.

Orientational energy. Let us start with the orientational energy. If we take the differential of the to be chosen rotational energy, we get:

$$dV_o(R_i^j) = V_o(R_i^j + dR_i^j) - V_o(R_i^j). \tag{16.28}$$

Since we have for any two vectors $v, w \in \mathbb{R}^3$ that:

$$v^T w = \frac{1}{2}\,\mathrm{tr}(\tilde{v}\tilde{w}),$$

because this energy function is only a function of a rotation, we would like to be able to express this differential as

$$dV_o(R_i^j) = m_o^T \delta\theta_j^i = \frac{1}{2}\,\mathrm{tr}(\tilde{m}_o \tilde{\delta\theta}_j^i) \tag{16.29}$$

where m_o would correspond to the rotational part of the wrench $\bar{W}_j^i$ that body j would apply to the spring expressed in frame Ψ_i. Nevertheless, due to the nodicity of a spring, $\bar{W}_j^i = -\bar{W}_i^{i,j}$, and therefore m_o is also equal to the torque that the spring applies to body i expressed in Ψ_i.

This implies that $\tilde{m}_o$ can be anything expressed by:

$$\tilde{m}_o(R_i^j) = 2\,\mathrm{as}(x(R_i^j))$$

where x is any matrix dependent on R_i^j which should be found. Furthermore, we want the differential of this torque to be such that in the neighborhood of the identity it satisfies the chosen local behavior expressed by Eq. (16.23). This implies that:

$$\tilde{m}_o(I + \tilde{\delta\theta}_j^i) = 2\,\mathrm{as}(G_o \tilde{\delta\theta}_j^i).$$

A straightforward function realizing this latter equation is

$$\tilde{m}_o(R_i^j) = -2\,\mathrm{as}(G_o R_i^j) \tag{16.30}$$

where we need a minus sign so that we create not a maximum but a minimum of the potential energy at the identity. This implies with Eq. (16.29) that:

$$dV_o(R_i^j) = -\,\mathrm{tr}(\mathrm{as}(G_o R_i^j)\tilde{\delta\theta}_j^i) = -\,\mathrm{tr}(G_o R_i^j \tilde{\delta\theta}_j^i) \tag{16.31}$$

where the last equality results from the following matrix identity[4]:

$$\mathrm{tr}(AB) = \mathrm{tr}(\mathrm{sy}(A)\,\mathrm{sy}(B)) + \mathrm{tr}(\mathrm{as}(A)\,\mathrm{as}(B)). \tag{16.32}$$

Eventually, since:

$$R_i^j \tilde{\delta\theta}_j^i = dR_i^j = (R_i^j + dR_i^j) - R_i^j$$

we can infer that a $V_o(R_i^j)$ satisfying Eq. (16.28) and Eq. (16.31) is:

$$V_o(R_i^j) = -\,\mathrm{tr}(G_o R_i^j). \tag{16.33}$$

[4] The operator sy(·) indicates the symmetric part of a matrix and as(·) the anti-symmetric one.

Remark 16.6.2. Note that the previous energy function exactly corresponds to the one presented in [2], but here we deduced it from the local behavior around the identity which can be specified by a proper choice of K_o. Note also that the element (k,l) of the matrix R_i^j can be interpreted as the scalar product of the k-th axis of Ψ_j with the l-th axis of Ψ_i. This means that when Ψ_i and Ψ_j coincide, R_i^j is the identity which corresponds to collineation of the coordinates frames. The element (k,l) of G_o can therefore be seen as a weight for the collineation of the axis k of Ψ_i and the axis l of Ψ_j. This has been the starting point for getting this energy function in [5]. With this interpretation, the necessity of the minus sign of Eq. (16.30) should become clear: the scalar product of two vectors is maximum when the vectors are colinear; to let it become a minimum we need to invert its sign.

We have therefore proven the following:

Theorem 16.6.1 (Orientational Elastic Wrenches). *A spring with elastic energy that is given by Eq. (16.33) and connecting body i and j would apply in a relative position* (p_i^j, R_i^j) *a wrench* $\tilde{W}^i = \left[m_o^T \; f_o^T\right]^T$ *to body i and expressed in frame* Ψ_i *such that:*

$$\tilde{m}_o = -2\,\mathrm{as}(G_o R_i^j) \tag{16.34}$$

$$\tilde{f}_o = 0 \tag{16.35}$$

Translational energy. Strangely enough, the translational energy is more involved than the orientational one. This is due to the fact that it is not physically meaningful to describe a purely translational anisotropic spring. This means that either we define a spring which generates a force proportional to the distance of the origins of Ψ_i and Ψ_j using a scalar constant which is independent from the direction, or any directional dependence also implies the generation of a torque between the two bodies.

Theorem 16.6.2 (Anisotropic translational springs). *Purely translational anisotropic springs do not exist: either a spring is isotropic*[5] *or it also generates a torque.*

Proof. First consider the following relation between wrenches which is the consequence of the nodicity of a spring:

$$w_j^i = -Ad^*_{h_i^j} w_i^j. \tag{16.36}$$

where w_j^i is the wrench that a spring connecting body B_i to B_j applies to body i expressed in $\mathcal{E}_i$. Where the wrench $w_j^i \in se_i^*(n)$ indicates the wrench that a spring connecting body B_i and body B_j applies to body B_i.

[5] Isotropic comes from the Greek *isos* (equal) and *tropos* (rotation) which means direction independent.

After the choice of coordinates Ψ_i and Ψ_j we can express this latter relation as:

$$\bar{W}_j^i = -Ad_{H_i^j}^T \bar{W}_i^j \tag{16.37}$$

where we have

$$Ad_{H_i^j} = \begin{bmatrix} R_i^j & 0 \\ \tilde{p}_i^j R_i^j & R_i^j \end{bmatrix} \Rightarrow Ad_{H_i^j}^T = \begin{bmatrix} R_j^i & \tilde{p}_j^i R_j^i \\ 0 & R_j^i \end{bmatrix} \tag{16.38}$$

and $\bar{W}_i^j = ((m_i^j)^T \ (f_i^j)^T)^T \in Re^6$. In order to obtain a purely translational force, the following should go:

$$\begin{bmatrix} 0 \\ f_j^i \end{bmatrix} = - \begin{bmatrix} R_j^i & \tilde{p}_j^i R_j^i \\ 0 & R_j^i \end{bmatrix} \begin{bmatrix} 0 \\ f_i^j \end{bmatrix}$$

which can be true for any force f_i^j if and only if:

$$\tilde{p}_j^i \, f_i^{i,j} = 0 \Leftrightarrow \tilde{p}_j^i f_j^i = 0 \Leftrightarrow p_j^i \wedge f_j^i = 0 \Leftrightarrow \exists k \in \mathbb{R} \, ; \, f_j^i = k \, p_j^i$$

for a scalar k where $f_i^{i,j}$ indicates the linear force applied by the spring to body B_j, but expressed in the frame Psi_i. The scalar k corresponds to the constant describing the stiffness of the isotropical translational spring.

Th. 16.6.2 implies that in the formulation of translational energy we should also consider the relative orientation of i and j so that it is defined correctly for the general anisotropic case. The usual potential energy we would use for a translational spring, and that would result in the proper contribution for Eq. (16.21), would be a quadratic form of the position difference of the origins of Ψ_i and Ψ_j, namely:

$$V_t(p_i^j) = \frac{1}{2}(p_i^j)^T K_t(p_i^j) \tag{16.39}$$

where K_t can be interpreted as being described with the coordinates of Ψ_j at all times. As it is also shown in [3], the choice of translational potential energy of Eq. (16.39) creates the problems described above since there is no dependence from a relative rotation. The problems are due to an asymmetry in the energy function which describes the stiffness only in reference Ψ_j. To solve these difficulties, we can "symmetrize" this energy, as [3], and choose:

$$V_t(p_i^j, p_j^i) = \frac{1}{4}(p_i^j)^T K_t(p_i^j) + \frac{1}{4}(p_j^i)^T K_t(p_j^i) \tag{16.40}$$

where the equality of the K_t in the two frames creates the desired symmetry. Note that at equilibrium, $\Psi_i = \Psi_j$ and we have an energy equivalent to Eq. (16.39). The energy of Eq. (16.40) does depend on the relative orientation since it is equal to:

$$V_t(p_i^j, R_i^j) = \frac{1}{4}(p_i^j)^T K_t(p_i^j) + \frac{1}{4}((R_i^j)^T p_i^j)^T K_t(R_i^j)^T p_i^j \tag{16.41}$$

where the identity $p_j^i = -R_j^i p_i^j$ can be shown when considering that $H_j^i = (H_i^j)^{-1}$. It can be proven that $p^T K_t p = -\operatorname{tr}(\tilde{p} G_t \tilde{p})$ for any $p \in \mathbb{R}^3$ and therefore Eq. (16.41) is equal to:

$$V_t(p_i^j, R_i^j) = -\frac{1}{4}\operatorname{tr}(\tilde{p}_i^j G_t \tilde{p}_i^j) - \frac{1}{4}\operatorname{tr}(\tilde{p}_i^j R_i^j G_t R_j^i \tilde{p}_i^j) \tag{16.42}$$

where we used the matrix identity $\operatorname{tr}(R^T A R) = \operatorname{tr}(A)$. The following result gives an expression for the wrenches generated by an elastic energy with such an energy function.

Theorem 16.6.3 (Translational Elastic Wrenches). *A spring with elastic energy given by Eq. (16.42) and connecting body i and j would apply in a relative position (p_i^j, R_i^j) a wrench $\bar{W}^i = \begin{bmatrix} m_t^T & f_t^T \end{bmatrix}^T$ to body i, and such that this wrench expressed in frame Ψ_i is:*

$$\tilde{m}_t = -\operatorname{as}(G_t R_j^i \tilde{p}_i^j \tilde{p}_i^j R_i^j) \tag{16.43}$$

$$\tilde{f}_t = -R_j^i \operatorname{as}(G_t \tilde{p}_i^j) R_i^j - \operatorname{as}(G_t R_j^i \tilde{p}_i^j R_i^j) \tag{16.44}$$

Proof. First of all, we should consider that in a certain configuration (p_i^j, R_i^j), the differential of the energy function should be such that:

$$dV_t(p_i^j, R_i^j) = (m_j^i)^T \delta\theta_j^i + (f_j^i)^T \delta p_j^i = \frac{1}{2}\operatorname{tr}(\tilde{m}_j^i \tilde{\delta\theta}_j^i) + \frac{1}{2}\operatorname{tr}(\tilde{f}_j^i \tilde{\delta p}_j^i) \tag{16.45}$$

where $\delta T_j^i = \begin{bmatrix} \delta\theta_j^{i\,T} & \delta p_j^{i\,T} \end{bmatrix}^T$ is an infinitesimal twist of body j with respect to i and expressed in Ψ_i. Furthermore, in [12] can be shown that we have $m_t := -m_i^{i,j} = m_j^i$ and $f_t := -f_i^{i,j} = f_j^i$. This implies that if we calculate the differential of the energy function and we shape it to a form like Eq. (16.45), we immediately obtain m_t and f_t. We will therefore calculate this differential. We get:

$$dV_t(p_i^j, R_i^j) = V_t(p_i^j + dp_i^j, R_i^j + dR_i^j) - V_t(p_i^j, R_i^j) \tag{16.46}$$

where

$$dH_i^j = \begin{bmatrix} dR_i^j & dp_i^j \\ 0 & 0 \end{bmatrix} = H_i^j \delta T_j^i = \begin{bmatrix} R_i^j & p_i^j \\ 0 & 1 \end{bmatrix} \begin{bmatrix} \tilde{\delta\theta}_j^i & \delta p_j^i \\ 0 & 0 \end{bmatrix} = \begin{bmatrix} R_i^j \tilde{\delta\theta}_j^i & R_i^j \delta p_j^i \\ 0 & 0 \end{bmatrix}. \tag{16.47}$$

We can split the energy in the two parts reported in Eq. (16.42) such that:

$$dV_t(p_i^j, R_i^j) = dV_t^{(1)}(p_i^j, R_i^j) + dV_t^{(2)}(p_i^j, R_i^j) \tag{16.48}$$

where

$$\begin{aligned}
dV_t^{(1)}(p_i^j, R_i^j) &= V_t^{(1)}(p_i^j + R_i^j \delta p_j^i, R_i^j + R_i^j \tilde{\delta\theta}_j^i) - V_t^{(1)}(p_i^j, R_i^j) = \\
&- \frac{1}{4}\operatorname{tr}((p_i^j + R_i^j \delta p_j^i)^{\tilde{}}\, G_t (p_i^j + R_i^j \delta p_j^i)^{\tilde{}}\,) - \frac{1}{4}\operatorname{tr}(\tilde{p}_i^j G_t \tilde{p}_i^j) = \\
&- \frac{1}{4}\operatorname{tr}(\tilde{\delta p}_j^i R_j^i G_t R_i^j \tilde{\delta p}_j^i) - \frac{1}{4}\operatorname{tr}(\tilde{p}_i^j G_t R_i^j \tilde{\delta p}_j^i R_j^i) - \frac{1}{4}\operatorname{tr}(R_i^j \tilde{\delta p}_j^i R_j^i G_t \tilde{p}_i^j).
\end{aligned} \tag{16.49}$$

Above we again used the identity $(Rp)^{\tilde{}} = R\tilde{p}R^T$. The first of the previous three terms is a second-order term one which we can therefore discard. By applying the identity $\operatorname{tr}(AB) = \operatorname{tr}(BA)$ we get:

$$\begin{aligned}
dV_t^{(1)}(p_i^j, R_i^j) &= -\frac{1}{2}\operatorname{tr}\left(\frac{R_j^i \tilde{p}_i^j G_t R_i^j - (R_j^i \tilde{p}_i^j G_t R_i^j)^T}{2}\,\tilde{\delta p}_j^i\right) = \\
&\qquad \frac{1}{2}\operatorname{tr}(-\operatorname{as}(R_j^i \tilde{p}_i^j G_t R_i^j)\tilde{\delta p}_j^i)
\end{aligned} \tag{16.50}$$

and since

$$-\operatorname{as}(R_j^i \tilde{p}_i^j G_t R_i^j) = -R_j^i \operatorname{as}(\tilde{p}_i^j G_t) R_i^j = -R_j^i \operatorname{as}(G_t \tilde{p}_i^j) R_i^j$$

this clearly gives the first term of Eq. (16.44). We can now analyse the differential of the second term $V_t^{(2)}(\cdot)$ of Eq. (16.42). We get:

$$\begin{aligned}
dV_t^{(2)}(p_i^j, R_i^j) &= -\frac{1}{4}\operatorname{tr}((p_i^j + R_i^j \delta p_j^i)^{\tilde{}}\,(R_i^j + R_i^j \tilde{\delta\theta}_j^i) G_t (R_j^i - \tilde{\delta\theta}_j^i R_j^i)(p_i^j + R_i^j \delta p_j^i)^{\tilde{}}\,) \\
&\qquad + \frac{1}{4}\operatorname{tr}(\tilde{p}_i^j R_i^j G_t R_j^i \tilde{p}_i^j).
\end{aligned} \tag{16.51}$$

Discarding second-order terms from the previous equation, we obtain:

$$\begin{aligned}
&dV_t^{(2)}(p_i^j, R_i^j) = \\
&- \frac{1}{4}\operatorname{tr}(\tilde{p}_i^j R_i^j G_t R_j^i R_i^j \tilde{\delta p}_j^i R_j^i - \tilde{p}_i^j R_i^j G_t \tilde{\delta\theta}_j^i R_j^i \tilde{p}_i^j + \\
&\tilde{p}_i^j R_i^j \tilde{\delta\theta}_j^i G_t R_j^i \tilde{p}_i^j + R_i^j \tilde{\delta p}_j^i R_j^i R_i^j G_t R_j^i \tilde{p}_i^j) = \\
&-\frac{1}{4}\operatorname{tr}(R_j^i \tilde{p}_i^j R_i^j G_t \tilde{\delta p}_j^i + \tilde{\delta p}_j^i G_t R_j^i \tilde{p}_i^j R_i^j) - \frac{1}{4}\operatorname{tr}(\tilde{p}_i^j R_i^j (\tilde{\delta\theta}_j^i G_t - G_t \tilde{\delta\theta}_j^i) R_j^i \tilde{p}_i^j) = \\
&-\frac{1}{2}\operatorname{tr}\left(\frac{R_j^i \tilde{p}_i^j R_i^j G_t - (R_j^i \tilde{p}_i^j R_i^j G_t)^T}{2}\,\tilde{\delta p}_j^i\right) - \frac{1}{4}\operatorname{tr}(\tilde{p}_i^j \tilde{p}_i^j R_i^j (\tilde{\delta\theta}_j^i G_t - G_t \tilde{\delta\theta}_j^i) R_j^i)
\end{aligned} \tag{16.52}$$

and since we have:

$$
\begin{aligned}
&-\frac{1}{4}\operatorname{tr}(\tilde{p}_i^j\tilde{p}_i^j R_i^j(\tilde{\delta\theta}_j^i G_t - G_t\tilde{\delta\theta}_j^i)R_j^i) = \\
&-\frac{1}{4}\operatorname{tr}(\tilde{p}_i^j\tilde{p}_i^j R_i^j\tilde{\delta\theta}_j^i G_t R_j^i) + \frac{1}{4}\operatorname{tr}(\tilde{p}_i^j\tilde{p}_i^j R_i^j G_t\tilde{\delta\theta}_j^i R_j^i) = \\
&-\frac{1}{4}\operatorname{tr}(G_t R_j^i\tilde{p}_i^j\tilde{p}_i^j R_i^j\tilde{\delta\theta}_j^i) + \frac{1}{4}\operatorname{tr}(R_j^i\tilde{p}_i^j\tilde{p}_i^j R_i^j G_t\tilde{\delta\theta}_j^i) = \\
&-\frac{1}{2}\operatorname{tr}\left(\frac{G_t R_j^i\tilde{p}_i^j\tilde{p}_i^j R_i^j - R_j^i\tilde{p}_i^j\tilde{p}_i^j R_i^j G_t}{2}\tilde{\delta\theta}_j^i\right) = \\
&\qquad\qquad -\frac{1}{2}\operatorname{tr}(\operatorname{as}(G_t R_j^i\tilde{p}_i^j\tilde{p}_i^j R_i^j)\tilde{\delta\theta}_j^i)
\end{aligned}
\tag{16.53}
$$

we finally obtain:

$$
dV_t^{(2)}(p_i^j, R_i^j) = \frac{1}{2}\operatorname{tr}(-\operatorname{as}(R_j^i\tilde{p}_i^j R_i^j G_t)\tilde{\delta p}_j^i) + \frac{1}{2}\operatorname{tr}(-\operatorname{as}(G_t R_j^i\tilde{p}_i^j\tilde{p}_i^j R_i^j)\tilde{\delta\theta}_j^i))
\tag{16.54}
$$

and considering that

$$
-\operatorname{as}(R_j^i\tilde{p}_i^j R_i^j G_t) = -\operatorname{as}(G_t R_j^i\tilde{p}_i^j R_i^j)
$$

this gives the other terms of Eq. (16.43) and Eq. (16.44).

Clearly, the wrench which the spring applies to the body j is opposite to the one applied to i. Observe also that around the identity the computed elastic wrench gives the desired behavior specified by K_t and expressed in Eq. (16.23).

Coupling energy. From Eq. (16.23), the coupling energy should be such that the corresponding torque and force linearization at the identity should satisfy:

$$
m_c(0 + R_i^j\delta p_j^i, I + I\tilde{\delta\theta}_j^i) = -2\operatorname{as}(G_c\tilde{\delta\theta}_j^i)
\tag{16.55}
$$

$$
f_c(0 + R_i^j\delta p_j^i, I + I\tilde{\delta\theta}_j^i) = -2\operatorname{as}(G_c\tilde{\delta p}_j^i).
\tag{16.56}
$$

It is possible to see that a simple energy function satisfying the previous relations at the identity and reported in [2] is

$$
V_t(p_i^j, R_i^j) = \operatorname{tr}(G_c R_j^i\tilde{p}_i^j).
\tag{16.57}
$$

It is then possible to give the following result:

Theorem 16.6.4 (Coupling Elastic Wrenches). *A spring with elastic energy given by Eq. (16.57) and connecting body i and j would apply in a relative position (p_i^j, R_i^j) a wrench $\bar{W}^i = \left[m_c^T \; f_c^T\right]^T$ to body i and expressed in frame Ψ_i such that:*

$$\tilde{m}_c = -2\,\mathrm{as}(G_c \tilde{p}_i^j R_i^j) \tag{16.58}$$
$$\tilde{f}_c = -2\,\mathrm{as}(G_c R_i^j) \tag{16.59}$$

Proof. Reasoned along the same line as for the translational energy we get:

$$\begin{aligned}
& dV_c(p_i^j, R_i^j) = V_c(p_i^j + dp_i^j, R_i^j + dR_i^j) - V_c(p_i^j, R_i^j) = \\
& \mathrm{tr}(G_c(R_j^i - \tilde{\delta\theta}_j^i R_j^i)(p_i^j + R_i^j \delta p_j^i)^{\tilde{}}) - \mathrm{tr}(G_c R_j^i \tilde{p}_j^i) = \\
& = \mathrm{tr}(G_c R_j^i R_i^j \tilde{\delta p}_j^i R_j^i - G_c \tilde{\delta\theta}_j^i R_j^i \tilde{p}_i^j) = \\
& \mathrm{tr}(R_j^i G_c \tilde{\delta p}_j^i) - \mathrm{tr}(R_j^i \tilde{p}_i^j G_c \tilde{\delta\theta}_j^i) = \\
& \qquad \frac{1}{2}\mathrm{tr}(2\,\mathrm{as}(R_j^i G_c)\tilde{\delta p}_j^i) + \frac{1}{2}\mathrm{tr}(-2\,\mathrm{as}(R_j^i \tilde{p}_i^j G_c)\tilde{\delta\theta}_j^i)
\end{aligned} \tag{16.60}$$

which after applying some properties of the antisymmetric part of the product of two matrices gives the result to be proven.

16.6.2 Storing positions

The position of each of the frames Ψ_x in Fig. 16.4 can be associated to the necessary changes of coordinates and relative positions from Ψ_x to a fixed frame Ψ_0 that is attached to the inertial space. We indicate such a matrix with H_x. More precisely we have that:

$$\begin{aligned}
H_i &:= H_i^0 = \Psi_0 \circ h_i^0 \circ \Psi_i^{-1} \\
H_{c(i)} &:= H_i H_{c(i)}^i \\
H_b &:= H_b^0 = \Psi_0 \circ h_b^0 \circ \Psi_b^{-1} \\
H_{c(b)} &:= H_b H_{c(b)}^b \\
H_{v(i)} &:= H_{v(i)}^0 = \Psi_0 \circ h_{v(i)}^0 \circ \Psi_{v(i)}^{-1} \\
H_{b(i)} &:= H_{v(i)} H_{b(i)}^{v(i)} \\
H_{v(b)} &:= H_{v(b)}^0 = \Psi_0 \circ h_{v(b)}^0 \circ \Psi_{v(b)}^{-1}
\end{aligned}$$

where $H_{c(i)}^i$, $H_{c(b)}^b$, $H_{b(i)}^{v(i)}$ are fixed changes of coordinates within the same space.

It is then easy to calculate the matrices which are needed in order to compute the elastic wrenches, namely:

$$H_{c(i)}^{v(i)} = (H_{v(i)})^{-1} H_{c(i)} \text{ and } H_{c(b)}^{v(b)} = (H_{v(b)})^{-1} H_{c(b)}. \tag{16.61}$$

The direct kinematics of the robotic hand should then be available as a map of the following form:

$$L(q) : Q \to \underbrace{SE(3) \times \ldots \times SE(3)}_{n \text{ times}} \; ; \; q \mapsto (H_1(q), \ldots, H_n(q)) \tag{16.62}$$

and the body position H_b is a state of the controller. The coordinates changes represented by $H^i_{c(i)}$, $H^b_{c(b)}$, $H^{v(i)}_{b(i)}$ are fixed, known, and chosen at the beginning, as will be shown later.

16.6.3 The wrenches of the system

If we use the proposed energy functions and we want to create an elastic force between two bodies i and j by means of control, we should proceed as follows.

Algorithm 1 (Calculation of elastic wrenches) *1. Choose a relative position r^j_i of minimal potential energy.*

2. *In this relative position, choose a common point which will be the center of stiffness.*
3. *Choose two coordinate systems Ψ_i and Ψ_j for i and j respectively, which have their origin in the center of stiffness and coincide at the equilibrium relative position r^j_i.*
4. *Choose the desired K_t, K_o, K_c which are expressed at equilibrium in the frames $\Psi_i = \Psi_j$.*
5. *Calculate the corresponding G_t, G_o, G_c with Eq. (16.26).*
6. *With H^j_i, the total wrench generated by the spring on body i and expressed in Ψ_i is the sum of the wrenches of the orientational, translational and coupling energies $\bar{W}^i = \left[(m^i)^T \; (f^i)^T\right]^T$ with:*

$$\tilde{m}^i = -2\,\mathrm{as}(G_o R^j_i) - \mathrm{as}(G_t R^i_j \tilde{p}^j_i \tilde{p}^j_i R^j_i) - 2\,\mathrm{as}(G_c \tilde{p}^j_i R^j_i) \tag{16.63}$$

$$\tilde{f}^i = -R^i_j\,\mathrm{as}(G_t \tilde{p}^j_i) R^j_i - \mathrm{as}(G_t R^i_j \tilde{p}^j_i R^j_i) - 2\,\mathrm{as}(G_c R^j_i) \tag{16.64}$$

7. *The wrench $\bar{W}^j$ that the spring applies to body j will be $\bar{W}^j = -Ad^T_{H^i_j} \bar{W}^i$, which implies that:*

$$m^j = -R^j_i m^i - \tilde{p}^j_i R^j_i f^i \tag{16.65}$$

$$f^j = -R^j_i f^i \tag{16.66}$$

If we apply Algorithm 1 to the pair of bodies $(i, v(i))$ for the springs connecting the tips to the virtual body, and to the pair $(b, v(b))$ for the hand spring, we can compute the elastic wrenches in the system. If we choose the $\Psi_{c(i)}$ as coordinates for the finger tips, $\Psi_{v(i)}$ as coordinates for the supporting bodies

$v(i)$, $\Psi_{c(b)}$ for the virtual body and $\Psi_{v(b)}$ for the hand virtual position, we have as representative matrices $H_{c(i)}^{v(i)}$ for the tips springs and $H_{c(b)}^{v(b)}$ for the hand.

We can then choose the desired stiffnesses for each of these springs, which we indicate as[6]:

$$K^{(i)} = \begin{bmatrix} K_o^{(i)} & K_c^{(i)} \\ K_c^{(i)} & K_t^{(i)} \end{bmatrix} \quad i = 1 \ldots n \quad \text{and} \quad K^b = \begin{bmatrix} K_o^b & K_c^b \\ K_c^b & K_t^b \end{bmatrix}. \tag{16.67}$$

With Algorithm 1 we can calculate therefore for each $H_{c(i)}^{v(i)}$ and $H_{c(b)}^{v(b)}$ the elastic wrenches applied to the tips $\bar{W}^{c(i)}$, the elastic wrenches applied to the supporting spaces $\bar{W}^{v(i)}$ and the direct elastic wrenches applied to the virtual body $\bar{W}^{c(b)}$.

Wrenches on the tips and control torques. Once we obtain the wrenches $\bar{W}^{c(i)}$ for each time step, we can directly calculate the torques we have to apply to the robot in order to get the same effect that these virtual springs would have.

Use a numerical representation of the Jacobian of Eq. (16.14) where the twists t_i^0 are expressed in a common fixed frame Ψ_0, and indicate it with $\bar{J}_{\text{tips}}(q)$.

Since the wrenches $\bar{W}^{c(i)}$ of each fingertip are expressed in the frame $\Psi_{c(i)}$, we need to transform them first to the fixed space in the frame Ψ_0 using:

$$\bar{W}^{0,c(i)} = Ad^T_{H_0^{c(i)}} \bar{W}^{c(i)} \tag{16.68}$$

the torques to set to the actuators will therefore be:

$$\tau = \bar{J}^T_{\text{tips}}(q) \begin{bmatrix} Ad^T_{H_0^{c(1)}} & \cdots & 0 \\ \vdots & \ddots & \vdots \\ 0 & \cdots & Ad^T_{H_0^{c(n)}} \end{bmatrix} \begin{bmatrix} \bar{W}^{c(1)} \\ \vdots \\ \bar{W}^{c(n)} \end{bmatrix}. \tag{16.69}$$

Elastic wrench on the virtual object. The total elastic wrench applied to the virtual object is the sum of all the wrenches generated by the springs attached to it.

We considered each tip spring energy as a function of the relative position of each tip i and the supporting space $v(i)$. As shown in Eq. (16.6), this formulation is actually used to describe a spring which connects each tip to the virtual object. When we look at the dual equation of Eq. (16.5), we see that the wrench applied to the body and the wrench applied to the supporting

---

[6] Note that at the center of stiffness, the coupling stiffness matrices $K_c^{(i)}$ and K_c^b are symmetric.

space $v(i)$ are the same. This can be thought of intuitively by considering the space $v(i)$ rigidly connected to b at a certain instant. We therefore only need to express these wrenches in the right coordinates. If we express all the wrenches in the coordinate Ψ_b, we get:

$$\bar{W}^b_{\text{elastic}} = \sum_{i=1}^{n} Ad^T_{H_b^{v(i)}} \bar{W}^{v(i)} + Ad^T_{H_b^{c(b)}} \bar{W}^{c(b)} \tag{16.70}$$

where $\bar{W}^{v(i)}$ and $\bar{W}^{c(b)}$ are calculated using Algorithm 1.

16.6.4 Simulation of the virtual object dynamics

So far we have completely analyzed how to calculate the torques to supply to the robot as a function of the positions of the finger tips and of H_b. The matrix H_b is a time-varying matrix describing the dynamic evolution of the virtual object which is the dynamic extension of the controller.

We have chosen the virtual object with a very simple inertial structure on purpose and its dynamics can be easily simulated in real time within the controller. Considering Eq. (16.3) and the coordinates systems used, we can give a coordinate expression for the virtual object dynamics.

Assuming the virtual body in a gravitational-less environment, it does not have potential energy and its Hamiltonian is:

$$H_B(H_b, P^b) = \frac{1}{2}(P^b)^T \begin{bmatrix} \frac{1}{j}I_3 & 0 \\ 0 & \frac{1}{m}I_3 \end{bmatrix} (P^b) \quad m, j \in \mathbb{R}. \tag{16.71}$$

The map h_b^0 of Eq. (16.3) is represented by the matrix $H_b^0 = H_b$. and therefore, the first Hamiltonian equation is:

$$\dot{H}_b = H_b \tilde{T}_b^{b,0} \quad \text{with} \quad \bar{T}_b^{b,0} = \frac{\partial H_B}{\partial P^b} = \begin{bmatrix} \frac{1}{j}I_3 & 0 \\ 0 & \frac{1}{m}I_3 \end{bmatrix} (P^b) \tag{16.72}$$

and the second equation is:

$$\dot{P}^b = P^b \wedge \bar{T}_b^{b,0} + \bar{W}^b_{\text{tot}} \tag{16.73}$$

where $\bar{W}^b_{\text{tot}}$ is the total wrench applied to the body and $(P^b \wedge)$ corresponds to the Lie-Poisson bracket and can be represented as a 6×6 matrix of the following form:

$$(P^b \wedge) := \begin{pmatrix} \tilde{P}^b_\omega & \tilde{P}^b_v \\ \tilde{P}^b_v & 0 \end{pmatrix}$$

where $P^b = \left(P_\omega^T \; P_v^T\right)^T$ and the "tilde" operator is defined in Eq. (16.20). If we assume the R_h of Sect. 16.4.4 equal to zero, the dissipating wrench is

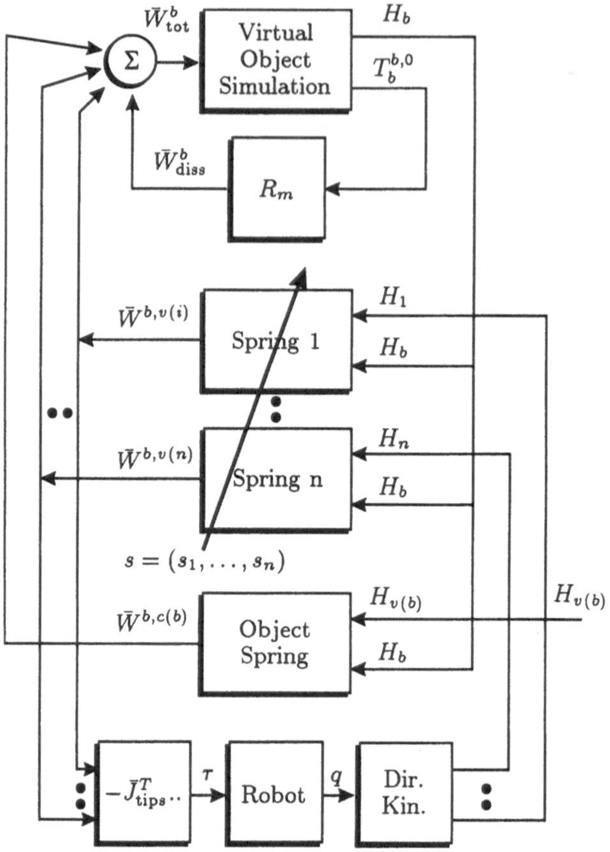

Fig. 16.6. The simplified grasping control scheme.

$\bar{W}^b_{\text{diss}} = R_m \bar{T}_b^{b,0}$ and if we use Eq. (16.70), the total wrench applied to the virtual body is:

$$\bar{W}^b_{\text{tot}} = \bar{W}^b_{\text{elastic}} + \bar{W}^b_{\text{diss}} \tag{16.74}$$

With the previous equations it is therefore possible to calculate $\dot{H}_b$ and $\dot{P}^b$, which are the state rates of the controller. The integration of $\dot{H}_b$ should be done with care since the matrix H_b should remain in $SE(3)$ during all steps. In order to do that, the orthonormal matrix R_b^0 within H_b^0 should be normalized after each step in order to keep it within $SO(3)$. With the integration of H_b, the position of the body is known at each step and the elastic wrenches can be calculated again closing the control loop. A simplified representation of the control scheme is reported in Fig. 16.6

16.6.5 Varying the length of springs

The variation of the length of a spring, can be easily taken into account by considering the supervisory control twist $t^b_{v(i)}$ reported in Fig. 16.1. If we indicate the expression of this twist in the base Ψ_b with $T^b_{b(i)}$, we can express the effective state for the variable springs as:

$$H^{v(i)}_{c(i)} = H^{v(i)}_{b(i)} H^{b(i)}_b H^b_{c(i)} \quad \text{with} \quad H^{b(i)}_b = e^{-T^b_{b(i)}} \tag{16.75}$$

where we choose $T^{b(i)}_b = s_i T_i$ with T_i a constant twist, and $e^T \in SE(3)$ indicates the matrix exponential of the twist $T \in se(3)$. The reason of this choice is explained in Sect. 16.7. The effective state will be therefore changed either by a change of $H^{b(i)}_b$ due to a supervisory control or by a change of $H^b_{c(i)} = (H_b)^{-1} H_{c(i)}$ corresponding to a change of the relative position of the virtual body with respect to the tip i.

It is now possible to explain the sequence of actions that can be used to grasp an object and to interact with it.

16.7 Grasping strategies

Suppose we have an object to be grasped and we want to plan a tip grasp configuration. We can describe this grasp configuration by defining the desired positions of the tips with respect to a "grasp focus" positioned somewhere between the fingers [10]. If we represent the focus by a frame H_f, the desired grasp configuration is described by n relative configurations $H^1_f, \ldots, H^n_f$.

If we position the virtual body at the focus ($H_b = H_f$) and consider it as a representative position for the whole grasp, to achieve the desired configuration it should be $H^i_b = H^i_f \; \forall i$.

If the hand would just touch the object to be grasped without exerting any force, the energy of the controller's springs should be zero, which implies that in such a situation $H_{c(i)} = H_{v(i)}$. By compositions we would therefore have:

$$H^i_{c(i)} H^{v(i)}_{b(i)} H^{b(i)}_b = H^i_b = H^i_f \Rightarrow H^{v(i)}_{b(i)} = H^{c(i)}_i H^i_f H^b_{b(i)} \tag{16.76}$$

The transformation $H^i_{c(i)}$ is chosen and expresses the position of the center of stiffness [2] for the tip i. We can then choose $H^b_{b(i)}$ in such a way that the translational part of $H^i_{b(i)}$ is zero so that the origins of Ψ_i and of $\Psi_{b(i)}$ coincide for the nominal grasp configuration. The rotational part of $H^i_{b(i)}$ can be chosen when we consider that we control the hand configuration by actually choosing a twist $\bar{T}_i \in se(3)$ such that $H_b^{b(i)}(s_i) = e^{s_i \bar{T}_i}$ and such that for $s_i = 1$, $H_b^{b(i)}$ would satisfy Eq. (16.76). We have therefore:

$$H_{b(i)}^i = \begin{bmatrix} R_{b(i)}^i & 0 \\ 0 & 1 \end{bmatrix} = H_f^i e^{-s_i \bar{T}_i}. \tag{16.77}$$

Which can be solved for $\bar{T}_i$ once $s_i = 1$ and $R_{b(i)}^i$ has been chosen. The choice of $R_{b(i)}^i$ is important because it expresses the way in which the hand is opened and closed by respectively increasing or decreasing the scalars s_i.

It we then set $H_{b(i)}^{v(i)} := H_i^{c(i)} H_f^i e^{-\bar{T}_i}$, we control the position of the extreme of each tip spring connected on the side of the virtual body by changing s_i, and get:

$$H_b^{v(i)}(s_i) = H_{b(i)}^{v(i)} e^{s_i \bar{T}_i}.$$

In order to perform a tip grasp, we have to proceed as follows:

Algorithm 2 (Tip grasping algorithm) *1. Choose the location of the center of stiffness for the object by choosing $H_b^{c(b)}$.*

2. *Choose a $H_{v(b)}$ far from the object to be grasped and in the neighborhood of the hand.*
3. *For each i, choose $H_b^{v(i)}(s_i)$ with s_i sufficiently larger than 1 (Opening Hand).*
4. *Move the virtual position of the object, $H_{v(b)}$, along a proper trajectory to $H_f H_{c(b)}^b = H_f (H_b^{c(b)})^{-1}$.*
5. *Decrease the s_i to a proper value smaller than 1 which represents a sufficiently high grasping internal energy (Closing Hand).*
6. *Change $H_{v(b)}$ to move the object and do what is needed.*

Proper choices of the center of stiffness for each tip would then specify how the robot reacts to small unexpected motions of the object due to external forces. In this way, it would also be possible to handle rolling contacts by exploiting the contact friction, but this requires further research.

The presented control system can not only be used to implement tips grasps but in the same way as for tips grasp, it is possible to open the hand and move the virtual position $H_{v(b)}$ over an object to be grasped rather than at its center. At the same time we can close the hand by decreasing the s_i and the result is that the hand folds around the object to be grasped implementing a full grasp.

16.8 Conclusions

In this chapter, an intrinsically passive algorithm for controlling multi-limbed robotic systems based on physical concepts has been presented. A major advantage is the ensured stability in any interactive situation for any environment. This because Liapunov stability is ensured choosing as Liapunov function the equivalent energy of the robot-controller coupled system and their power-continuous interconnection. This equivalent energy is composed of the potential energy of the controller springs, the kinetic energy of the robot and of the virtual object. The general Hamiltonian structure of the controller has been shown as the spatial interconnection of generalised Hamiltonian systems.

References

1. P.C. Breedveld. *Physical Systems Theory in Terms of Bond Graphs.* PhD thesis, Technische Hogeschool Twente, Enschede, The Netherlands, February 1984. ISBN 90-90005999-4.

2. Ernest D. Fasse. On the spatial compliance of robotic manipulators. *ASME J.of Dynamic Systems, Measurement and Control*, 119:839–844, 1997.

3. Ernest D. Fasse and Peter C. Breedveld. Modelling of elastically coupled bodies: Part i: General theory and geometric potential function method. *Accepted for publication in ASME J. of Dynamic Systems, Measurement and Control*, 1997.

4. Ernest D. Fasse and Peter C. Breedveld. Modelling of elastically coupled bodies: Part ii: Exponential- and generalized-coordinate methods. *Accepted for publications in ASME J. of Dynamic Systems, Measurement and Control*, 1997.

5. Ernest D. Fasse and Jan F. Broenink. A spatial impedance controller for robotic manipulation. *IEEE Trans. on Robotics and Automation*, 13:546–556, 1997.

6. Neville Hogan. Impedance control: An approach to manipulation: Part I-Theory. *ASME J. of Dynamic Systems, Measurement and Control*, 107:1–7, March 1985.

7. Josip Lončarić . *Geometrical Analysis of Compliant Mechanisms in Robotics.* PhD thesis, Harvard University, Cambridge (MA), April 3 1985.

8. Richard M. Murray, Zexiang Li, and S.Shankar Sastry. *A Methematical Introduction to Robotic Manipulation.* CRC Press, March 1994. ISBN 0-8493-7981-4.

9. Romeo Ortega, Antonio Loria, Rafael Kelly, and Laurent Praly. On passivity-based output feedback global stabilization of euler-lagrange systems. In *Conference on Decision and Control*, 1994.

10. Nancy S. Polland and Tomás Lozano-Pérez. Grasp stability and feasibility for an arm with an articulated hand. *IEEE*, 1990.

11. Stefano Stramigioli. Creating artificial damping by means of damping injection. In K.Danai, editor, *Proceedings of the ASME Dynamic Systems and Control Division*, volume DSC.58, pages 601–606, Atlanta, (GE), 1996.

12. Stefano Stramigioli. *From Manifolds to Interactive Robot Control*. PhD thesis, Delft University of Technology, Delft, The Netherlands, December 4 1998. ISBN 90-9011974-4, `http://lcewww.et.tudelft.nl/~stramigi`.

13. Stefano Stramigioli. A novel impedance grasping strategy based on the virtual object concept. In *IEEE Mediterranean Conference 1998*, Alghero, Italy, June 1998.

14. Stefano Stramigioli, Bernhard Maschke, and Arjan van der Schaft. Passive output feedback and port interconnection. In *Proc. Nonlinear Control Systems Design Symposium1998*, Enschede, The Netherlands, July 1998.

17. A nonsmooth hybrid maximum principle

Héctor J. Sussmann[1,2]

Department of Mathematics
Rutgers, the State University of New Jersey
Hill Center—Busch Campus
110 Frelinghuysen Road
Piscataway, NJ 08854-8019, USA
sussmann@hilbert.rutgers.edu

Summary.

We present two versions of the maximum principle for nonsmooth hybrid optimal control problems, the first one of which requires differentiability along the reference trajectory and yields an adjoint equation of the usual kind, while the second one only requires approximability to first order by Lipschitz maps, and yields an adjoint differential inclusion involving a generalized gradient of the approximating Hamiltonian.

17.1 Introduction

In this paper we present a version of the maximum principle for nonsmooth hybrid optimal control problems, under weak regularity conditions. The class of hybrid problems to be considered is defined in §17.2. The maximum principle is stated in §17.3 as a general assertion involving terms that are not yet precisely defined, and without a detailed specification of technical assumptions. One version of the principle, where the terms are precisely defined and the appropriate technical requirements are completely specified, is stated in §17.4 for problems where all the basic objects—the dynamics, the Lagrangian, and the cost functions for the switchings and the endpoint constraints—are differentiable along the reference arc, but a considerable amount of nonsmoothness is allowed away from the reference control. A fairly detailed outline of the proof of the result of §17.4 is presented in §17.5. Another version

[1] Research supported in part by NSF Grant DMS-9803411 and AFOSR Grant 0923.
[2] Most of this work was done in the Netherlands, during a three-month visit at the University of Groningen, to which the author is immensely grateful for its generous hospitality and exciting intellectual atmosphere.

of the hybrid maximum principle, allowing nondifferentiability along the reference control—and requiring only "first-order approximability by Lipschitz maps"—is stated without proof in §17.6.

Our results are stronger than the usual versions of the finite-dimensional maximum principle. For example, even the theorem for classical differentials applies to situations where the maps are not of class C^1, and can fail to be Lipschitz continuous. The "nonsmooth" result applies to maps that are neither Lipschitz continuous nor differentiable in the classical sense. This has been shown for the non-hybrid case in several of our earlier papers on the maximum principle (cf. Sussmann [5, 6]), and in each case it would be trivial to construct hybrid examples of a similar nature.

On the other hand, the results presented here are much weaker than what can actually be proved by our methods. More general versions, involving systems of differential inclusions, and "flows" that do not arise from vector fields, will be discussed in subsequent papers.

A simpler version of our results, dealing only with autonomous systems, was announced—without proof—in the conference paper [7].

17.2 Hybrid optimal control problems

Throughout this paper, the expression "smooth manifold"—or, simply, the word "manifold"—means "finite-dimensional Hausdorff manifold of class C^1 without boundary." If M is a manifold, and $x \in M$, then T_xM, T_x^*M, TM, T^*M denote, respectively, the tangent and cotangent spaces of M at x, and the tangent and cotangent bundles of M.

Definition 17.2.1. *A* ***finite family of state spaces*** *is a pair* $(\mathcal{Q}, \mathcal{M})$ *such that*

FFSS1. $\mathcal{Q}$ *is a finite set;*

FFSS2. $\mathcal{M} = \{M_q\}_{q\in\mathcal{Q}}$ *is a family of smooth manifolds, indexed by* $\mathcal{Q}$. ◊

If $(\mathcal{Q}, \mathcal{M})$ is a finite family of state spaces, then for each pair $(q, q') \in \mathcal{Q} \times \mathcal{Q}$ we use $\mathcal{M}_{q,q'}$ to denote the product $M_q \times M_{q'} \times \mathbb{R} \times \mathbb{R}$.

Definition 17.2.2. *A* ***switching constraint*** *for a finite family of state spaces* $(\mathcal{Q}, \mathcal{M})$ *is a family* $\mathcal{S} = \{S_{q,q'}\}_{(q,q')\in\mathcal{Q}\times\mathcal{Q}}$ *such that* $S_{q,q'}$ *is a subset of* $\mathcal{M}_{q,q'}$ *for every pair* $(q, q') \in \mathcal{Q} \times \mathcal{Q}$. ◊

The following is the definition of "hybrid control system" that will be adopted for the purposes of this paper.

Definition 17.2.3. *A **hybrid control system** is a 6-tuple*

$$\Sigma = (\mathcal{Q}, \mathcal{M}, \mathcal{U}, f, \boldsymbol{\mathcal{U}}, \mathcal{S})$$

such that

HCS1. $(\mathcal{Q}, \mathcal{M})$ *is a finite family of state spaces;*

HCS2. $\mathcal{U} = \{U_q\}_{q\in\mathcal{Q}}$ *is a family of sets;*

HCS3. $f = \{f_q\}_{q\in\mathcal{Q}}$ *is a family such that* f_q *is, for each* q, *a partially defined map from* $M_q \times U_q \times \mathbb{R}$ *to* TM_q, *having the property that* $f_q(x,u,t)$ *belongs to* T_xM_q *for every* $(x,u,t) \in M_q \times U_q \times \mathbb{R}$ *for which* $f_q(x,u,t)$ *is defined;*

HCS4. $\boldsymbol{\mathcal{U}} = \{\boldsymbol{U}_q\}_{q\in\mathcal{Q}}$ *is a family consisting, for each* q, *of a set* $\boldsymbol{U}_q$, *each of whose members is a map* $\eta : I_\eta \to U_q$ *defined on some subinterval* I_η *of* $\mathbb{R}$;

HCS5. $\mathcal{S} = \{S_{q,q'}\}_{(q,q')\in\mathcal{Q}\times\mathcal{Q}}$ *is a switching constraint for* $(\mathcal{Q}, \mathcal{M})$. ◇

The sets $S_{q,q'}$ are the *switching sets* of Σ, and are allowed to be empty. One should think of $S_{q,q'}$ as the set of all 4-tuples (x, x', t, t') such that $x \in M_q$, $x' \in M_{q'}$, and a switching (or "jump") from state $x \in M_q$ to state $x' \in M_{q'}$ is permitted at time t, with a resetting of the clock to time t'. Usually, one does not want to permit clock resetting, but for mathematical reasons it is better to allow it in principle, and exclude it, when desired, by just taking the switching sets $S_{q,q'}$ to consist only of points of the form (x, x', t, t).

The members of $\mathcal{Q}$ are called *locations*. The families $\mathcal{M}$, $\mathcal{U}$, are, respectively, the *family of state spaces* and the *family of control spaces* of Σ. For each q, the manifold M_q, the set U_q, the map f_q, and the set $\boldsymbol{U}_q$ are, respectively, the *state space*, the *control space*, the *dynamical law*, and the *class of admissible controls* at location q. Usually, $\mathcal{Q}$ will be the set of states of some finite automaton.

Definition 17.2.4. *A **control** for a hybrid system* Σ *as above is a triple* $\zeta = (\mathbf{q}, \mathbf{I}, \boldsymbol{\eta})$ *such that*

- $\mathbf{q} = (q_1, \ldots, q_\nu)$ *is a finite sequence of locations;*
- $\mathbf{I} = (I_1, \ldots, I_\nu)$ *is a finite sequence of compact intervals;*
- $\boldsymbol{\eta} = (\eta_1, \ldots, \eta_\nu)$ *is a finite sequence such that* η_j *belongs to* $\boldsymbol{U}_{q_j}$ *and* $I_{\eta_j} = I_j$ *for* $j = 1, \ldots, \nu$.

If $\zeta = (\mathbf{q}, \mathbf{I}, \boldsymbol{\eta})$ *is a control, and* $\mathbf{I} = (I_1, \ldots, I_\nu)$ *for* $j = 1, \ldots, \nu$, *we use* $\mathbf{q}(\zeta)$, $\mathbf{I}(\zeta)$, $\boldsymbol{\eta}(\zeta)$, $\nu(\zeta)$, *to denote, respectively, the finite sequences* $\mathbf{q}$, $\mathbf{I}$, $\boldsymbol{\eta}$, *and the natural number* ν. *If* $I_j = [t_j, \tau_j]$, *we use* $\mathbf{t}(\zeta)$, $\boldsymbol{\tau}(\zeta)$ *to denote the sequences* $(t_1,\ldots,t_\nu)$ *and* $(\tau_1,\ldots,\tau_\nu)$, *and we let* $a_\zeta = t_1$, $b_\zeta = \tau_\nu$. *Then* a_ζ, b_ζ, $\nu(\zeta)-1$, *and* $\mathbf{q}(\zeta)$ *are, respectively, the* ***initial time***, *the* ***terminal time***, *the* ***number of switchings***, *and the* ***switching strategy*** *of* ζ. ◇

Definition 17.2.5. *If* $\Sigma = (\mathcal{Q},\mathcal{M},\mathcal{U},f,\boldsymbol{\mathcal{U}},S)$ *is a hybrid system as above,* ζ *is a control for* Σ, *and* $\nu = \nu(\zeta)$, *then a* ***pretrajectory*** *for* ζ *is a* ν*-tuple* $\boldsymbol{\xi} = (\xi_1,\ldots,\xi_\nu)$ *such that, if*

$$\mathbf{I}(\zeta) = (I_1,\ldots,I_\nu)\,,\ I_j = [t_j,\tau_j],\ \mathbf{q}(\zeta) = (q_1,\ldots,q_\nu)\,,\ \boldsymbol{\eta}(\zeta) = (\eta_1,\ldots,\eta_\nu)\,,$$

then, for each $j \in \{1,\ldots,\nu\}$, ξ_j *is an absolutely continuous map from* I_j *to the manifold* M_{q_j}, *having the property that* $f_{q_j}(\xi_j(t),\eta_j(t),t)$ *is defined and* $\dot{\xi}_j(t) = f_{q_j}(\xi_j(t),\eta_j(t),t)$ *for almost all* $t \in I_j$. ◇

Definition 17.2.6. *If* Σ *is a hybrid system as above, a* ***pretrajectory-control pair*** *for* Σ *is a pair* $(\boldsymbol{\xi},\zeta)$ *such that* ζ *is a control for* Σ *and* $\boldsymbol{\xi}$ *is a pretrajectory of* Σ *for* ζ.

We use $PTCP(\Sigma)$ *to denote the set of all pretrajectory-control pairs of the system* Σ. ◇

Definition 17.2.7. *An* ***endpoint constraint*** *for a finite family of state spaces* $(\mathcal{Q},\mathcal{M})$ *is a family* $\mathcal{E} = \{E_{q,q'}\}_{(q,q')\in\mathcal{Q}\times\mathcal{Q}}$ *of sets such that* $E_{q,q'}$ *is, for each* $(q,q') \in \mathcal{Q}\times\mathcal{Q}$, *a subset of* $\mathcal{M}_{q,q'}$. ◇

Notice that, mathematically, an endpoint constraint is exactly the same kind of object as a switching condition. This is why the part of the maximum principle that has to do with the switchings will have the same form as the transversality condition.

Definition 17.2.8. *Let* $\Sigma = (\mathcal{Q},\mathcal{M},\mathcal{U},f,\boldsymbol{\mathcal{U}},S)$ *be a hybrid control system as in the previous definitions, and let* $\Xi = (\boldsymbol{\xi},\zeta)$ *belong to* $PTCP(\Sigma)$. *Let* $\nu = \nu(\zeta)$, $\boldsymbol{\xi} = (\xi_1,\ldots,\xi_\nu)$, $\mathbf{q}(\zeta) = (q_1,\ldots,q_\nu)$, $\mathbf{t}(\zeta) = (t_1,\ldots,t_\nu)$, $\boldsymbol{\tau}(\zeta) = (\tau_1,\ldots,\tau_\nu)$, $\mathbf{I}(\zeta) = (I_1,\ldots,I_\nu)$, $\mathcal{S} = \{S_{q,q'}\}_{(q,q')\in\mathcal{Q}\times\mathcal{Q}}$. *Then*

- *The* ***endpoint condition*** *of* $\boldsymbol{\xi}$ *(or of* Ξ*) is the 4-tuple*

$$\partial\boldsymbol{\xi} \stackrel{\text{def}}{=} \partial\Xi \stackrel{\text{def}}{=} (\xi_\nu(b_\zeta),\xi_1(a_\zeta),b_\zeta,a_\zeta) \in \mathcal{M}_{q_\nu,q_1}\,. \tag{17.1}$$

- *If* $1 \le j < \nu$, *the "j-th jump" of* $\boldsymbol{\xi}$ *(or of* Ξ*) is the 4-tuple*

$$\partial_j\boldsymbol{\xi} \stackrel{\text{def}}{=} \partial_j\Xi \stackrel{\text{def}}{=} (\xi_j(\tau_j),\xi_{j+1}(t_{j+1}),\tau_j,t_{j+1}) \in \mathcal{M}_{q_j,q_{j+1}}\,. \tag{17.2}$$

- *If* $\mathcal{E} = \{E_{q,q'}\}_{(q,q')\in\mathcal{Q}\times\mathcal{Q}}$ *is an endpoint constraint for* $(\mathcal{Q},\mathcal{M})$, *we say that* Ξ ***satisfies the constraint*** $\mathcal{E}$ *if* $\partial\Xi$ *belongs to* E_{q_ν,q_1}.

- *We say that* $\boldsymbol{\xi}$ *(or* Ξ*)* ***satisfies the switching conditions*** *for* Σ *if* $\partial_j \Xi$ *belongs to* $S_{q_j,q_{j+1}}$ *whenever* $j \in \{1, \dots, \nu - 1\}$. ◇

Definition 17.2.9. *If* $\Sigma = (\mathcal{Q}, \mathcal{M}, \mathcal{U}, f, \boldsymbol{\mathcal{U}}, \mathcal{S})$ *is a hybrid system as above, then*

- *we say that a pretrajectory* $\boldsymbol{\xi}$ *of* Σ *is a* ***trajectory*** *of* Σ *if* $\boldsymbol{\xi}$ *satisfies the switching conditions for* Σ*;*
- *we use* $TCP(\Sigma)$ *to denote the set of all trajectory-control pairs of* Σ *(i.e., the set of all* $\Xi = (\boldsymbol{\xi}, \zeta) \in PTCP(\Sigma)$ *such that* $\boldsymbol{\xi}$ *is a trajectory of* Σ*), and* $TCP(\Sigma; \mathcal{E})$ *to denote the set of all* $\Xi \in TCP(\Sigma)$ *that satisfy the endpoint constraint* $\mathcal{E}$. ◇

Definition 17.2.10. *If* Σ *is a hybrid system as above, then a* ***Lagrangian*** *for* Σ *is a family* $L = \{L_q\}_{q \in \mathcal{Q}}$ *such that*

- L_q *is, for each* $q \in \mathcal{Q}$*, a partially defined real-valued function on the product* $M_q \times U_q \times \mathbb{R}$*,*
- *whenever* $q \in \mathcal{Q}$*,* $\eta \in \boldsymbol{U}_q$ *has domain* $[\alpha, \beta]$*, and* $\xi : [\alpha, \beta] \to M_q$ *is an absolutely continuous solution of* $\dot{\xi}(t) = f_q(\xi(t), \eta(t), t)$ *a.e., it follows that the function* $[\alpha, \beta] \ni t \to L(\xi(t), \eta(t), t)$ *is defined for almost every* t*, and is integrable.*

A ***switching cost function*** *for* Σ *is a family* $\Phi = \{\Phi_{q,q'}\}_{(q,q') \in \mathcal{Q} \times \mathcal{Q}}$ *such that each* $\Phi_{q,q'}$ *is an extended real-valued function on* $S_{q,q'}$ *that never takes the value* $-\infty$.

An ***endpoint cost function*** *for* Σ *is a family* $\varphi = \{\varphi_{q,q'}\}_{(q,q') \in \mathcal{Q} \times \mathcal{Q}}$ *such that each* $\varphi_{q,q'}$ *is an extended real-valued function on* $\mathcal{M}_{q,q'}$ *that never takes the value* $-\infty$. ◇

If $L = \{L_q\}_{q \in \mathcal{Q}}$ is a Lagrangian for the hybrid control system Σ, then we can define the corresponding *Lagrangian cost functional* $C_L : TCP(\Sigma) \to \mathbb{R}$, by letting

$$C_L(\boldsymbol{\xi}, \zeta) = \sum_{j=1}^{\nu} \int_{I_j} L_{q_j}(\xi_j(t), \eta_j(t), t)\, dt\,, \tag{17.3}$$

where $\nu = \nu(\zeta)$, $\mathbf{I}(\zeta) = (I_1, \dots, I_\nu)$, $\mathbf{q}(\zeta) = (q_1, \dots, q_\nu)$, $\boldsymbol{\eta}(\zeta) = (\eta_1, \dots, \eta_\nu)$, and $\boldsymbol{\xi} = (\xi_1, \dots, \xi_\nu)$.

If Φ is a switching cost function for Σ, and φ is an endpoint cost function, then we associate with Φ and φ the functional $\hat{C}_{\Phi,\varphi} : TCP(\Sigma) \to \mathbb{R} \cup \{+\infty\}$ that assigns to each $\Xi = (\boldsymbol{\xi}, \zeta) \in TCP(\Sigma)$ the number

$$\hat{C}_{\Phi,\varphi}(\boldsymbol{\xi},\zeta) = \varphi_{q_\nu,q_1}(\partial\Xi) + \sum_{j=1}^{\nu-1} \Phi_{q_j,q_{j+1}}(\partial_j\Xi), \tag{17.4}$$

where $\nu = \nu(\zeta)$, and $(q_1,\ldots,q_\nu)$ is the switching strategy of ζ.

Definition 17.2.11. *A* ***hybrid Bolza cost functional*** *for Σ is an extended real-valued functional $C : TCP(\Sigma) \to \mathbb{R} \cup \{+\infty\}$ such that $C = C_L + \hat{C}_{\Phi,\varphi}$ for some L, Φ, φ that are, respectively, a Lagrangian, a switching cost function, and an endpoint cost function for Σ.* ◇

Given a hybrid control system Σ, a Bolza cost functional C for Σ, and an endpoint constraint $\mathcal{E}$, we will consider the *optimal control problem* $\mathcal{P}(\Sigma,C,\mathcal{E})$, whose objective is to minimize $C(\boldsymbol{\xi},\zeta)$ in the class $TCP(\Sigma;\mathcal{E})$. We observe that the endpoint constraint sets $E_{q,q'}$ could all be of the special form $E^0_{q,q'} \times \{b\} \times \{a\}$, where a, b are fixed real numbers, independent of q, q', and each $E^0_{q,q'}$ is a subset of $M_q \times M_{q'}$. In that special case, all the members $\Xi = (\boldsymbol{\xi},\zeta)$ of $TCP(\Sigma;\mathcal{E})$ satisfy $a_\zeta = a$, $b_\zeta = b$, so we have a *problem with fixed initial and terminal times.* In addition, the switching sets $S_{q,q'}$ could be of the form $S_{q,q'} = S^0_{q,q'} \times \{\bar{t}_{q,q'}\} \times \{\bar{t}_{q,q'}\}$, where $S^0_{q,q'} \subseteq M_q \times M_{q'}$ and the $\bar{t}_{q,q'}$ are fixed real numbers, in which case we would be dealing with a *problem with fixed switching times and no clock resetting.*

17.3 The general form of the maximum principle

Let us assume that

A1. $\Sigma = (\mathcal{Q},\mathcal{M},\mathcal{U},f,\boldsymbol{\mathcal{U}},\mathcal{S})$ is a hybrid control system;

A2. $C = C_L + \hat{C}_{\Phi,\varphi}$ is a hybrid Bolza cost functional for Σ;

A3. $\mathcal{E}$ is an endpoint constraint for $(\mathcal{Q},\mathcal{M})$;

A4. $\Xi^\#$ (the "reference trajectory-control pair") belongs to $TCP(\Sigma;\mathcal{E})$, and

$$\begin{aligned} \Xi^\# &= (\boldsymbol{\xi}^\#,\zeta^\#), & \boldsymbol{\xi}^\# &= (\xi_1^\#,\ldots,\xi_{\nu^\#}^\#), \\ \zeta^\# &= (\mathbf{q}^\#,\mathbf{I}^\#,\boldsymbol{\eta}^\#), & \mathbf{q}^\# &= (q_1^\#,\ldots,q_{\nu^\#}^\#), \\ \mathbf{I}^\# &= (I_1^\#,\ldots,I_{\nu^\#}^\#), & \boldsymbol{\eta}^\# &= (\eta_1^\#,\ldots,\eta_{\nu^\#}^\#). \end{aligned}$$

The *maximum principle* gives a necessary condition for $\Xi^\#$ to be a solution of $\mathcal{P}(\Sigma,C,\mathcal{E})$. The result only depends on comparing trajectories with the same switching strategy, and does not require the candidate arc $\Xi^\#$ to be a true solution. Moreover, even within the class of arcs corresponding to a fixed switching strategy, only arcs that are close to $\Xi^\#$ are compared with $\Xi^\#$. So we introduce the following definition.

Definition 17.3.1. *A **local solution** of a problem $\mathcal{P}(\Sigma, \mathcal{C}, \mathcal{E})$ is a trajectory-control pair $\Xi^\# = (\boldsymbol{\xi}^\#, \zeta^\#) = (\xi_1^\#, \ldots, \xi_{\nu^\#}^\#, \zeta^\#)$ such that there exist neighborhoods $\mathcal{N}_1, \ldots, \mathcal{N}_{\nu^\#}$ of the graphs of $\xi_1^\#, \ldots, \xi_{\nu^\#}^\#$ in $M_{q_1} \times \mathbb{R}, \ldots, M_{q_{\nu}^\#} \times \mathbb{R}$ having the property that $\Xi^\#$ minimizes the cost $\mathcal{C}(\Xi)$ in the class of all the trajectory-control pairs $\Xi = (\boldsymbol{\xi}, \zeta) = (\xi_1, \ldots, \xi_\nu, \zeta) \in TCP(\Sigma, \mathcal{E})$ such that $\mathbf{q}(\zeta) = \mathbf{q}(\zeta^\#)$ (so that, in particular, $\nu = \nu^\#$) and the graph $G(\xi_j)$ of ξ_j is contained in $\mathcal{N}_j$ for $j = 1, \ldots, \nu^\#$. (Here the "graph" of ξ_j is the set*

$$G(\xi_j) \stackrel{\text{def}}{=} \{(\xi_j(t), t) : t \in \text{Domain}(\xi_j)\}, \tag{17.5}$$

so $G(\xi_j) \subseteq M_{q_j} \times \mathbb{R}$.) ◇

We now present the maximum principle for hybrid systems as a true "principle," that is, a not very precise mathematical statement that can be rendered precise in various ways, giving rise to different "versions" of the principle. Two such versions—both completely precise and rigorous—will be stated in subsequent sections of the paper.

The maximum principle. *Assume that A1-A4 hold, and $\Xi^\#$ is a local solution of $\mathcal{P}(\Sigma, \mathcal{C}, \mathcal{E})$. Then there exists an adjoint pair $(\boldsymbol{\psi}, \psi_0)$ along $\Xi^\#$ that satisfies the weak Hamiltonian maximization, nontriviality, and transversality conditions for $\mathcal{P}(\Sigma, \mathcal{C}, \mathcal{E})$ along $\Xi^\#$.* ◇

To turn the above statement into a theorem, we have to specify technical assumptions on the 12-tuple of data $(\mathcal{Q}, \mathcal{M}, \mathcal{U}, f, \boldsymbol{\mathcal{U}}, S, L, \Phi, \varphi, \mathcal{E}, \boldsymbol{\xi}^\#, \zeta^\#)$, and assign a precise meaning to the notions of "adjoint pair," "weak Hamiltonian maximization," "nontriviality," and "transversality." This will be done in detail in the following section for problems whose reference vector fields and Lagrangians are differentiable along the reference trajectory. The changes needed for the nondifferentiable case will be sketched in §17.6.

17.4 A version involving classical differentials

We now make the maximum principle precise in the setting of maps having classical differentials.

First, we let

$$I_0^\# = I_{\nu^\#}^\#, \qquad I_{\nu^\#+1}^\# = I_1^\#, \qquad q_0^\# = q_{\nu^\#}^\#, \qquad q_{\nu^\#+1}^\# = q_1^\#,$$

and write

$$
\begin{array}{lll}
I_j^\# & = [t_j^\#, \tau_j^\#] & \text{for} \quad j = 0, 1, \ldots, \nu^\# + 1\,, \\
a^\# & = t_1^\#\,, & \\
b^\# & = \tau_{\nu^\#}^\#\,, & , \\
M_j^\# & = M_{q_j^\#} & \text{for} \quad j = 0, 1, \ldots, \nu^\# + 1\,, \\
\mathcal{M}_j^\# & = \mathcal{M}_{q_j^\#, q_{j+1}^\#} & \text{for} \quad j = 1, \ldots, \nu^\#\,, \\
S_j^\# & = S_{q_j^\#, q_{j+1}^\#} & \text{for} \quad j = 1, \ldots, \nu^\# - 1\,, \\
S_j^\# & = E_{q_{\nu^\#}^\#, q_1^\#} & \text{for} \quad j = \nu^\#\,, \\
S_0^\# & = S_{\nu^\#}^\#\,, & \\
\Phi_j^\# & = \Phi_{q_j^\#, q_{j+1}^\#} & \text{for} \quad j = 1, \ldots, \nu^\# - 1\,, \\
\Phi_j^\# & = \varphi_{q_{\nu^\#}^\#, q_1^\#} & \text{for} \quad j = \nu^\#\,.
\end{array}
$$

Next, we drop the superscript $^\#$ in our discussion (except in the statement of our hypotheses A5-A15, where the notation of the previous sections will be maintained), and write ν, Ξ, $\boldsymbol{\xi}$, ξ_j, $\mathbf{I}$, I_j, ζ, $\mathbf{q}$, q_j, $\boldsymbol{\eta}$, η_j, t_j, τ_j, a, b, M_j, $\mathcal{M}_j$, S_j, Φ_j instead of $\nu^\#$, $\Xi^\#$, $\boldsymbol{\xi}^\#$, $\xi_j^\#$, $\mathbf{I}^\#$, $I_j^\#$, $\zeta^\#$, $\mathbf{q}^\#$, $q_j^\#$, $\boldsymbol{\eta}^\#$, $\eta_j^\#$, $t_j^\#$, $\tau_j^\#$, $a^\#$, $b^\#$, $M_j^\#$, $\mathcal{M}_j^\#$, $S_j^\#$, $\Phi_j^\#$. Then

$$
\begin{array}{lll}
\Xi = (\boldsymbol{\xi}, \zeta)\,, & \boldsymbol{\xi} = (\xi_1, \ldots, \xi_\nu)\,, & \zeta = (\mathbf{q}, \mathbf{I}, \boldsymbol{\eta})\,, \\
\mathbf{I} = (I_1, \ldots, I_\nu)\,, & \mathbf{q} = (q_1, \ldots, q_\nu)\,, & \boldsymbol{\eta} = (\eta_1, \ldots, \eta_\nu)\,.
\end{array}
$$

In order to state the transversality condition, and to define the notion of adjoint pair, we need a concept of "tangent cone" to a set. There are many nonequivalent definitions of tangent cone, and we choose for our purposes the notion of a "Boltyanskii approximating cone":

Definition 17.4.1. *Let S be a subset of a smooth manifold X, and let $\bar{s} \in S$. A* ***Boltyanskii approximating cone*** *to S at $\bar{s}$ is a closed convex cone K in the tangent space $T_{\bar{s}}X$ to X at $\bar{s}$ such that there exist a neighborhood V of 0 in $T_{\bar{s}}X$ and a continuous map $\mu : V \cap K \to X$ with the property that $\mu(V \cap K) \subseteq S$, $\mu(0) = \bar{s}$, and $\mu(v) = \bar{s} + v + o(\|v\|)$ as $v \to 0$ via values in $V \cap K$.* ◇

(The condition "$\mu(v) = \bar{s} + v + o(\|v\|)$ as $v \to 0$" appears to depend, in principle, on a choice of local coordinates about $\bar{s}$, but is in fact independent of that choice.)

We now define the notion of an "adjoint pair" along Ξ, and what it means for such a pair to be "Hamiltonian-maximizing." For that purpose, we first stipulate that

S1. *"tangent cone" means "Boltyanskii approximating cone."*

We then make the following assumptions.

A5. For $j \in \{1, \ldots, \nu^\# - 1\}$, K_j is a tangent cone to the set $S_{q_j^\#, q_{j+1}^\#}$ at $\partial_j \Xi^\#$.

A6. K_e is a tangent cone to the set $E_{q_1^\#, q_{\nu^\#}^\#}$ at $\partial \Xi^\#$.

(In particular, K_j is a subset of the tangent space $T_{\partial_j \Xi} \mathcal{M}_{q_j, q_{j+1}}$, which is canonically identified with the product $T_{\xi_j(\tau_j)} M_{q_j} \times T_{\xi_{j+1}(t_{j+1})} M_{q_{j+1}} \times \mathbb{R} \times \mathbb{R}$. Similarly, the cone K_e is a subset of $T_{\partial \Xi} \mathcal{M}_{q_\nu, q_1}$, which is canonically identified with $T_{\xi_\nu(b)} M_{q_\nu} \times T_{\xi_1(a)} M_{q_1} \times \mathbb{R} \times \mathbb{R}$.)

We will also write K_ν for K_e and $\partial_\nu \Xi$ for $\partial \Xi$, so K_j *is a tangent cone to* S_j *at* $\partial_j \Xi$ *for* $j = 1, \ldots, \nu$.

We recall that if V is a finite-dimensional real linear space and $S \subseteq V$, then the *polar* of S is the set

$$S^\perp = \{w \in V^\dagger : w \cdot v \le 1 \text{ for all } v \in S\},$$

where $V^\dagger$ is the dual of V. If S is a cone (that is, S is nonempty and such that $r \cdot v \in S$ whenever $v \in S$ and $r \ge 0$), then

$$S^\perp = \{w \in V^\dagger : w \cdot v \le 0 \text{ for all } v \in S\}.$$

It is well known that if S is a cone then $S^\perp$ is a closed convex cone, and $S^{\perp\perp}$ is the closed convex hull of S (using the standard identification of $V^{\dagger\dagger}$ with V), so $S^{\perp\perp} = S$ if and only if S is closed and convex.

Next, we assume

A7. If $j \in \{1, \ldots, \nu^\#\}$ then for almost every time $t \in I_j^\#$ the maps $x \to f_{q_j^\#}(x, \eta_j^\#(t), t)$ and $x \to L_{q_j^\#}(x, \eta_j^\#(t), t)$ are defined on a neighborhood of $\xi_j^\#(t)$ and are differentiable at $\xi_j^\#(t)$.

A8. If $j \in \{1, \ldots, \nu^\# - 1\}$, then $\Phi_{q_j^\#, q_{j+1}^\#}$ is differentiable at $\partial_j \Xi^\#$ along $S_{q_j^\#, q_{j+1}^\#}$.

A9. $\varphi_{q_{\nu^\#}^\#, q_1^\#}$ is differentiable at $\partial \Xi^\#$ along $E_{q_{\nu^\#}^\#, q_1^\#}$.

Remark 17.4.1. The precise meaning of "differentiability along a set" in A8 and A9 is as follows: a partially defined function h on a smooth manifold X is *differentiable along* S at a point $\bar{s}$ of a subset S of X if (a) for some neighborhood V of $\bar{s}$, the inclusion $V \cap S \subseteq \text{Domain}(h)$ holds, and (b) there exists a linear functional λ on the tangent space $T_{\bar{s}} X$ such that

$$\lim_{s \to \bar{s}, s \in S} \frac{h(s) - h(\bar{s}) - \lambda \cdot (s - \bar{s})}{\|s - \bar{s}\|} = 0. \tag{17.6}$$

Naturally, (17.6) makes sense relative to a choice of local coordinates about $\bar{s}$, but it is easy to verify that the validity of (17.6) does not depend on that choice. The linear functional λ need not be unique but, if K is a tangent cone to S at $\bar{s}$ (in the sense of S1), then the restriction of λ to K is unique, since $\lambda(v) = \lim_{\varepsilon\downarrow 0} \varepsilon^{-1}(h(\mu(\varepsilon v)) - h(\bar{s}))$ whenever $v \in K$, if μ is any map that satisfies the conditions of Definition 17.4.1. In particular, a condition such as "$\theta + \rho\nabla h(\bar{s}) \in K^{\perp}$" makes sense intrinsically, if θ is a given covector at $\bar{s}$ and ρ is a number. Indeed, it suffices to interpret the condition to mean "$\theta + \rho\lambda \in K^{\perp}$ for any λ such that (17.6) holds." ◇

For each $q \in Q$, we introduce a partially defined real-valued function H_q on $T^*M_q \times U_q \times \mathbb{R} \times \mathbb{R}$ by letting

$$H_q(x, p, u, p_0, t) = p \cdot f_q(x, u, t) - p_0 L_q(x, u, t) \tag{17.7}$$

for $(x, p, u, p_0, t) \in T^*M_q \times U_q \times \mathbb{R} \times \mathbb{R}$ such that $f_q(x, u, t)$ and $L_q(x, u, t)$ are defined.

Definition 17.4.2. *If S1 and A1-A9 hold, then an* ***adjoint pair along*** Ξ *is a pair* $(\boldsymbol{\psi}, \psi_0)$ *with the property that:*

- $\boldsymbol{\psi}$ *is a* ν*-tuple* $(\psi_1, \ldots, \psi_\nu)$ *such that each* ψ_j *is a field of covectors along* ξ_j *(that is,* ψ_j *is a map from the interval* I_j *to the cotangent bundle* $T^*M_{q_j}$*, with the property that* $\psi_j(t) \in T^*_{\xi_j(t)}M_{q_j}$ *for every* $t \in I_j$*);*
- *each* ψ_j *is absolutely continuous;*
- $\psi_0 \in \mathbb{R}$ *and* $\psi_0 \geq 0$*;*
- *each* ψ_j *satisfies the* ***adjoint equation***

$$\dot{\psi}_j(t) = -\frac{\partial H_{q_j}}{\partial x}(\xi_j(t), \psi_j(t), \eta_j(t), \psi_0, t) \text{ for a.e. } t \in I_j; \tag{17.8}$$

- *for each* $j \in \{1, \ldots, \nu - 1\}$*, the* ***switching condition***

$$(-\psi_j(\tau_j), \psi_{j+1}(t_{j+1}), h_j^+, -h_{j+1}^-) - \psi_0\omega_j \in K_j^{\perp} \tag{17.9}$$

holds, where $\omega_j = \nabla\Phi_{q_j,q_{j+1}}(\partial_j\Xi)$*, and*

$$h_j^+ = \begin{cases} \lim_{s\downarrow 0} \frac{1}{s}\int_{\tau_j - s}^{\tau_j} H_{q_j}(\xi_j(t), \psi_j(t), \eta_j(t), \psi_0, t)\, dt & \text{if the limit exists}, \\ 0 & \text{otherwise}, \end{cases}$$

$$h_j^- = \begin{cases} \lim_{s\downarrow 0} \frac{1}{s}\int_{t_j}^{t_j + s} H_{q_j}(\xi_j(t), \psi_j(t), \eta_j(t), \psi_0, t)\, dt & \text{if the limit exists}, \\ 0 & \text{otherwise}. \end{cases}$$

Definition 17.4.3. *If S1 and A1-A9 hold, $(\boldsymbol{\psi}, \psi_0)$ is an adjoint pair along Ξ, and $j \in \{1, \dots, \nu\}$, we say that $(\boldsymbol{\psi}, \psi_0)$ satisfies the* ***weak Hamiltonian maximization condition on the j-th interval*** *if for every $u \in U_{q_j}$ the inequality*

$$H_{q_j}(\xi_j(t), \psi_j(t), u, \psi_0, t) \leq H_{q_j}(\xi_j(t), \psi_j(t), \eta_j(t), \psi_0, t) \tag{17.10}$$

holds for almost all $t \in I_j$. We say that $(\boldsymbol{\psi}, \psi_0)$ satisfies the ***strong Hamiltonian maximization condition on the j-th interval*** *if the identity*

$$H_{q_j}(\xi_j(t), \psi_j(t), \eta_j(t), \psi_0, t) = \max\{H_{q_j}(\xi_j(t), \psi_j(t), u, \psi_0, t) : u \in U_{q_j}\} \tag{17.11}$$

holds for almost all $t \in I_j$. ◇

Notice that the only difference between "weak" and "strong" Hamiltonian maximization is that weak maximization says that (17.10) holds, for each $u \in U_{q_j}$, at all points in the complement of a "bad" null set $B_j(u)$ that could depend on u, whereas strong maximization says that $B_j(u)$ can be chosen independently of u.

Weak maximization implies strong maximization in several important cases, such as, for example, (a) when the control sets are separable topological spaces and the dynamics and Lagrangian are continuous with respect to the control, and (b) when the dynamics and Lagrangian do not depend explicitly on time.

Proposition 17.4.1. *Let $j \in \{1, \dots, \nu\}$. Assume that either*

(1) U_{q_j} is a separable topological space and the maps

$$U_{q_j} \ni u \rightarrow (f_{q_j}(\xi_j(t), u, t), L_{q_j}(\xi_j(t), u, t)) \in TM_{q_j} \times \mathbb{R}$$

are continuous for almost every $t \in I_j$,

or

(2) there exists a compact subset X_j of the set space M_{q_j}, containing the set $\xi_j(I_j) \stackrel{\text{def}}{=} \{\xi_j(t) : t \in I_j\}$ and such that $f_{q_j}(x, u, t)$ and $L_{q_j}(x, u, t)$ are defined for all $(x, u, t) \in X_j \times U_{q_j} \times I_j$ and do not depend on t.

Then weak Hamiltonian maximization on the j-th interval implies strong Hamiltonian maximization on the j-th interval.

Proof. If V is a countable dense subset of U_{q_j}, then weak maximization implies that there exists a null subset B of I_j such that (17.10) holds whenever

$t \in I_j \backslash B$ and $u \in V$. If the continuity hypotheses of (1) hold, then (17.10) holds whenever $t \in I_j \backslash B$, and $u \in U_{q_j}$, so (17.11) holds for all $t \in I_j \backslash B$.

If (2) holds, then for each $u \in U_{q_j}$ the map

$$X_j \ni x \to f_{q_j}(x, u, t) \stackrel{\text{def}}{=} F^u(x) \in TM_{q_j}$$

is a continuous vector field on X_j, and the scalar function

$$X_j \ni x \to L_{q_j}(x, u, t) \stackrel{\text{def}}{=} L^u(x) \in \mathbb{R}$$

is continuous. Let $VF(X_j)$ denote the space of all continuous vector fields on X_j, endowed with the topology of uniform convergence. Then $VF(X_j)$ is metrizable and separable. Clearly, the space $C^0(X_j, \mathbb{R})$ of continuous real-valued functions on X_j is metric and separable. It follows that the set $\mathcal{W} = \{(F^u, L^u) : u \in U_{q_j}\}$ is a subset of the separable metrizable space $VF(X_j) \times C^0(X_j, \mathbb{R})$. Hence $\mathcal{W}$ is a metrizable separable space. So there exists a countable subset V of U_{q_j} such that the set $\mathcal{W}^V \stackrel{\text{def}}{=} \{(F^u, L^u) : u \in V\}$ is dense in $\mathcal{W}$. For each $u \in V$, let B^u be a null subset of I_j such that (17.10) holds for all $t \in I_j \backslash B^u$. Let $B = \cup_{u \in V} B^u$. Then B is a null set, and (17.10) holds whenever $u \in V$ and $t \in I_j \backslash B$. If $u \in U_{q_j}$ is arbitrary, then there exists a sequence $\{u_k\}_{k=1}^{\infty}$ in U_{q_j} such that $F^{u_k} \to F^u$ and $L^{u_k} \to L^u$ uniformly on X_j as $k \to \infty$. Then (17.11) holds whenever $t \in I_j \backslash B$. ◊

Definition 17.4.4. *If S1 and A1-A9 hold, and* $(\boldsymbol{\psi}, \psi_0)$ *is an adjoint pair along* $\Xi^{\#}$, *we say that* $(\boldsymbol{\psi}, \psi_0)$ *satisfies the* ***transversality condition*** *for* $\mathcal{E}$ *and* φ *if*

$$(-\psi_\nu(b), \psi_1(a), h_\nu^+, -h_1^-) - \psi_0 \omega_e \in K_e^\perp, \tag{17.12}$$

where $\omega_e = \nabla \varphi_{q_\nu, q_1}(\partial \Xi)$, *and the numbers* $h_j^\pm$ *are those that were introduced in Definition 17.4.2.* ◊

Definition 17.4.5. *If* $(\boldsymbol{\psi}, \psi_0)$ *is an adjoint pair along* Ξ, *we say that* $(\boldsymbol{\psi}, \psi_0)$ *satisfies the* ***nontriviality condition*** *if either* $\psi_0 \neq 0$ *or at least one of the functions* ψ_j *is not identically zero.* ◊

We have now completed the list of definitions needed to make the statement of the maximum principle meaningful, but in order to make the result true we need additional conditions. In order to state these conditions, we first give some preliminary definitions. For this purpose, we first define the "augmented dynamical laws"

$$\tilde{f}_q(x, u, t) \stackrel{\text{def}}{=} (f_q(x, u, t), L_q(x, u, t)), \tag{17.13}$$

so $\tilde{f}_q(x, u, t) \in T_x M_q \times \mathbb{R}$ whenever $x \in M_q$, $u \in U_q$, $t \in \mathbb{R}$ are such that both $f_q(x, u, t)$ and $L_q(x, u, t)$ are defined.

Definition 17.4.6. *We say that Σ is **autonomous at location** q if the set of those $(x,u) \in M_q \times U_q$ for which $\tilde{f}_q(x,u,t)$ is defined is independent of t and, for (x,u) in that set, the value of $\tilde{f}_q(x,u,t)$ is independent of t.* ◇

Next, we define various types of "control variations." In the following discussion, "$\lceil$" denotes "restriction." If U is a set, a,b,c are real numbers such that $a \leq b \leq c$, and $\eta_1 : [a,b] \to U$, $\eta_2 : [b,c] \to U$ are maps, then the ***concatenation of*** η_1 ***and*** η_2 is the map $\eta_2 * \eta_1 : [a,c] \to U$ given by

$$(\eta_2 * \eta_1)(t) = \begin{cases} \eta_1(t) & \text{if} \quad a \leq t \leq b\,, \\ \eta_2(t) & \text{if} \quad b < t \leq c\,. \end{cases}$$

So $\eta_2 * \eta_1$ is defined only when the domains I_{η_i} of η_i are compact intervals such that $\max(I_{\eta_1}) = \min(I_{\eta_2})$. Clearly, concatenation is an associative operation, in the sense that $(\eta_3 * \eta_2) * \eta_1$ is defined if and only if $\eta_3 * (\eta_2 * \eta_1)$ is defined, and in that case $(\eta_3 * \eta_2) * \eta_1 = \eta_3 * (\eta_2 * \eta_1)$.

Definition 17.4.7. *Let U be a set, let a,b be real numbers such that $a \leq b$, and let $\eta : [a,b] \to U$ be a map. Then*

1. *if $r \geq 0$, the **right r-translation** and **left r-translation of** η **by** r are the maps $T_r^+(\eta) : [a+r, b+r] \to U$, $T_r^-(\eta) : [a-r, b-r] \to U$, given by*

$$T_r^+(\eta)(t) = \eta(t-r) \quad \text{for} \quad t \in [a+r, b+r]\,,$$
$$T_r^-(\eta)(t) = \eta(t+r) \quad \text{for} \quad t \in [a-r, b-r]\,;$$

2. *if $a \leq c \leq b$, $r \geq 0$, and η' is a U-valued map whose domain contains the interval $[c, c+r]$, then the **right r-expansion of the map** η **by** η' **at time** c is the map $E^+_{r,c,\eta'}(\eta) : [a, b+r] \to U$ given by*

$$E^+_{r,c,\eta'}(\eta) = T_r^+(\eta\lceil[c,b]) * \eta'\lceil[c,c+r] * \eta\lceil[a,c]\,;$$

if $c = b$, then we omit the subscript c, so $E^+_{r,\eta'}(\eta) \stackrel{\text{def}}{=} \eta'\lceil[b, b+r] * \eta$;

3. *if $a \leq c \leq b$, $r \geq 0$, and η' is a U-valued map whose domain contains the interval $[c-r, c]$, then the **left r-expansion of the map** η **by** η' **at time** c is the map $E^-_{r,c,\eta'}(\eta) : [a-r, b] \to U$ given by*

$$E^-_{r,c,\eta'}(\eta) = \eta\lceil[c,b] * \eta'\lceil[c-r,c] * T_r^-(\eta\lceil[a,c])\,;$$

if $c = a$, then we omit the subscript c, so $E^-_{r,\eta'}(\eta) \stackrel{\text{def}}{=} \eta * \eta'\lceil[a-r, a] * \eta\lceil[a,c]$;

3. *if $a \leq c_1 \leq c_2 \leq b$, then*

   a. *the **right shortening of** η **at** $[c_1, c_2]$ is the map*

$$S^+_{c_1,c_2}(\eta) = \eta\lceil[c_2, b] * T^+_{c_2-c_1}(\eta\lceil[a, c_1])\,;$$

b. the ***left shortening of*** η ***at*** $[c_1, c_2]$ *is the map*

$$S^-_{c_1,c_2}(\eta) = T^-_{c_2-c_1}(\eta\lceil[c_2, b]) * \eta\lceil[a, c_1]\,;$$

4. if A *is a subset of* $[a, b]$, *and* θ *is a* U*-valued map whose domain contains* A, *then the* ***replacement of*** η ***by*** θ ***on*** A *is the map* $R_{A,\theta}(\eta) : [a, b] \to U$ *given by*

$$R_{A,\theta}(\eta) = \begin{cases} \eta(t) & \text{if} \quad t \in [a, b]\backslash A\,, \\ \theta(t) & \text{if} \quad t \in A\,. \end{cases} \qquad \diamondsuit$$

We let $\mu(U)$ denote the set of all U-valued maps defined on compact intervals, so

$$\mu(U) \stackrel{\text{def}}{=} \bigcup_{-\infty < a \leq b < \infty} U^{[a,b]}\,, \tag{17.14}$$

where $U^{[a,b]}$ is the set of all mappings from $[a, b]$ to U. Then the transformations T^+_r, T^-_r, $R_{A,\theta}$, $S^+_{c_1,c_2}$, $S^-_{c_1,c_2}$, $E^+_{r,c,\eta'}$, $E^-_{r,c,\eta'}$, $E^+_{r,\eta'}$, $E^-_{r,\eta'}$, are partially defined maps from $\mu(U)$ to $\mu(U)$. (For example, $R_{A,\theta}(\eta)$ is defined if and only if $A \subset \text{Domain}(\eta) \cap \text{Domain}(\theta)$.) So we can define *iterates* of these maps. For example, if $\mathbf{A} = (A_1, \ldots, A_m)$ is an m-tuple of subsets of $\mathbb{R}$, and $\mathbf{u} = (u_1, \ldots, u_m)$ is an m-tuple of members of U (regarded as constant U-valued maps) then $R_{\mathbf{A},\mathbf{u}}$ is the composite map

$$R_{\mathbf{A},\mathbf{u}} \stackrel{\text{def}}{=} R_{A_1,u_1} \circ R_{A_2,u_2} \circ \ldots \circ R_{A_m,u_m}\,. \tag{17.15}$$

Similarly, $E^\sigma_{\mathbf{r},\mathbf{c},\theta}$ and $E^\sigma_{\mathbf{r},\theta}$ are the composite maps

$$E^\sigma_{\mathbf{r},\mathbf{c},\theta} \stackrel{\text{def}}{=} E^{\sigma_1}_{r_1,c_1,\theta_1} \circ E^{\sigma_2}_{r_2,c_2,\theta_2} \circ \ldots \circ E^{\sigma_m}_{r_m,c_m,\theta_m}\,,$$
$$E^\sigma_{\mathbf{r},\theta} \stackrel{\text{def}}{=} E^{\sigma_1}_{r_1,\theta_1} \circ E^{\sigma_2}_{r_2,\theta_2} \circ \ldots \circ E^{\sigma_m}_{r_m,\theta_m}\,,$$

if $\mathbf{r} = (r_1, \ldots, r_m)$ and $\mathbf{c} = (c_1, \ldots, c_m)$ are m-tuples of nonnegative real numbers, $\theta = (\theta_1, \ldots, \theta_m)$ is an m-tuple of members of $\mu(U)$, and $\sigma = (\sigma_1, \ldots, \sigma_m)$ is a sequence of members of $\{-, +\}$.

Definition 17.4.8. *We say that* $\mathbf{U}_{q_j}$ *is a* ***fixed-time measurable weak variational neighborhood of the control*** η_j ***if***

(FTMVN) For every positive integer m *and every* m*-tuple* $\mathbf{u} = (u_1, \ldots, u_m)$ *of members of* U_{q_j} *there exists a positive number* δ *such that, whenever* $\mathbf{A} = (A_1, \ldots, A_m)$ *is an* m*-tuple of pairwise disjoint measurable subsets of* I_j *for which* $\text{meas}(A_1 \cup \ldots \cup A_m) \leq \delta$, *it follows that the function* $R_{\mathbf{A},\mathbf{u}}(\eta)$ *belongs to* $\mathbf{U}_{q_j}$. $\diamondsuit$

Definition 17.4.9. *We say that* U_{q_j} *is a **weak variable-time measurable variational neighborhood of the control*** η_j *if*

(WVTMVN) *For every positive integer* m *and every triple* $(\mathbf{u},\mathbf{v},\mathbf{w})$ *of* m*-tuples* $\mathbf{u}=(u_1,\ldots,u_m)$, $\mathbf{v}=(v_1,\ldots,v_m)$, $\mathbf{w}=(w_1,\ldots,w_m)$ *of members of* U_{q_j} *there exists a positive number* δ *with the property that, if* $\alpha,\beta,r_1,\ldots,r_m,s_1,\ldots,s_m$ *are nonnegative real numbers whose sum does not exceed* δ, *and* $\mathbf{A}=(A_1,\ldots,A_m)$ *is an arbitrary* m*-tuple of pairwise disjoint measurable subsets of* $[t_j+\alpha,\tau_j-\beta]$ *such that* $\mathrm{meas}(A_1\cup\ldots\cup A_m)\le\delta$, *then the function*

$$V_{\mathbf{r},\mathbf{v},\alpha,\mathbf{s},\mathbf{w},\beta,\mathbf{A},\mathbf{u}}(\eta)\stackrel{\mathrm{def}}{=}E^-_{\mathbf{r},\mathbf{v}}(E^+_{\mathbf{s},\mathbf{w}}(S^+_{t_j,t_j+\alpha}(S^-_{\tau_j-\beta,\tau_j}(R_{\mathbf{A},\mathbf{u}}(\eta)))))\tag{17.16}$$

where $\mathbf{r}=(r_1,\ldots,r_m)$ *and* $\mathbf{s}=(s_1,\ldots,s_m)$) *belongs to* U_{q_j}. ◇

Definition 17.4.10. *We say that* U_{q_j} *is a **strong variable-time measurable variational neighborhood of the control*** η_j *if*

(SVTMVN) *For every positive integer* m *and every* m*-tuple* $\mathbf{u}=(u_1,\ldots,u_m)$ *of members of* U_{q_j} *there exists a positive number* δ *with the property that, if* $\mathbf{c}=(c_1,\ldots,c_m)$, $\mathbf{d}=(d_1,\ldots,d_m)$ *are* m*-tuples of members of* I_j *such that* $t_j<c_1<\ldots<c_m<\tau_j$, $t_j<d_1<\ldots<d_m<\tau_j$, *and* $c_i\ne d_k$ *for all* i,k, (σ,σ') *is a pair of sequences* $\sigma=(\sigma_1,\ldots,\sigma_m)$, $\sigma'=(\sigma'_1,\ldots,\sigma'_m)$ *of members of* $\{-,+\}$, $\alpha,\beta,r_1,\ldots,r_m,s_1,\ldots,s_m$ *are nonnegative real numbers such that*

$$\alpha+\beta+r_1+\cdots+r_m+s_1+\cdots+s_m\le\delta\,,$$

and $\mathbf{A}=(A_1,\ldots,A_m)$ *is an arbitrary* m*-tuple of pairwise disjoint measurable subsets of the interval* $[t_j+\alpha,\tau_j-\beta]$ *satisfying* $\mathrm{meas}(A_1\cup\ldots\cup A_m)\le\delta$, *then the function*

$$W_{\mathbf{r},\mathbf{d},\sigma',\mathbf{s},\mathbf{c},\mathbf{r},\sigma,\mathbf{A},\mathbf{u}}(\eta)\stackrel{\mathrm{def}}{=}E^{\sigma'}_{\mathbf{s},\mathbf{d},\eta}(S^{\sigma}_{\mathbf{c},\mathbf{r}}(R_{\mathbf{A},\mathbf{u}}(\eta)))\tag{17.17}$$

where $\mathbf{r}=(r_1,\ldots,r_m)$ *and* $\mathbf{s}=(s_1,\ldots,s_m)$) *belongs to* U_{q_j}. ◇

Definition 17.4.11. *For each of the three types (fixed-time, weak variable-time, and strong variable-time) of "measurable variational neighborhoods" introduced above, the corresponding notion of an **interval variational neighborhood** is defined in exactly the same way, except that the sets* A_k *are required to be intervals rather than general measurable subsets.* ◇

Thus, we have a total of six different types of "variational neighborhoods" of a control η.

Remark 17.4.2. Our definitions of the various types of variational neighborhoods say, in all six cases, that $\boldsymbol{U}_{q_j}$ is a "variational neighborhood" of η_j if it contains all the controls obtained from η_j by making "sufficiently small" variations in an appropriate class $\mathcal{V}$. Where the six definitions differ is, naturally, in the choice of $\mathcal{V}$. The fixed-time variational neighborhood corresponds to choosing $\mathcal{V}$ to consist of *substitutions of finitely many constant controls for the control η, without changing the total time interval.* Weak variable-time variational neighborhoods are obtained when, in addition, we allow *shortenings at the endpoints and expansions by constant controls at the endpoints.* Finally, strong variable-time variational neighborhoods arise when we allow *shortenings and expansions by constant controls not only at the endpoints but at all points of I_j.*

Those readers who prefer neighborhoods to be derived from a topology can easily verify that ***for each of our six definitions of "neighborhood," if we define an "open subset" of $\mu(U)$ to be a subset which is a neighborhood of each of its points, then the set of open subsets constitutes a topology, and the corresponding notion of neighborhood is exactly the one of the original definition.*** ◇

Definition 17.4.12. *If $q \in \mathcal{Q}$, J is an interval, $\eta : J \to \boldsymbol{U}_q$ is a function, and $x \in M_q$, we say that (x,t) is a* ***forward regular point for*** *η if*

(1) there exists a positive $\bar{\delta}$ such that $\tilde{f}_q(x', \eta(s), s)$ is defined whenever x' belongs to M_q, $\operatorname{dist}(x', x) \leq \bar{\delta}$, and $t \leq s \leq t+\bar{\delta}$, and depends continuously on x' and measurably on s,

and

(2) if $\sigma^{\eta,q}_{x,t,\delta}$ is the modulus of continuity defined by

$$\sigma^{\eta,q}_{x,t,\delta}(s) \stackrel{\text{def}}{=} \sup\{\|\tilde{f}_q(x', \eta(s), s) - \tilde{f}_q(x, \eta(t), t)\| : x' \in M_q, d(x', x) \leq \delta\}, \tag{17.18}$$

then the number

$$\nu(\delta) \stackrel{\text{def}}{=} \frac{1}{\delta} \int_t^{t+\delta} \sigma^{\eta,q}_{x,t,\delta}(s)\, ds \tag{17.19}$$

goes to zero as $\delta \downarrow 0$. ◇

Remark 17.4.3. In Definition 17.4.12, "dist" is the distance function corresponding to any Riemannian metric g on a neighborhood of x in M_q. It is easy to see that the forward regularity property does not depend on the choice of g. ◇

"Backward regularity" is defined in the same way, except that in (1) the condition "$t \leq s \leq t + \bar{\delta}$" is replaced by "$t - \bar{\delta} \leq s \leq t$," and in (2) the integral of (17.19) is taken over the interval $[t-\delta, t]$ rather than over $[t, t+\delta]$.

We are now ready to state our technical conditions. In the statements, we will use

$$\tilde{Y}_j(t) \stackrel{\text{def}}{=} \frac{\partial \tilde{f}_{q_j^\#}}{\partial x}(\xi_j^\#(t), \eta_j^\#(t), t)\,. \tag{17.20}$$

We recall that the existence of the right-hand side of (17.20) for almost every $t \in I_j$ is guaranteed by A7.

A10. For each $j \in \{1, \dots, \nu^\#\}$ and each $u \in U_{q_j^\#}$

1. $\tilde{f}_{q_j^\#}(x, u, t)$ is defined for all (x, t) belonging to some neighborhood $\mathcal{N}_j(u)$ in $M_{q_j^\#} \times \mathbb{R}$ of the graph $G(\xi_j^\#)$ (cf. (17.5);
2. for each $t \in \mathbb{R}$, the map $x \to \tilde{f}_{q_j^\#}(x, u, t)$ is continuous on the set $\{x : (x, t) \in \mathcal{N}_j(u)\}$,
3. for each $x \in M_{q_j^\#}$, the map $t \to \tilde{f}_{q_j^\#}(x, u, t)$ is Lebesgue measurable on the set $\{t : (x, t) \in \mathcal{N}_j(u)\}$.

A11. The maps $I_j^\# \ni t \to \tilde{f}_{q_j^\#}(x, \eta_j^\#(t), t)$ are Lebesgue measurable for every $j \in \{1, \dots \nu^\#\}$, $x \in M_{q_j^\#}$.

A12. For every $j \in \{1, \dots, \nu^\#\}$, the map $\tilde{Y}_j$ is Lebesgue integrable on $I_j^\#$.

A13. $\Phi_{q,q'}$ and $\varphi_{q,q'}$ are finite-valued and continuous on the sets $S_{q,q'}$, $E_{q,q'}$, for all $(q, q') \in \mathcal{Q} \times \mathcal{Q}$.

A14. The differentiability at the point $x = \xi_j^\#(t)$ of the augmented vector fields $M_{q_j^\#} \ni x \to \tilde{f}_{q_j^\#}(x, \eta_j^\#(t), t)$ has the following Carathéodory-type uniformity: there exist measurable functions $k_j^\delta : I_j^\# \to [0, +\infty]$ such that

$$\lim_{\delta \downarrow 0} \int_{I_j^\#} k_j^\delta(t)\, dt = 0\,,$$

having the property that, for almost all $t \in I_j^\#$, the inequality

$$\left\| \tilde{f}_{q_j^\#}(x, \eta_j^\#(t), t) - \tilde{f}_{q_j^\#}(\xi_j^\#(t), \eta_j^\#(t), t) - \tilde{Y}_j(t) \cdot (x - \xi^\#(t)) \right\| \\ \leq k_j^\delta(t) \|x - \xi_j^\#(t)\| \qquad (17.21)$$

holds whenever $\|x - \xi_j^\#(t)\| \leq \delta$.

A15. For each $j \in \{1, \dots, \nu^\#\}$, either

I. one of the following two conditions holds:

I.1. $\boldsymbol{U}_{q_j^\#}$ is a weak variable-time measurable variational neighborhood of $\eta_j^\#$, $(\xi_j^\#(t_j^\#), t_j^\#)$ is a backward regular point for all constant controls and a forward regular point for $\eta_j^\#$, and $(\xi_j^\#(\tau_j^\#), \tau_j^\#)$ is forward regular for all constant controls and backward regular for $\eta_j^\#$.

I.2. $\boldsymbol{U}_{q_j^\#}$ is a fixed-time measurable variational neighborhood of $\eta_j^\#$, the switching set $S_j^\#$ is a subset of the Cartesian product $M_j^\# \times M_{j+1}^\# \times \{\tau_j^\#\} \times \mathbb{R}$, and the switching set $S_{j-1}^\#$ is a subset of $M_{j-1}^\# \times M_j^\# \times \mathbb{R} \times \{t_j^\#\}$.

or

II. the following two conditions hold:

a. the "constant control augmented vector fields" satisfy integral bounds

$$\|\tilde{f}_{q_j^\#}(x, u, t)\| \leq k_{u,j}(t) \quad \text{for all} \quad (x, u, t) \in \mathcal{N}_j(u)\,, \qquad (17.22)$$

where each $k_{u,j}$ is a nonnegative Lebesgue integrable function on $\mathbb{R}$,

b. one of conditions I.1, I.2 holds, with "measurable" replaced by "interval,"

or

III. Σ is autonomous at location $q_j^\#$, and $\boldsymbol{U}_{q_j^\#}$ is a strong variable-time interval variational neighborhood of $\eta_j^\#$.

Theorem 17.4.1. *If S1 and A5-A15 hold, then the maximum principle is true. Moreover, the adjoint vector can be chosen in such a way that, in addition, the Hamiltonian function* $t \to H_{q_j^\#}(\xi_j^\#(t), \psi_j(t), \eta_j^\#(t), \psi_0, t)$ *is almost everywhere constant on* $I_j^\#$ *for every* j *for which A15.III holds.* ◊

Theorem 17.4.1 is stronger than the classical versions of the maximum principle given, e.g., in Pontryagin *et al.* [3] and Berkovitz [1], because it does not require the maps to be of class C^1, or even Lipschitz continuous, and only assumes that they are differentiable along the reference trajectory.

17.5 Outline of the proof

For $n \in \mathbb{N}$, we use $\mathbb{R}^n$, $\mathbb{R}_n$, $\mathbb{R}^{n\times n}$ to denote, respectively, the spaces of n-dimensional real column vectors, n-dimensional real row vectors, and $n \times n$ real matrices.

For simplicity, we will assume that the state spaces M_q are open subsets of Euclidean spaces $\mathbb{R}^{n_q}$.

Remark 17.5.1. The proof for the general manifold case is identical, except that more care has to be exercised to define intrinsically various objects such as $\frac{\partial f_q}{\partial x}$. Alternatively, one could prove the theorem on manifolds by an extra "hybridization" of our problem, in which each curve ξ_j is covered by domains Ω_j^k of coordinate patches (Ω_j^k, κ_j^k), and these coordinate patches are treated as new locations, with state spaces $\kappa_j^k(\Omega_j^k) \subseteq \mathbb{R}^{\dim M_j}$, while the changes of coordinates are treated as switchings. ◊

Then the vector fields f_q are $\mathbb{R}^{n_q}$-valued maps, and the differentials $\frac{\partial f_q}{\partial x}$ are $\mathbb{R}^{n_q\times n_q}$-valued. We let $n_j = n_{q_j}$, and define

$$Y_j(t) = \frac{\partial f_{q_j}}{\partial x}(\xi_j(t), \eta_j(t), t)\,, \quad y_j(t) = \frac{\partial L_{q_j}}{\partial x}(\xi_j(t), \eta_j(t), t)\,, \tag{17.23}$$

so Y_j and y_j are integrable functions, with values in the spaces $\mathbb{R}^{n_j\times n_j}$ and $\mathbb{R}_{n_j}$, respectively.

The proof of Theorem 17.4.1 is carried out by means of "needle variations." Our variations will be *set-valued maps*, because they are constructed by associating to each value of the variation parameter "the solution" of a certain initial value problem with a continuous but non-Lipschitz right-hand side. Since the solution need not be unique, the object associated to the parameter value will in fact be a *set* of curves rather than a single curve. We introduce the following notation: "$F : A \longrightarrow\!\!\!\rightarrow B$" means "$F$ is a set-valued map from A to B." If $F : A \longrightarrow\!\!\!\rightarrow B$, then the *graph* of F is the set $\{(a,b) : a \in A, b \in F(a)\}$. (Often, one defines a set-valued map from A to B to be a subset of $A \times B$. In that case, the graph of F is just F.)

We first remark that for every $j \in \{1, \dots, \nu\}$ there exists an integrable function $\hat{k}_j$ on I_j such that the integral bound

$$\|\tilde{f}_{q_j}(x,\eta_j(t),t)\| \leq \hat{k}_j(t) \quad \text{for all} \quad (x,t) \in \mathcal{N}_j \tag{17.24}$$

holds for some neighborhood $\mathcal{N}_j$ of the graph of ξ_j in $M_{q_j} \times I_j$. (This follows from A12, A14, the absolute continuity of ξ_j, and the integrability of the function $t \to L_{q_j}(\xi_j(t), \eta_j(t), t)$.)

We will begin by doing the proof for the case corresponding to Part II of condition A15, assuming the the "interval" analogue of I.2 holds. We will then discuss how to modify the proof to take care of the other cases.

So we assume that the integral bounds of A15.II.a hold for all j, so that for every j and every $u \in U_{q_j}$ there exists an integrable function $k_{u,j}$ on $\mathbb{R}$ such that (17.22) holds. We also assume that every U_{q_j} is a fixed-time interval variational neighborhood of η_j, and that each switching set S_j is contained in the product $M_j \times M_{j+1} \times \{\tau_j\} \times \{t_{j+1}\}$. (As explained before, this includes in particular the set S_ν, which is defined to be E_{q_ν,q_1}. Recall that $M_{\nu+1} = M_1$ and $t_{\nu+1} = t_1$.) It follows that $S_j = S_j^0 \times \{\tau_j\} \times \{t_{j+1}\}$ for some subset S_j^0 of $M_j \times M_{j+1}$.

Fix a positive integer m and then choose, for each index $j \in \{1,\ldots,\nu\}$, an m-tuple $\mathbf{u}_j = (u_j^1, u_j^2, \ldots, u_j^m) \in U_{q_j}^m$ of control values. Let

$$\mathcal{N}_j(\mathbf{u}) = \mathcal{N}_j(u_j^1) \cap \ldots \cap \mathcal{N}_j(u_j^m).$$

Then the Scorza-Dragoni theorem together with the integral bounds implies that for each j there exists a null subset $B_j(\mathbf{u})$ of $\mathbb{R}$ such that if $(x,t) \in \mathcal{N}_j(\mathbf{u})$ and $t \notin B_j(\mathbf{u})$ then (x,t) is a forward and backward regular point for η_j and for each of the constant controls $u_j^1, \ldots, u_j^m$.

Pick, for each j, m distinct points $s_j^1, \ldots, s_j^m$ of I_j, not belonging to $B_j(\mathbf{u})$, and such that $s_j^1 < s_j^2 < \ldots < s_j^m < \tau_j$. Fix a small positive real number $\bar{\varepsilon}$. Let $\mathbf{E} = ([0,\bar{\varepsilon}]^m)^\nu$, so $\mathbf{E}$ is the set of all ν-tuples $\boldsymbol{\varepsilon} = (\boldsymbol{\varepsilon}_1, \ldots, \boldsymbol{\varepsilon}_\nu)$ of m-tuples $\boldsymbol{\varepsilon}_j = (\varepsilon_j^1, \ldots, \varepsilon_j^m)$ of numbers ε_j^k such that $0 \leq \varepsilon_j^k \leq \bar{\varepsilon}$ for all j,k. Then for each $\boldsymbol{\varepsilon} \in \mathbf{E}$ we can construct the modified control

$$\zeta^{\mathbf{u},\mathbf{s},\boldsymbol{\varepsilon}} \stackrel{\text{def}}{=} (\mathbf{q}, \mathbf{I}, \boldsymbol{\eta}^{\mathbf{u},\mathbf{s},\boldsymbol{\varepsilon}}),$$

having the same switching strategy $\mathbf{q}$ and the same sequence $\mathbf{I}$ of intervals as ζ, such that

$$\boldsymbol{\eta}^{\mathbf{u},\mathbf{s},\boldsymbol{\varepsilon}} = (\eta_1^{\mathbf{u},\mathbf{s},\boldsymbol{\varepsilon}}, \ldots, \eta_\nu^{\mathbf{u},\mathbf{s},\boldsymbol{\varepsilon}}),$$

$$\eta_j^{\mathbf{u},\mathbf{s},\boldsymbol{\varepsilon}}(t) = \begin{cases} \eta_j(t) & \text{if} \quad t \in I_j \backslash (\cup_{k=1}^m [s_j^k, s_j^k + \varepsilon_j^k]), \\ u_j^k & \text{if} \quad t \in [s_j^k, s_j^k + \varepsilon_j^k]. \end{cases}$$

Fix small balls $D_1, \ldots, D_m$ in the tangent spaces $T_{\xi_1(t_1)}M_1, \ldots, T_{\xi_\nu(t_\nu)}M_\nu$, centered at 0, and smooth maps $\Delta_j : D_j \to M_j$ such that $\Delta_j(0) = \xi_j(t_j)$ and the differential of Δ_j at 0 is the identity map of $T_{\xi_j(t_j)}M_j$.

Let $\mathbf{P} = \mathbf{E} \times \mathbf{D}$. For each $(\boldsymbol{\varepsilon}, \mathbf{v}) \in \mathbf{P}$ let $\xi^{\mathbf{u},\mathbf{s}}(\boldsymbol{\varepsilon}, \mathbf{v})$ be the set of all ν-tuples $\tilde{\xi} = (\tilde{\xi}_1, \ldots, \tilde{\xi}_\nu)$ such that each $\tilde{\xi}_j$ is an absolutely continuous curve in M_j, defined on I_j, which is a solution of the initial value problem

$$\begin{cases} \dot{x} &= f_{q_j}(x, \eta_j^{\mathbf{u},\mathbf{s},\boldsymbol{\varepsilon}}(t), t) \quad \text{for} \quad \text{a.e.} \quad t \in I_j\,, \\ x(t_j) &= \Delta_j(v_j)\,. \end{cases} \tag{17.25}$$

Notice that in general $\xi^{\mathbf{u},\mathbf{s}}(\boldsymbol{\varepsilon}, \mathbf{v})$ will not consist of a single element, because the right-hand of the differential equation of (17.25) is just continuous with respect to x, so uniqueness of solutions of (17.25) is not guaranteed. On the other hand, global existence of solutions on I_j follows— if $\bar{\varepsilon}$ and the balls D_j are small enough—because (a) the differential equation of (17.25) has local solutions —thanks to the continuity of the right-hand side with respect to x, the measurability with respect to t, and the integral bounds (17.22) and (17.24), and (b) a solution of (17.25), as long as it exists, will stay in any prespecified neighborhood of the graph of ξ_j —provided, once again, that $\bar{\varepsilon}$ and the D_j are small enough—due to A12 and A14 (which imply bounds

$$\|\tilde{f}_{q_j}(x, \eta_j(t), t) - \tilde{f}_{q_j}(\xi_j(t), \eta_j(t), t)\| \leq \kappa_j(t)\|x - \xi_j(t)\|\,,$$

with κ_j integrable, valid whenever $\|x - \xi_j(t)\| \leq \delta$ and δ is small enough), and Gronwall's inequality.

We let $\mathcal{V}^{\mathbf{u},\mathbf{s}}$ be the set-valued map $\mathbf{P} \ni (\boldsymbol{\varepsilon}, \mathbf{v}) \to \xi^{\mathbf{u},\mathbf{s}}(\boldsymbol{\varepsilon}, \mathbf{v})$. It then follows by a standard application of the Ascoli-Arzelà theorem that the graph of $\mathcal{V}^{\mathbf{u},\mathbf{s}}$— i.e., the set of all pairs $((\boldsymbol{\varepsilon}, \mathbf{v}), \tilde{\xi})$ such that $\tilde{\xi} \in \xi^{\mathbf{u},\mathbf{s}}(\boldsymbol{\varepsilon}, \mathbf{v})$—is a compact subset of the product $\mathbf{P} \times \mathbf{C}^0(\mathbf{I}, \mathbf{M})$, where $\mathbf{C}^0(\mathbf{I}, \mathbf{M}) \stackrel{\text{def}}{=} C^0(I_1, M_1) \times \ldots \times C^0(I_\nu, M_\nu)$. So $\mathcal{V}^{\mathbf{u},\mathbf{s}}$ *is an upper semicontinuous set-valued map from* $\mathbf{P}$ *to* $\mathbf{C}^0(\mathbf{I}, \mathbf{M})$, *with nonempty compact values.*

To the variation $\mathcal{V}^{\mathbf{u},\mathbf{s}}$ we associate the set-valued "endpoint-cost map"

$$\mathcal{EC}^{\mathbf{u},\mathbf{s}} : \mathbf{P} \longrightarrow\!\!\!\!\to \mathcal{M}_1^0 \times \ldots \times \mathcal{M}_\nu^0 \times \mathbb{R}$$

given by

$$\mathcal{EC}^{\mathbf{u},\mathbf{s}}(\boldsymbol{\varepsilon}, \mathbf{v}) = \left\{ \left(\boldsymbol{\partial}^0 \tilde{\xi}, C_L(\tilde{\xi}, \zeta^{\mathbf{u},\mathbf{s},\boldsymbol{\varepsilon}}) \right) : \tilde{\xi} \in \xi^{\mathbf{u},\mathbf{s}}(\boldsymbol{\varepsilon}, \mathbf{v}) \right\}, \tag{17.26}$$

where

$$\begin{aligned} \boldsymbol{\partial}^0 \tilde{\xi} &\stackrel{\text{def}}{=} (\partial_1^0 \tilde{\xi}, \partial_2^0 \tilde{\xi}, \ldots, \partial_{\nu-1}^0 \tilde{\xi}, \partial_\nu^0 \tilde{\xi})\,, \\ \mathcal{M}_j^0 &\stackrel{\text{def}}{=} M_j \times M_{j+1}\,, \\ \partial_j^0 \tilde{\xi} &\stackrel{\text{def}}{=} (\tilde{\xi}_j(\tau_j), \tilde{\xi}_{j+1}(t_j))\,. \end{aligned}$$

Then $\mathcal{EC}^{\mathbf{u},\mathbf{s}}$ *is an upper semicontinuous set-valued map from* $\mathbf{P}$ *to the product* $\mathcal{M}_1^0 \times \ldots \mathcal{M}_\nu^0 \times \mathbb{R}$, *with nonempty compact values.*

Moreover, the maps $\mathcal{V}^{\mathbf{u},\mathbf{s}}$ and $\mathcal{EC}^{\mathbf{u},\mathbf{s}}$ are "regular" in the sense of [4], [5], or [6]. (This follows because the vector fields that occur in (17.25) can be regularized, i.e., approximated by smooth vector fields, for which the corresponding initial value problems have unique solutions. The variation maps $\mathcal{V}_k^{\mathbf{u},\mathbf{s}}$ and endpoint-cost maps $\mathcal{EC}_k^{\mathbf{u},\mathbf{s}}$ arising from these regularized vector fields are then single-valued and continuous, and approximate the set-valued maps $\mathcal{V}^{\mathbf{u},\mathbf{s}}$, $\mathcal{EC}^{\mathbf{u},\mathbf{s}}$ in the "graph convergence" sense used in [4] to define regularity.)

The map $\mathcal{EC}^{\mathbf{u},\mathbf{s}}$ is differentiable at 0, and its differential $D\mathcal{EC}^{\mathbf{u},\mathbf{s}}(0)$ can be explicitly computed. The result is, of course, a linear map

$$D\mathcal{EC}^{\mathbf{u},\mathbf{s}}(0) : (\mathbb{R}^m)^\nu \times T_{\xi_1(t_1)}M_1 \times \ldots \times T_{\xi_\nu(t_\nu)}M_\nu \to T_{\partial_1^0\Xi}\mathcal{M}_1^0 \times \ldots \times T_{\partial_\nu^0\Xi}\mathcal{M}_\nu^0 \times \mathbb{R} ,$$

and is given by the formula

$$D\mathcal{EC}^{\mathbf{u},\mathbf{s}}(0)(\boldsymbol{\varepsilon}, \mathbf{v}) = ((w_1, v_2), (w_2, v_3), \ldots, (w_\nu, v_1), \alpha(\boldsymbol{\varepsilon}, \mathbf{v})) , \qquad (17.27)$$

where

$$\alpha(\boldsymbol{\varepsilon}, \mathbf{v}) = \sum_{j=1}^{\nu} \int_{t_j}^{\tau_j} y_j(t) \cdot D\Theta^j_{t,t_j}(\xi_j(t_j)) \cdot v_j \, dt$$
$$+ \sum_{j=1}^{\nu} \sum_{k=1}^{m} \varepsilon_j^k \left(\lambda_j^k + \int_{s_j^k}^{\tau_j} y_j(t) \cdot D\Theta^j_{t,s_j^k}(\xi_j(s_j^k)) \cdot \tilde{w}_j^k \, dt \right) ,$$

$$w_j = D\Theta^j_{\tau_j,t_j}(\xi_j(t_j)) \cdot v_j + \sum_{k=1}^{m} \varepsilon_j^k D\Theta^j_{\tau_j,s_j^k}(\xi_j(s_j^k)) \cdot \tilde{w}_j^k , \qquad (17.28)$$

$$\tilde{w}_j^k = f_{q_j}(\xi_j(s_j^k), u_j^k, s_j^k) - f_{q_j}(\xi_j(s_j^k), \eta_j(s_j^k), s_j^k) , \qquad (17.29)$$

$$\lambda_j^k = L_{q_j}(\xi_j(s_j^k), u_j^k, s_j^k) - L_{q_j}(\xi_j(s_j^k), \eta_j(s_j^k), s_j^k) , \qquad (17.30)$$

and $\{\Theta^j_{t,s}\}_{s,t\in I_j}$ is the family of flow maps corresponding to the reference vector field $(x, t) \to f_{q_j}(x, \eta_j(t), t)$. (Each differential $D\Theta^j_{t,s}(\xi_j(s))$ is then a linear map from $T_{\xi_j(s)}M_j$ to $T_{\xi_j(t)}M_j$. The flow maps $\Theta^j_{t,s}$ are of course set-valued, but our technical hypotheses imply that $\Theta^j_{t,s}$ is differentiable at $\xi_j(s)$ in the ordinary sense.)

Let $\Omega_j(\cdot,\cdot)$ denote the fundamental matrix solution of the linear time-varying differential equation $\dot{X} = Y_j(t) \cdot X$, so $I_j \times I_j \ni (t, s) \to \Omega_j(t, s) \in \mathbb{R}^{n_j \times n_j}$ is a continuous map having the property that

$$\Omega_j(t, s) = \text{identity}_{\mathbb{R}^n_j} + \int_s^t Y_j(r) \cdot \Omega_j(r, s) \, dr . \qquad (17.31)$$

Then

$$\alpha(\boldsymbol{\varepsilon},\mathbf{v}) = \sum_{j=1}^{\nu}\int_{t_j}^{\tau_j} y_j(t)\cdot\Omega_j(t,t_j)\cdot v_j\,dt$$
$$+\sum_{j=1}^{\nu}\sum_{k=1}^{m}\varepsilon_j^k\Big(\lambda_j^k+\int_{s_j^k}^{\tau_j} y_j(t)\cdot\Omega_j(t,s_j^k)\cdot\tilde{w}_j^k\,dt\Big),$$
$$w_j = \Omega_j(\tau_j,t_j)\cdot v_j+\sum_{k=1}^{m}\varepsilon_j^k\Omega_j(\tau_j,s_j^k)\cdot\tilde{w}_j^k\,.$$

Every pair $(\tilde{\xi},\zeta^{\mathbf{u},\mathbf{s},\boldsymbol{\mathcal{E}}})$, for $\tilde{\xi}\in\xi^{\mathbf{u},\mathbf{s}}(\boldsymbol{\varepsilon},\mathbf{v})$, is a pretrajectory-control pair of Σ. Moreover, $(\tilde{\xi},\zeta^{\mathbf{u},\mathbf{s},\boldsymbol{\mathcal{E}}})$ belongs to $TCP(\Sigma;\mathcal{E})$ if and only if

$$\partial_j^0\tilde{\xi}\in S_j^0 \quad \text{for} \quad j=1,\ldots,\nu\,. \tag{17.32}$$

If (17.32) holds, then it is clear that the total cost $\mathcal{C}(\tilde{\xi},\zeta^{\mathbf{u},\mathbf{s},\boldsymbol{\mathcal{E}}})$ is equal to the sum $C_L(\tilde{\xi},\zeta^{\mathbf{u},\mathbf{s},\boldsymbol{\mathcal{E}}})+c(\tilde{\xi},\zeta^{\mathbf{u},\mathbf{s},\boldsymbol{\mathcal{E}}})$, where

$$c(\tilde{\xi},\zeta^{\mathbf{u},\mathbf{s},\boldsymbol{\mathcal{E}}}) = \sum_{j=1}^{\nu}\Phi_j^0(\partial_j^0\tilde{\xi})\,, \tag{17.33}$$

and Φ_j^0 is the function $S_j^0\ni(x,x')\to\Phi_j(x,x',\tau_j,t_{j+1})$.

The assumption that Ξ is optimal implies that

$$\mathcal{C}(\tilde{\xi},\zeta^{\mathbf{u},\mathbf{s},\boldsymbol{\mathcal{E}}})\geq\mathcal{C}(\Xi) \quad \text{whenever} \quad (17.32) \quad \text{holds}\,. \tag{17.34}$$

Now, fix smooth functions $\sigma_j:\mathcal{M}_j^0\to\mathbb{R}$ such that $\sigma_j(\partial_j^0\Xi)=0$ and $\sigma_j(x,x')>0$ whenever $(x,x')\neq\partial_j^0\Xi$. Then let $\mathcal{G}$ be the set of all points $((x_1,x_1'),\ldots,(x_\nu,x_\nu'),r)$ of $\mathcal{M}_1^0\times\ldots\times\mathcal{M}_\nu^0\times\mathbb{R}$ having the property that $(x_j,x_j')\in S_j^0$ for $j=1,\ldots,\nu$ and

$$r\leq\mathcal{C}(\Xi)-\sum_{j=1}^{\nu}\Phi_j^0(x_j,x_j')-\sum_{j=1}^{\nu}\sigma_j(x_j,x_j')\,.$$

Then

$$\mathcal{E}C^{\mathbf{u},\mathbf{s}}(\mathbf{P})\cap\mathcal{G}=\{G_*\}\,, \tag{17.35}$$

where

$$G_*\stackrel{\text{def}}{=}(\partial_1^0\Xi,\ldots,\partial_\nu^0\Xi,C_L(\Xi))\,. \tag{17.36}$$

We now determine a Boltyanskii approximating cone $\mathcal{K}$ for the set $\mathcal{G}$ at G_*. For this purpose, we first observe that each cone K_j must be equal to the product $K_j^0 \times \{0\} \times \{0\}$, where K_j^0 is a closed convex cone in $T_{\partial_j^0 \Xi}\mathcal{M}_j^0$ which is a Boltyanskii approximating cone to S_j^0 at $\partial_j^0 \Xi$. (Proof: let V_j be a neighborhood of 0 in the tangent space $T_{\partial_j \Xi}\mathcal{M}_j$, and let $\alpha_j : V_j \cap K_j \to S_j$ be a continuous map such that

$$\alpha_j(z) = \partial_j \Xi + z + o(\|z\|) \quad \text{as} \quad z \to_{K_j} 0 .$$

Write $\alpha_j(z) = (\alpha_j^1(z), \alpha_j^2(z), \alpha_j^3(z), \alpha_j^4(z))$, so $\alpha_j^1(z) \in M_j$, $\alpha_j^2(z) \in M_{j+1}$, $\alpha_j^3(z) \in \mathbb{R}$, and $\alpha_j^4(z) \in \mathbb{R}$. Then $\alpha_j^3(z) \equiv \tau_j$ and $\alpha_j^4(z) \equiv t_{j+1}$. If we let $z = (z_1, z_2, z_3, z_4)$, we see that $\alpha_j^3(z) = \alpha_j^3(0)$, while on the other hand $\alpha_j^3(z) = \alpha_j^3(0) + z_3 + o(\|z\|)$. Therefore $\alpha_j^3(\rho z) = \alpha_j^3(0) + \rho z_3 + o(\rho)$ as $\rho \downarrow 0$, and then $\rho z_3 = o(\rho)$, so $z_3 = 0$. A similar argument proves that $z_4 = 0$. So $K_j = K_j^0 \times \{0\} \times \{0\}$, as stated. Moreover, if we define a map β_j by letting $\beta_j(w) = (\alpha_j^1(w,0,0), \alpha_j^2(w,0,0))$, then β is a continuous map from a neighborhood of 0 in K_j^0 to S_j^0, and $\beta_j(w) = \partial_j^0 \Xi + w + o(\|w\|)$ as $w \to_{K_j^0} 0$.) We then define $\mathcal{K}$ to be the set of all $((z_1, z_1'), \ldots, (z_\nu, z_\nu'), r)$ such that $(z_j, z_j') \in K_j^0$ for $j = 1, \ldots, \nu$, and

$$r \leq -\sum_{j=1}^{\nu} \nabla \Phi_j^0(\partial_j^0 \Xi) \cdot (z_j, z_j') . \tag{17.37}$$

Then $\mathcal{K}$ is a closed convex cone in $T_{G_*}(\mathcal{M}_1^0 \times \ldots \times \mathcal{M}_\nu^0 \times \mathbb{R})$, and it is easy to see that $\mathcal{K}$ is a Boltyanskii approximating cone to $\mathcal{G}$ at G_*. Moreover, $\mathcal{K}$ is not a linear subspace of $T_{G_*}(\mathcal{M}_1^0 \times \ldots \times \mathcal{M}_\nu^0 \times \mathbb{R})$, because $(0,-1) \in \mathcal{K}$ but $(0,1) \notin \mathcal{K}$.

It then follows from the general separation theorem of [4] that there exists a nonzero linear functional Ψ on $T_{G_*}(\mathcal{M}_1^0 \times \ldots \times \mathcal{M}_\nu^0 \times \mathbb{R})$ which is nonnegative on $\mathcal{K}$ and nonpositive on $\hat{\mathcal{K}}$, where

$$\hat{\mathcal{K}} \stackrel{\text{def}}{=} D\mathcal{EC}^{\mathbf{u},\mathbf{s}}(0)\Big(([0,\infty[^m)^\nu \times T_{\xi_1(t_1)}M_1 \times \ldots \times T_{\xi_\nu(t_\nu)}M_\nu\Big) . \tag{17.38}$$

Write

$$\Psi = (\psi_1^+, \psi_2^-, \psi_2^+, \psi_3^-, \ldots, \psi_\nu^+, \psi_1^-, -\psi_0) , \tag{17.39}$$

where each ψ_j^+ is a linear functional on $T_{\xi_j(\tau_j)}M_j$, each ψ_j^- is a linear functional on $T_{\xi_j(t_j)}M_j$, and $\psi_0 \in \mathbb{R}$. Since the vector $(0,-1)$ belongs to $\mathcal{K}$, the fact that Ψ is nonnegative on $\mathcal{K}$ implies that $\psi_0 \geq 0$. Next, fix a j and a vector $v \in T_{\xi_j(t_j)}M_j$, and let $\boldsymbol{\varepsilon} = 0$, $\mathbf{v} = (v_1, \ldots, v_\nu)$, where $v_\ell = 0$ if $\ell \neq j$, $v_j = v$. With this choice of $\boldsymbol{\varepsilon}$ and $\mathbf{v}$, the vector $D\mathcal{EC}^{\mathbf{u},\mathbf{s}}(0)(\boldsymbol{\varepsilon}, \mathbf{v})$ is equal to $((w_1, v_2), \ldots, (w_\nu, v_1), r)$, where the v_ℓ and w_ℓ vanish when $\ell \neq j$, $v_j = v$, $w_j = \Omega_j(\tau_j, t_j) \cdot v$, and $r = \int_{t_j}^{\tau_j} y_j(t) \cdot \Omega_j(t, t_j) \cdot v \, dt$. The fact that Ψ is nonpositive on $\hat{\mathcal{K}}$ implies that

$$\psi_j^- \cdot v + \psi_j^+ \cdot \Omega_j(\tau_j, t_j) \cdot v - \psi_0 \int_{t_j}^{\tau_j} y_j(t) \cdot \Omega_j(t, t_j) \cdot v \, dt \leq 0 . \tag{17.40}$$

Since this is true for all v in the linear space $T_{\xi_j(t_j)}M_j$, it follows that

$$\psi_j^- + \psi_j^+ \cdot \Omega_j(\tau_j, t_j) - \psi_0 \int_{t_j}^{\tau_j} y_j(t) \cdot \Omega_j(t, t_j) \, dt = 0 . \tag{17.41}$$

Let ψ_j be the unique solution of the adjoint equation (17.8) that satisfies the terminal condition $\psi_j(\tau_j) = \psi_j^+$. Then $\dot{\psi}_j(t) = -\psi_j(t) \cdot Y_j(t) + \psi_0 \cdot y_j(t)$, from which it follows that

$$\psi_j(t) = \psi_j^+ \cdot \Omega_j(\tau_j, t) - \psi_0 \int_t^{\tau_j} y_j(s) \cdot \Omega_j(s, t) \, ds . \tag{17.42}$$

(Proof: $\Omega_j(t,s) \cdot \Omega_j(s,t) =$ identity, so

$$\frac{\partial \Omega_j(t,s)}{\partial s} \cdot \Omega_j(s,t) + \Omega_j(t,s) \cdot \frac{\partial \Omega_j(s,t)}{\partial s} = 0 ,$$

and then

$$\frac{\partial \Omega_j(t,s)}{\partial s} \cdot \Omega_j(s,t) + \Omega_j(t,s) \cdot Y_j(s) \cdot \Omega_j(s,t) = 0 ,$$

so

$$\frac{\partial \Omega_j(t,s)}{\partial s} = -\Omega_j(t,s) \cdot Y_j(s) .$$

Using this, if we define $\psi_j^*(t)$ to be the right-hand side of (17.42), we see that

$$\dot{\psi}_j^*(t) = -\psi_j^*(t) \cdot Y_j(t) + \psi_0 \cdot y_j(t) .$$

Since $\psi_j^*(\tau_j) = \psi_j^+ = \psi_j(\tau_j)$, the conclusion follows.)

The identities (17.41) and (17.42) imply that

$$\psi_j^- = -\psi_j(t_j) . \tag{17.43}$$

So we have shown that (17.43) holds for all j.

Now fix j, pick a point (z, z') in K_j^0, and define $\mathbf{z} = ((z_1, z_1'), \ldots, (z_\nu, z_\nu'), r)$, where $z_\ell = 0$ and $z_\ell' = 0$ if $\ell \neq j$, $z_j = z$, $z_j' = z'$, and $r = -\nabla \Phi_j^0(\partial_j^0 \Xi) \cdot (z, z')$. Then $\mathbf{z}$ clearly belongs to $\mathcal{K}$, so the nonnegativity of Ψ on $\mathcal{K}$ implies that $\psi_j^+ \cdot z + \psi_{j+1}^- \cdot z' - \psi_0 r \geq 0$, that is,

$$\Big((\psi_j(\tau_j), -\psi_{j+1}(t_{j+1})) + \psi_0 \nabla \Phi_j^0(\partial_j^0 \Xi) \Big) \cdot (z, z') \geq 0 ,$$

or, equivalently,

$$\Big((-\psi_j(\tau_j), \psi_{j+1}(t_{j+1})) - \psi_0 \nabla \Phi_j^0(\partial_j^0 \Xi)\Big) \cdot (z, z') \le 0\,.$$

Since this is true for all $(z, z') \in K_j^0$, and $K_j = K_j^0 \times \{0\} \times \{0\}$, the conclusion that

$$(-\psi_j(\tau_j), \psi_{j+1}(t_{j+1}), h_j^+, -h_{j+1}^-) - \psi_0 \nabla \Phi_j(\partial_j \Xi) \in K_j^\perp$$

follows trivially. For $j = 1, \ldots, \nu - 1$, this yields the switching conditions, and for $j = \nu$ this is the transversality condition.

Next, let us fix j, and also fix $k \in \{1, \ldots, m\}$. Choose $\boldsymbol{\varepsilon} = (\boldsymbol{\varepsilon}_1, \ldots, \boldsymbol{\varepsilon}_\nu)$, where we let $\boldsymbol{\varepsilon}_\ell = 0$ if $\ell \ne j$, and $\boldsymbol{\varepsilon}_j = (\varepsilon_j^1, \ldots, \varepsilon_j^m)$, where $\varepsilon_j^i = 0$ if $i \ne k$, and $\varepsilon_j^k = 1$. Also, choose $\mathbf{v} = 0$. Then $D\mathcal{E}\mathcal{C}^{\mathbf{u},\mathbf{s}}(0)(\boldsymbol{\varepsilon}, \mathbf{v})$ is equal to $((w_1, 0), (w_2, 0), \ldots, (w_\nu, 0), \alpha(\boldsymbol{\varepsilon}, 0))$, where

$$\begin{aligned} w_i &= 0 \qquad \text{if} \quad i \ne j\,, \\ w_j &= \Omega_j(\tau_j, s_j^k) \cdot \tilde{w}_j^k\,, \\ \alpha(\boldsymbol{\varepsilon}, 0) &= \lambda_j^k + \int_{s_j^k}^{\tau_j} y_j(t) \cdot \Omega_j(t, s_j^k) \cdot \tilde{w}_j^k \, dt\,. \end{aligned}$$

The nonpositivity of Ψ on $\hat{\mathcal{K}}$ then implies that $\psi_j^+ \cdot w_j - \psi_0 \alpha(\boldsymbol{\varepsilon}, 0) \le 0$, i.e., that

$$\psi_j^+ \cdot \Omega_j(\tau_j, s_j^k) \cdot \tilde{w}_j^k - \psi_0 \int_{s_j^k}^{\tau_j} y_j(t) \cdot \Omega_j(t, s_j^k) \cdot \tilde{w}_j^k \, dt - \psi_0 \lambda_j^k \le 0\,,$$

that is

$$\psi_j(s_j^k) \cdot \tilde{w}_j^k - \psi_0 \lambda_j^k \le 0\,. \tag{17.44}$$

In view of (17.29) and (17.30), this says that

$$\begin{aligned} &\psi_j(s_j^k) \cdot f_{q_j}(\xi_j(s_j^k), u_j^k, s_j^k) - \psi_0 L_{q_j}(\xi_j(s_j^k), u_j^k, s_j^k) \\ &\le \psi_j(s_j^k) \cdot f_{q_j}(\xi_j(s_j^k), \eta_j(s_j^k), s_j^k)\,, -\psi_0 L_{q_j}(\xi_j(s_j^k), \eta_j(s_j^k), s_j^k)\,, \end{aligned}$$

i.e., that

$$H_{q_j}(\xi_j(s_j^k), \psi_j(s_j^k), u_j^k, \psi_0, s_j^k) \le H_{q_j}(\xi_j(s_j^k), \psi_j(s_j^k), \eta_j(s_j^k), \psi_0, s_j^k)\,. \tag{17.45}$$

The inequality (17.45) is precisely the Hamiltonian maximization condition, except that we have only established it for finitely many times and finitely many control values. The complete Hamiltonian maximization condition then follows by a standard compactness argument that we omit.

This completes the proof under the special assumptions described earlier in this section. We now outline how to prove the result in the general case, still assuming that A15.II.a holds. Assume that, for a particular j, U_{q_j} is a weak variable-time interval variational neighborhood of η_j, and the regularity conditions of A15.I.1 are satisfied. This implies, to begin with, that the limits that occur in the right-hand sides of the definitions of $h_j^{\pm}$ (cf. Def. 17.4.2) exist, and are given by

$$h_j^+ = H_{q_j}(\xi_j(\tau_j), \psi_j(\tau_j), \eta_j(\tau_j), \psi_0, \tau_j)\,,$$
$$h_j^- = H_{q_j}(\xi_j(t_j), \psi_j(t_j), \eta_j(t_j), \psi_0, t_j)\,.$$

In this case, we can make some extra variations in addition to those considered in the previous argument. Specifically, we can do "shortenings" at the endpoints t_j, τ_j (that is, we can restrict η_j to a subinterval $[t_j+\alpha, \tau_j-\beta]$, with $\alpha \geq 0$ and $\beta \geq 0$. In the definition of the endpoint-cost map given in (17.26), we substitute $\partial_j\tilde{\xi}$ for $\partial_j^0\tilde{\xi}$, that is, we add two extra time variables, corresponding to variations of t_j and of τ_j. Corresponding to these variables, the functional Ψ will now have two new components $\hat{\psi}_j^-$, $\hat{\psi}_j^+$, and the cone $\hat{\mathcal{K}}$ will have two new coordinates δt_j, $\delta\tau_j$. The linear part of the effect on $\mathcal{EC}^{\mathbf{u},\mathbf{s}}$ of a "shortening at the right endpoint" will involve $-\beta f_{q_j}(\xi_j(\tau_j), \eta_j(\tau_j), \tau_j)$ (variation of $\xi_j(t_j)$), $-\beta$ (variation of τ_j), and $-\beta L_{q_j}(\xi_j(\tau_j), \eta_j(\tau_j), \tau_j)$ (variation of the Lagrangian cost). This will give the inequality

$$-\psi_j^+ \cdot f_{q_j}(\xi_j(\tau_j), \eta_j(\tau_j), \tau_j) - \hat{\psi}_j^+ + \psi_0 L_{q_j}(\xi_j(\tau_j), \eta_j(\tau_j), \tau_j) \leq 0\,,$$

that is,

$$H_{q_j}(\xi_j(\tau_j), \psi_j(\tau_j), \eta_j(\tau_j), \psi_0, \tau_j) \geq -\hat{\psi}_j^+\,.$$

On the other hand, the linear part of the effect on $\mathcal{EC}^{\mathbf{u},\mathbf{s}}$ of an "expansion at the right endpoint" using the control value u will involve $\beta f_{q_j}(\xi_j(\tau_j), u, \tau_j)$ (variation of $\xi_j(t_j)$), β (variation of τ_j), and $\beta L_{q_j}(\xi_j(\tau_j), u, \tau_j)$ (variation of the Lagrangian cost). This will give the inequality

$$\psi_j^+ \cdot f_{q_j}(\xi_j(\tau_j), u, \tau_j) + \hat{\psi}_j^+ - \psi_0 L_{q_j}(\xi_j(\tau_j), u, \tau_j) \leq 0\,,$$

that is,

$$H_{q_j}(\xi_j(\tau_j), \psi_j(\tau_j), u, \psi_0, \tau_j) \leq -\hat{\psi}_j^+\,.$$

Since this happens for all values of u, we can specialize to $u = \eta_j(\tau_j)$ and conclude that

$$\hat{\psi}_j^+ = -H_{q_j}(\xi_j(\tau_j), \psi_j(\tau_j), \eta_j(\tau_j), \psi_0, \tau_j) = -h_j^+\,.$$

A similar argument yields the identity

$$\hat{\psi}_j^- = H_{q_j}(\xi_j(t_j), \psi_j(t_j), \eta_j(t_j), \psi_0, t_j) = h_j^-\,.$$

Now, if the same conditions are also satisfied for $j+1$, then we use the cone K_j instead of K_j^0. If $(z, z', \delta\tau_j, \delta t_{j+1})$ belongs to K_j, we will get a member $(z, z', \delta\tau_j, \delta t_{j+1}, r)$ of $\mathcal{K}$ by choosing $r = -\nabla\Phi_j(\partial_j\Xi).(z, z', \delta\tau_j, \delta t_{j+1})$. This will then yield the inequality

$$\psi_j^+ \cdot z + \psi_{j+1}^- \cdot z' + \hat{\psi}_j^+ \delta\tau_j + \hat{\psi}_{j+1}^- \delta t_{j+1} - \psi_0 r \geq 0\,,$$

that is,

$$\begin{aligned} 0 \leq\ & \psi_j(\tau_j) \cdot z - \psi_{j+1}(t_{j+1}) \cdot z' - h_j^+ \delta\tau_j + h_{j+1}^- \delta t_{j+1} \\ & + \psi_0 \nabla \Phi_j(\partial_j \Xi).(z, z', \delta\tau_j, \delta t_{j+1}) \end{aligned}$$

which implies

$$(-\psi_j(\tau_j), \psi_{j+1}(t_j), h_j^+, -h_{j+1}^-) - \psi_0 \nabla \Phi_j(\partial_j \Xi) \in K_j^\perp\,, \tag{17.46}$$

as desired.

The intermediate case, when the interval analogue of A15.I.2 holds for $j+1$, is easily handled, and we omit the argument.

Finally, in the autonomous case we observe, first of all, that the integral bounds are automatically satisfied. In this case, one can make shortening and expansion variations at almost all times. Once again, these variations change the times t_j and τ_j, so Ψ has corresponding components $\hat{\psi}_j^-$, $\hat{\psi}_j^+$. The identities

$$-\hat{\psi}_j^+ = H_{q_j}(\xi_j(t), \psi_j(t), \eta_j(t), \psi_0) = \hat{\psi}_j^-$$

(where we have explicitly spelled out the fact that in H_{q_j} there is no direct dependence on t) now follow for almost all $t \in I_j$. This clearly implies that the function $I_j \ni t \to H_{q_j}(\xi_j(t), \psi_j(t), \eta_j(t), \psi_0)$ is a. e. constant. It then follows, once again, that the limits occurring in the definition of $h_j^\pm$ exist, and in addition

$$H_{q_j}(\xi_j(t), \psi_j(t), \eta_j(t), \psi_0) = h_j^+ = -h_j^-$$

for almost all t. The switching condition (17.46) is now established exactly as before.

To conclude, we sketch how to remove the assumption on the integral bounds, if the "measurable" conditions A15.I.1 or A15.I.2 hold for a particular j, instead of their weaker "interval" counterparts A15.II.1, A15.II.2.

In this case, we replace the vector field system f_{q_j} and the corresponding Lagrangian by a new system $\hat{f}_{q_j}$ and a new Lagrangian $\hat{L}_{q_j}$, with a different control space $\hat{U}_{q_j}$. We take $\hat{U}_{q_j}$ to be $U_{q_j} \times \mathbb{N}$, so a control value for the new system is now a pair (u, N) where $u \in U_{q_j}$ and $N \in \mathbb{N}$. We fix a compact neighborhood $\mathcal{N}$ of the graph of ξ_j and define, for each $u \in U_{q_j}$, the upper bound function

$$\theta_u(t) = \sup\{\|f_{q_j}(x, u, t)\| + |L_{q_j}(x, u, t)| : x \in \mathcal{N}(t)\}\,,$$

where $\mathcal{N}(t) = \{x : (x, t) \in \mathcal{N}\}$. Then θ_u is measurable and almost everywhere finite. We then define

$$\hat{f}_{q_j}(x,(u,N),t) = \begin{cases} f_{q_j}(x,u,t) & \text{if} \quad \sigma_u(t) \le N\,, \\ f_{q_j}(x,\eta_j(t),t) & \text{if} \quad \sigma_u(t) > N\,, \end{cases}$$

$$\hat{L}_{q_j}(x,(u,N),t) = \begin{cases} L_{q_j}(x,u,t) & \text{if} \quad \sigma_u(t) \le N\,, \\ L_{q_j}(x,\eta_j(t),t) & \text{if} \quad \sigma_u(t) > N\,. \end{cases}$$

It is then easy to see that the integral bounds are automatically satisfied for the new system. Moreover, the substitution of a constant control (u, N) for the reference control on a subinterval J of I_j is equivalent to the substitution of the control value u for $\eta_j(t)$ for t in the measurable set $\{s \in J : \sigma_u(s) \le N\}$. In view of conditions A15.I or A15.II, this results in an admissible control for the original system. So the theorem under condition A15.II applies.

17.6 The case of a nondifferentiable reference vector field and Lagrangian

We now outline—without proof—how to remove the differentiability assumptions. For simplicity, we will only consider the removal of the differentiability requirement on the reference vector field and Lagrangian, even though it is also possible to weaken the differentiability requirement on the functions Φ_j.

Instead of A7, A12 and A14, we should now assume

A'. For each j belonging to the index set $\{1,\ldots,\nu^\#\}$ there exist maps $(x,t) \to \hat{f}_j(x,t)$ and $(x,t) \to \hat{L}_j(x,t)$ such that

A'.1 for almost every $t \in I_j^\#$ the maps $x \to \hat{f}_j(x,t)$ and $x \to \hat{L}_j(x,t)$ are Lipschitz with a Lipschitz constant $c_j(t)$ such that $\int_{t_j}^{\tau_j} c_j(t)\,dt < \infty$,

A'.2. for every x the maps $t \to \hat{f}_j(x,t)$ and $t \to \hat{L}_j(x,t)$ are measurable,

A'.3. there exist measurable functions $k_j^\delta : I_j^\# \to [0,+\infty]$ such that

$$\lim_{\delta \downarrow 0} \int_{I_j^\#} k_j^\delta(t)\,dt = 0\,,$$

having the property that, for almost all $t \in I_j^\#$, the inequality

$$\begin{aligned} &\|f_{q_j^\#}(x,\eta_j^\#(t),t) - \hat{f}_j(x,t)\| + |L_{q_j^\#}(x,\eta_j^\#(t),t) - \hat{L}_j(x,t)| \\ &\le k_j^\delta(t)\|x - \xi_j^\#(t)\| \end{aligned} \tag{17.47}$$

holds whenever $\|x - \xi_j^\#(t)\| \le \delta$.

Under these new conditions, all the terms that occur in the statement of the maximum principle are well defined, except only for the adjoint equation (17.8). We now reinterpret the adjoint equation to mean

$$-\dot{\psi}_j(t) \in \partial_x \hat{H}_j(\xi_j(t), \psi_j(t), \psi_0, t) \text{ for a.e. } t \in I_j, \tag{17.48}$$

where $\partial_x \hat{H}_j(x, p, p_0, t)$ is the generalized gradient (in the sense of Clarke [2]) of the function $x \to \hat{H}_j(x, p, p_0, t)$, where

$$\hat{H}_j(x, p, p_0, t) = p \cdot \hat{f}_j(x, t) - p_0 \hat{L}_j(x, t).$$

Theorem 17.6.1. *If S1, A5-6, A8-11, A13, A15 and A' hold, and in the definition of the adjoint equation (17.48) is substituted for (17.8), then the maximum principle is true.* ◇

References

1. L. D. Berkovitz, *Optimal Control Theory*, Springer-Verlag, New York, 1974.

2. F. H. Clarke, *Optimization and Nonsmooth Analysis*, Wiley Interscience, New York, 1983.

3. L.S. Pontryagin, V.G. Boltyanskii, R.V. Gamkrelidze and E.F. Mischenko, *The Mathematical Theory of Optimal Processes*, Wiley, New York, 1962.

4. H. J. Sussmann, Multidifferential calculus: chain rule, open mapping and transversal intersection theorems, in *Optimal Control: Theory, Algorithms, and Applications*, W. W. Hager and P. M. Pardalos, Editors, Kluwer Academic Publishers, 1998, pp. 436-487.

5. H. J. Sussmann, Geometry and optimal control, in *Mathematical Control Theory*, J. Baillieul and J. C. Willems, Eds., Springer-Verlag, New York, 1998, pp. 140-198.

6. H. J. Sussmann, Transversality conditions and a strong maximum principle for systems of differential inclusions, *Proc. 37th IEEE Conference on Decision and Control*, Tampa, FL, Dec. 1998. IEEE publications, 1998, pp. 1-6.

7. H. J. Sussmann, A Maximum Principle for hybrid optimal control problems, to appear in *Proc. 38th IEEE Conference on Decision and Control*, Phoenix, AZ, Dec. 1999. IEEE publications, 1999.

18. A converse Lyapunov theorem for robust exponential stochastic stability

John Tsinias and John Spiliotis

National Technical University
Department of Mathematics
Zografou Campus
15780, Athens, Greece
jtsin@math.ntua.gr

Summary.

Concepts of exponential global robust–stability for stochastic systems are introduced and analyzed in terms of Lyapunov functions. The main results of the paper are used to derive a Lyapunov like characterization for the concept of input–to–state–exponential stochastic stability introduced in earlier works by the first author.

18.1 Introduction

We consider stochastic systems:

$$dx = f(x,u)dt + g(x,u)dw, \quad x \in R^n, \quad u \in I \subset R \tag{18.1}$$

where u denotes the input and w is a Wiener process. Without any loss of generality we may assume that both u and w are single valued and u takes values on a compact interval I of the real line. We denote by $F(I)$ the convex set of all random functions $u(t) = u(\omega,t), (\omega,t) \in \Omega \times R$ taking values on I which are measurable in (t,ω) and F_t- adapted. Finally, assume that the dynamics $f,g : R^{n+1} \to R^n$ are everywhere continuous, $C^2((R^n \setminus \{0\}) \times I)$- without any loss of generality we may assume in the sequel that f and g are everywhere C^2- and satisfy the following properties:

- zero is an equilibrium for (18.1)

$$f(0,u) = g(0,u) = 0, \quad \forall\, u \in I \quad ; \tag{18.2a}$$

- there are constants $C_1, C_2 > 0$ such that

$$\left|\frac{\partial f}{\partial x}(x,u)\right| + \left|\frac{\partial g}{\partial x}(x,u)\right| \leq C_1, \tag{18.2b}$$

$$\left|\frac{\partial f}{\partial u}(x,u)\right| + \left|\frac{\partial g}{\partial u}(x,u)\right| \leq C_2|x|, \quad \forall\ (x,u) \in R^n \times I. \tag{18.2c}$$

It turns out by (1.2a,c) and boundedness of I that f and g satisfy the restriction on growth:

$$|f(x,u)| + |g(x,u)| \leq C_3|x|, \quad \forall (x,u) \in R^n \times I \tag{18.3}$$

for some $C_3 > 0$. It is known (see for instance [1, 5]) that, if f and g are locally Lipschitz, the restriction on growth (18.3) guarantees existence and uniqueness of solutions for (18.1), namely, for every input $u \in F(I)$, $x_0 \in R^n$ and $t_0 \in R$ there exists a unique solution $X(t, t_0, x_0, u)$ of (18.1) starting from x_0 at time $t = t_0$ which is defined for all t and almost all $\omega \in \Omega$.

Our purpose is to characterize concepts of robust exponential stochastic stability in terms of Lyapunov functions. Theorem 3.4 is the main result of the paper and establishes that, under certain hypothesis, exponential stability of zero for (18.1) is equivalent to existence of an appropriate Lyapunov function. This result partially extends the well known converse Lyapunov theorem of Lin–Sontag–Wang [7] concerning robust asymptotic stability for deterministic control systems and generalizes the converse stability theorems for deterministic and stochastic differential equations which establish existence of quadratic Lyapunov functions under the presence of exponential stability (see [2, 3]). The proof of our main theorem result is inspired from the analysis made in [7, 14] and is mainly based on some important technical results obtained by P. L. Lions in [8].

The results of robust stability enable us to derive a Lyapunov characterization of the **exponential input–to–state stability** (expISS) for stochastic systems (Proposition 4.1). The notions of exponential ISS we present constitute natural extensions of the well known deterministic ISS introduced by E. Sontag (see for instance [2, 10, 11]) plus exponential stability, and have been recently used in [12, 13] to derive sufficient conditions for global feedback stabilization for stochastic systems.

18.2 Notions of exponential robust stability

Definition 18.2.1. *We say that zero is* **exponentially uniformly globally asymptotically stable** *(expUGAS) for (18.1), if there exist constants*

$C, \quad \ell > 0$ and a bounded measurable function $\theta : R^n \times I \to R$ such that for every $x_0 \in R^n$, $u \in F(I)$ and almost all ω we have

$$|X(t,t_0,x_0,u)| \leq C|x_0| \exp\left(-\ell(t-t_0) + \int_{t_0}^{t} \theta(X(s,t_0,x_0,u),u(s))dw(s)\right) \tag{18.4}$$

$$\forall\ t \in [t_0, T), \quad T := T(\omega) \leq \infty$$

It should be noticed that boundedness of $\theta(\cdot)$ over $R^n \times I$ implies that

$$M_{t,t_0} := \int_{t_0}^{t} \theta(X(s,t_0,x_0,u),u(s))dw(s)$$

is a C^0 martingale with $E(M_{t,t_0} - M_{s,t_0}|F_s) = E(M_{t,s}|F_s) = 0$ for every $t \geq s \geq t_0$ and $E(M^2_{t,t_0}) \leq C(t-t_0)$, $\forall\ t \geq t_0$, $u \in F(I)$, where C is a positive constant being independent of u. Then (18.4) in conjunction with the supermartingale inequality:

$$P\left(\sup_{t \geq s} M^2_{t,t_0}(t-t_0)^{-2} > \varepsilon\right) \leq \varepsilon^{-1} E\left(M^2_{s,t_0}(s-t_0)^{-2}\right) \leq \varepsilon^{-1} C(s-t_0)^{-1},$$

$$\forall\ \varepsilon > 0, \quad s > t_0$$

implies:

I. "global uniform stability":

$$\lim_{x_0 \to 0} P\left[\sup\{|X(t,t_0,x_0,u)|, \quad t \geq t_0, \quad u \in F(I)\} > \varepsilon\right] = 0,$$

II. "global uniform attractivity":

$$P[X(t,t_0,x_0,u)| \to 0, \quad \text{as } t \to \infty, \quad \text{uniformly on } u \in F(I)] = 1$$

for all $x_0 \in R^n$.

The concepts above generalize the notions of stability and attractivity as given by Khasminskii and Kushner for stochastic differential equations (see [1, 3, 4, 6]) and both characterize the general concept of **uniform global asymptotic stability (UGAS)** for stochastic control systems. In the present work we limit ourselves to the study of the exponential UGAS, as well as of the notion of "exponential UGAS in the rth mean", whose precise definition is given below.

Definition 18.2.2. *We say that zero* **is exponentially UGAS in the rth mean** *(r-expUGAS) for some $r > 0$, if there exist constants $C, \ell > 0$ such that for every $x_0 \in R^n$ and $u \in F(I)$ it holds that*

$$E(|X(t,t_0,x_0,u)|^r) \le C|x_0|^r \exp(-\ell(t-t_0)) \tag{18.5}$$
$$\forall\, t \in [t_0,T), T = T(\omega) \le \infty$$

In addition to hypothesis (18.2) let us further assume that the following hold:

-
$$\frac{\partial^2 f}{\partial x_i \partial x_j}(x,u),\ \frac{\partial^2 g}{\partial x_i \partial x_j}(x,u),\quad 1 \le i,\ j \le n \ \text{ are bounded over } R^n \times I; \tag{18.6a}$$

- the noisy term g satisfies the following property ("non–degeneration condition"): for every nonzero x there is a constant $k_x > 0$ and a $u_0 \in I$ such that
$$g(x,u_0)g^T(x,u_0) \ge k_x I_n \tag{18.6b}$$
(^{T} stands for transpose and I_n is the n-dimensional unit matrix).

The main result of the paper (Theorem 3.4) establishes that under (18.2) and (18.6) r-expUGAS is equivalent to the existence of a Lyapunov function $V : R^n \to R^+$ which is $C^\infty(R^n \setminus \{0\}) \cap C^{[r]}(R^n)$ ($[r]$ is the integral part of r) and satisfies the following properties:

$$I. \quad c_1|x|^r \le V(x) \le c_2|x|^r \tag{18.7a}$$

$$II. \quad |DV(x)| \le c_3|x|^{r-1} \quad (x \ne 0) \tag{18.7b}$$

$$III. \quad L^u_{(1.1)}V(x) := DVf(x,u) + \frac{1}{2}\sum_{1\le i,j\le n} \frac{\partial^2 V}{\partial x_i \partial x_j}(g(x,u)g^T(x,u))_{i,j}$$
$$\le -c_4|x|^r, \quad \forall\,(x,u) \in R^n \times I \tag{18.7c}$$

for some positive constants $c_i (i = 1,2,3,4)$; (DV denotes the derivative of V).

Relationships between the two concepts of exponential robust stability given above and their characterization in terms of Lyapunov functions are established in next section. Among other things we prove that r-expUGAS is equivalent to expUGAS for r sufficiently small, provided that both (18.2) and (18.6) are fulfilled (Corollary 3.7). Finally, the results on robust stability are used in Section 4 to analyze notions of stochastic exponential ISS (see Definition 2.3 below) in terms of Lyapunov functions for stochastic systems (18.1) with C^2 dynamics under the assumptions:

$$f(0,0) = g(0,0) = 0 \tag{18.8a}$$

$$\left|\frac{\partial f}{\partial x}\right| + \left|\frac{\partial f}{\partial u}\right| + \left|\frac{\partial g}{\partial x}\right| + \left|\frac{\partial g}{\partial u}\right| \leq C, \quad \forall (x,u) \in R^n \times I \tag{18.8b}$$

for any compact subset $I \subset R$ and for some constant $C > 0$, whose choice depends on I.

Definition 18.2.3. *[12, 13]:*

• *We say that (18.1) satisfies the* **exponential input–to–state stability property** *(expISS), if there exist a positive definite function $\gamma : R^+ \to R^+$, a function $\theta(\cdot)$ and positive constants C and ℓ as in Definition 2.1, in such a way that (18.4) is satisfied for every random input u being measurable in (t,ω) and F_t- adapted with*

$$|u(t)| \leq \gamma(|X(t,t_0,x,u)|), \quad \forall\, t \in [t_0, T) := T(\omega) \leq \infty \tag{18.9}$$

provided that the corresponding solution $X(\cdot)$ of (18.1) exists, thus, is t-continuous, (t,ω)- measurable and F_t- adapted.

• *We say that (18.1) satisfies the* **exponential ISS in the rth-mean** *(r-expISS), if there exist a positive definite function $\gamma : R^+ \to R^+$ and positive constants C and ℓ such that (18.5) holds for every F_t- adapted input u for which (18.9) is fulfilled.*

18.3 Main results

The following propositions offer some links between expUGAS and r-expUGAS.

Proposition 18.3.1. *Consider the system (18.1) and assume that (18.2) are satisfied. Suppose that zero is expUGAS and let*

$$K := \sup_{x \neq 0, u \in I} \frac{|g(x,u)|^2}{|x|^2} \tag{18.10}$$

whose existence is guaranteed by (18.2b). Then for any

$$r \in (0, 2\ell K^{-1}) \tag{18.11}$$

where ℓ is the constant defined in (18.4), zero is r-expUGAS.

Proof: For simplicity let $X(t) := X(t,0,x,u)$, $x \neq 0$. Notice that $X(t) \neq 0$ for all t almost surely, since zero is an equilibrium. Condition (18.4) implies

$$\log|X(t)| \le \log|x| - \ell t + \int_0^t \theta(X(s), u(s))dw(s)$$

hence, for any $u \in F(I)$ we get:

$$L^u_{(1.1)} \log|x| = \lim_{t\to 0} \frac{E\log|X(t)| - \log|x|}{t} \le -\ell \tag{18.12}$$

Let r and c be a pair of positive constants with

$$\ell > c > \frac{rK}{2} \tag{18.13}$$

From (18.10), (18.12) and (18.13) we obtain

$$\begin{aligned}
\frac{1}{r|x|^r} L^u_{(1.1)}|x|^r - c = {} & \frac{x^T f(x,u)}{|x|^2} + \frac{1}{2|x|^2}\sum_{i=1}^n (g(x,u)g^T(x,u))_{i,i} \\
& + \left(\frac{r}{2} - 1\right)\frac{1}{|x|^4}\sum_{1\le i,j\le n} x_i x_j (g(x,u)g^T(x,u))_{i,j} - c \\
\le {} & \frac{x^T f(x,u)}{|x|^2} + \frac{1}{2|x|^2}\sum_{i=1}^n (g(x,u)g^T(x,u))_{i,i} \\
& - \frac{1}{|x|^4}\sum_{1\le i,j\le n} x_i x_j (g(x,u)g^T(x,u))_{i,j} \\
= {} & L^u_{(1.1)} \log|x| \le -\ell
\end{aligned} \tag{18.14}$$

It turns out from (18.14) that $L^u_{(1.1)}|x|^r \le -\bar{\ell}|x|^r$ with $\bar{\ell} := r(\ell - c) > 0$, therefore

$$\frac{d}{dt}E|X(t)|^r \le -\bar{\ell}E|X(t)|^r$$

which implies r-expUGAS.

We next establish that the existence of a Lyapunov function implies exponential robust stability.

Proposition 18.3.2. *Suppose that there exists a $C^2(R^n \setminus \{0\})$ function V satisfying all properties (18.7). Then*

(i) Zero is r-expUGAS

(ii) Zero is expUGAS assuming in addition that either

$$g(x,u)g^T(x,u) \ge 0, \quad \forall\ (x,u) \in R^n \times I \tag{18.15}$$

or

$$\frac{Kc_3^2}{2c_1^2} < \frac{c_4}{c_2} \tag{18.16}$$

c_1, c_2, c_3, c_4 and K being the constants defined in (18.7) and (18.10), respectively.

Proof: (i) Let $X(t) := X(t, t_0, x, u)$, $x \neq 0$, $u \in F(I)$. From (18.7) we get

$$\frac{d}{dt} EV(X(t)) = E(L^u_{(1.1)} V(X(t))) \leq -c_4 E(|X(t)|^r) \tag{18.17}$$

The desired (18.5) is a direct consequence of (18.17) and (18.7a).
(ii) Suppose for example that (18.16) holds. Taking into account (18.7) and applying Ito's theorem we estimate:

$$\log V(X(t)) = \log V(x) + \int_{t_0}^{t} \frac{1}{V(X(s))} \Bigg\{ L^u_{(1.1)} V(X(s)) - \frac{1}{2} \sum_{1 \leq i,j \leq n} \cdot \left(g(X(s), u(s)) g^T(X(s), u(s))\right)_{i,j} \left(\frac{\partial V}{\partial x_i} \frac{\partial V}{\partial x_j} \frac{1}{V}\right)(X(s)) \Bigg\} ds + M_{t,t_0} \leq \log V(x_0) - \left(\frac{c_4}{c_2} - \frac{K c_3^2}{2 c_1^2}\right)(t - t_0) + M_{t,t_0} \tag{18.18a}$$

where

$$M_{t,t_0} := \int_{t_0}^{t} \theta(X(s), u(s)) dw(s), \ \theta(x, u) := \frac{1}{V(x)} \nabla V(x) g(x, u), \ x \neq 0 \tag{18.18b}$$

It turns out from (18.2), (18.7) and (18.18) that zero is expUGAS, specifically, (18.4) holds with $\ell = c_4 c_2^{-1} - \frac{1}{2} K c_3^2 c_1^{-2}$ and $C = c_2 c_1^{-1}$. Likewise, we establish expUGAS under (18.15).

We illustrate the nature of Proposition 3.2 by the following simple example.

Example 18.3.1. Consider the system

$$dx = x\, dt + 5x(1 - u) dw, \quad u \in I := [-1/2, 1/2], \quad x \in R$$

Then it can be easily verified that the function $V(x) = |x|^r$, $0 < r < 1/5$ satisfies (2.4a,b,c) thus Proposition 3.2 asserts that zero is r-expUGAS, as well as expUGAS with respect to the above system.

Conversely, we have:

Theorem 18.3.1. *Consider the system (18.1) and assume that (18.2) as well (18.6) are fulfilled. If zero is r-expUGAS, then there exists a* $C^{[r]}(R^n) \cap C^\infty(R^n \setminus \{0\})$ *function V which satisfies all properties (18.7).*

In order to establish Theorem 3.4 we need the following important technical result proved by P. L. Lions in [8].

Lemma 18.3.1. (first statement of [Lemma 2.1,8]): *Let $\psi_\sigma : R^n \to R^+$ be a $C^\infty(R^n)$ map whose support is the unit disk with $\int_{R^n} \psi(s)ds = 1$ and let*

$$\psi_\sigma := \frac{1}{\sigma^n}\psi\left(\frac{\cdot}{\sigma}\right), \quad \sigma > 0 \tag{18.19}$$

Consider a function $V \in W^{1,\infty}_{loc}(R^n)$ and define

$$V_\sigma(x) := (V * \psi_\sigma)(x) = \int_{R^n} V(x + \sigma s)\psi(s)ds \tag{18.20}$$

Then under (18.2)

$$I_\sigma(x) := \sup_{u \in I} \left| L^u_{(1.1)}(V * \psi_\sigma) - (L^u_{(1.1)}V) * \psi_\sigma \right| \in L^\infty_{loc}(R^n) \tag{18.21}$$

i.e., for any bounded $Q \subset R^n$ there is a constant $K > 0$ (being independent of u and ε) such that

$$\left| L^u_{(1.1)}(V * \psi_\sigma) - (L^u_{(1.1)}V) * \psi_\sigma \right| \leq K, \quad \forall\, u \in I, \quad x \in Q \tag{18.22}$$

It should be noticed that, in addition to (18.22), [Lemma 2.1,8] establishes that $I_\sigma(x) \to 0$ as $\sigma \to 0$ for almost all x. The proof of Theorem 3.4 needs however the stronger type of convergence: $I_\sigma(x) \to 0$ as $s \to 0$ for all x. For deterministic systems (namely, when the noisy term $g(\cdot)$ identically vanishes) this is true (see [7, 14]). We next establish, by employing the same analysis made in [8], that the requirement above occurs for the stochastic case (18.1), provided that the boundedness condition (18.6a) for the second derivatives of f and g, as well as the non–degenation condition (18.6b) are fulfilled.

The proof of Theorem 3.4 also requires the following lemma, which is a well known result in stochastic control theory (see for instance [8]).

Lemma 18.3.2. *Let $h \in C^0(R^n)$ and $V \in W^{1,\infty}_{loc}(R^n)$ and assume that*

$$\frac{1}{t}(EV(X(t,0,x,u)) - V(x)) \leq \int_0^t Eh(X(s,0,x,u))ds$$

for all $x \in R^n$, $u \in F(I)$ and $t > 0$ near zero. Then under (18.2) it holds that

$$L^u_{(1.1)}V \leq h \ \ in\ D'(R^n)$$

namely, in the sense of distributions.

The proof of the previous lemma follows by repeating the same arguments used in [8, p.131].

Proof of Theorem 3.4

We distinguish two cases.

Case I: $r = 2$

We define

$$V(x) := \sup_{u \in F(I)} \int_0^T E|X(s,0,x,u)|^2 ds \tag{18.23}$$

where $T > 0$ yet to be specified. We first show that V is Lipschitz; specifically, there exists a constant $K > 0$ such that

$$|V(x) - V(y)| \leq K(|x| + |y|)(|x - y|), \quad \forall\, x, y \in R^n \tag{18.24}$$

Indeed, definition (18.23) of V implies that for any $\varepsilon > 0$ and $x \in R^n$ there corresponds an input $u \in F(I)$ with

$$V(x) \leq E\int_0^T |X(s)|^2 ds + \varepsilon; \quad -V(y) \leq -E\int_0^T |Y(s)|^2 ds \tag{18.25}$$

where for simplicity we denote $X(t) := X(t,0,x,u)$ and $Y(t) := X(t,0,y,u)$. By applying Ito's theorem and Gronwall's inequality and invoking our hypothesis (1.2a,b) we get

$$\sup_{0 \leq t \leq T} E^{1/2}\left(|X(t)|^2 + |Y(t)|^2\right) \leq C_1(|x| + |y|) \tag{18.26a}$$

$$\sup_{0 \leq t \leq T} E^{1/2}\left(|X(t) - Y(t)|^2\right) \leq C_2|x - y| \tag{18.26b}$$

for some positive constants C_1, C_2 being independent of x, y and u. By (18.25) and (18.26) it then follows

$$\begin{aligned} V(x) - V(y) &\leq \int_0^T E\left(|X(s)|^2 - |Y(s)|^2\right) ds + \varepsilon \\ &\leq \int_0^T E^{1/2}\left(|X(s) - Y(s)|^2\right) E^{1/2}\left(|X(s) + Y(s)|^2\right) ds + \varepsilon \\ &\leq K(|x| + |y|)|x - y| + \varepsilon, \quad \forall\, x, y \in R^n, \quad \varepsilon > 0 \end{aligned}$$

for some $K > 0$ independent of ε. Likewise, we obtain

$$V(y) - V(x) \leq K(|x| + |y|)|x - y| + \varepsilon$$

and this proves (18.24). From (18.24) we get

$$|DV(x)| \leq 2K|x| \tag{18.27}$$

for almost all x for which the derivative DV of V exists. We next show that V satisfies:

$$k_1|x|^2 \leq V(x) \leq k_2|x|^2, \quad \forall\, x \in R^n \tag{18.28}$$

for some $k_2 \geq k_1 > 0$. Indeed, invoking (1.2a,b) and using Ito's formula and Gronwall's inequality we can determine positive constants $k_2 \geq k_1 > 0$ such that

$$\sup_{0\leq t\leq T} E|X(t,0,x,u)|^2 \leq k_2|x|^2 \tag{18.29a}$$

$$k_1|x|^2 \leq \inf_{0\leq t\leq T} E|X(t,0,x,u)|^2 \tag{18.29b}$$

for all x and $u \in F(I)$. The desired (18.28) is consequence of (18.29) and definition (18.23) of V. We now prove that by appropriate selection of T the following property holds:

For every random input $u \in F(I)$ there is a time $t_0 > 0$ with

$$\frac{1}{t}(EV(X(t,0,x,u)) - V(x)) \leq -\frac{1}{2t}\int_0^t E|X(s,0,x,u)|^2 ds + t \tag{18.30}$$

$$\forall\, t \in (0, t_0]$$

In order to establish (18.30) we again recall our assumptions (1.2a,b,c) which guarantee the existence of a constant $C_1 > 0$ such that

$$\begin{aligned} &|f(x,u) - f(y,\overline{u})|\,|x-y| + \left||g(x,u)|^2 - |g(y,\overline{u})|^2\right| \\ &\leq C_1\left(|x-y|^2 + (|x|^2+|y|^2)|u-\overline{u}|^2\right) \end{aligned} \tag{18.31}$$

for every $x, y \in R^n$, $u, \overline{u} \in R$. By using Ito's formula and Gronwall's inequality and taking into account (18.26a) and (18.31) it follows that for every compact region $Q \subset R^n$ it holds that

$$\begin{aligned} &\sup_{0\leq t\leq T} \left|E(|X(t,0,x,u)|^2 - |X(t,0,y,\overline{u})|^2)\right| \\ &\leq \sup_{0\leq t\leq T} 2E^{1/2}\left(|X(t,0,x,u)|^2 + |X(t,0,y,\overline{u})|^2\right) \\ &\quad E^{1/2}\left(|X(t,0,x,u) - |X(t,0,y,\overline{u})|^2\right) \\ &\leq C_2(ess\sup\{|u-\overline{u}|,\ t\in[0,T],\ \omega\in\Omega\} + |x-y|), \\ &\quad \forall\, x,y \in Q,\ u,\overline{u} \in F(I), \end{aligned} \tag{18.32}$$

for certain $C_2 > 0$. Condition (18.32) enable us, by applying standard partition of unity arguments, to establish the existence of a family of maps:

$$U_t : \Omega \times R^n \times R^+ \to R, \quad t > 0$$

of the form

$$U_t(x, s) = \sum p_i(x) u_i(s), \quad u_i \in L(I) \tag{18.33a}$$

$$\sum p_i(x) = 1 \tag{18.33b}$$

where $p_i : R^n \to R^+$ are C^∞, have compact support, the summation $\sum$ above is finite and in such a way that

$$V(x) - t^2 < \int_0^T E|X(s, 0, x, U_t(x, s))|^2 ds < V(x) \tag{18.34}$$

Note that, according to (18.33) and convexivity of $F(I), U_t(\cdot)$ takes values in $I, U_t(x, \cdot)$ is of class $F(I)$ and, since the summation in (18.33) is finite, $U_t(\cdot, s)$ is locally Lipschitz. It turns out that for any vector $z \in R^n$ and $t > 0$ the control

$$u_t^z(s) := U_t(X(s, 0, z), s)$$

where $X(\cdot)$ is the solution of

$$dx = f(x, U_t(x, s))ds + g(x, U_t(x, s))dw$$

$$X(\cdot)|_{s=0} = z$$

is of class $F(I)$. Consider an input $u \in F(I)$, denote

$$X(\cdot) := X(\cdot, 0, x, u), u_t := u_t^z|_{z=X(t)} \tag{18.35}$$

define

$$\overline{u}_t(s) := \begin{cases} u_t(s), & t < s < \infty \\ u(s), & 0 \le s \le t \end{cases} \tag{18.36}$$

and let $\hat{u}_t(\cdot) \in F(I)$ with

$$\hat{u}_t(s) = \overline{u}_t(s - t) \quad \text{for } s \ge t \tag{18.37}$$

Using (18.34) and taking into account (18.35), (18.36), (18.37) and our hypothesis that (18.5) holds with $r = 2$ we get

$$\frac{1}{t}(EV(X(t)) - V(x)) \leq t + \frac{1}{t}E\left\{\int_0^T E\left|X(s,0,z,\overline{u}_t)\right|^2\Big|_{z=X(t)} ds\right.$$

$$\left. - \int_0^T E|X(s,0,x,\overline{u}_t)|^2 ds\right\}$$

$$= t + \frac{1}{t}\left\{E\int_0^T E\left(|X(s+t,0,x,\overline{u}_t)|^2\big| F_t\right) ds - \int_0^T E|X(s,0,x,\overline{u}_t)|^2 ds\right\}$$

$$\leq t + \frac{1}{t}\left\{\int_t^{T+t} E\left(|X(s,0,x,\hat{u}_t)|^2\big| F_t\right) ds - \int_0^T E|X(s,0,x,\overline{u}_t)|^2 ds\right\}$$

$$\leq t + \frac{C|x|^2}{t}\int_T^{T+t} \exp(-\ell s) ds - \frac{1}{t}\int_0^t E|X(s)|^2 ds \leq t - \frac{1}{2t}\int_0^t E|X(s)|^2 ds$$

for $t > 0$ sufficiently small, provided that T has been selected sufficiently large in such a way that $2Cexp(-\ell T) < 1$. This proves (18.30).

The map V as defined by (18.23) is in general nonsmooth. We build a $C^\infty(R^n \setminus \{0\})$ function in such a way that all properties (18.7) hold with $r = 2$. We proceed among the same lines in [7, 14]. We first need the following additional facts that play an important role to the rest procedure.

Fact I: *For any open bounded $Q \subset R^n$ there is a constant $C > 0$ such that*

$$D^2 V \geq -C \quad in\ D^{'}(Q) \tag{18.38}$$

($D^2 V$ mean the second derivative at V), provided that (18.6a) is satisfied.

Indeed, a well known result concerning differentiability of solutions for stochastic systems (see for instance [5]) asserts that under (18.2) and (18.6a) the processes $|X(\cdot)|^2$ is mean–square–differentiable with respect to initial state x and further it holds that

$$E \sup_{0\leq s\leq T} |D_x X(s,0,x,u)|^2 + E \sup_{0\leq s\leq T} |D_x^2 X(s,0,x,u)|^2 \leq C(1+|x|)^q,$$

$$\forall\, x \in R^n, \quad u \in I$$

for some appropriate constants C and q. Then, if we call

$$\phi^u(x) := \int_0^T E|X(s,0,x,u)|^2 ds$$

we can easily deduce

$$\sup_u |D^2\phi^u(x)| \leq \overline{C}(1+|x|)^{\overline{q}}$$

for some positive $\overline{C}$ and $\overline{q}$. Let e be a unitary vector in R^n, and let Q be an open subset of R^n. By making use of the mean valued theorem the previous inequality yields:

$$\begin{aligned}
&V(x+he)+V(x-he)-2V(x) \\
&\geq -\sup_u\left[\phi^u(x+he)+\phi^u(x-he)\right]-\sup_u 2\phi^u(x) \\
&\geq -\sup_u\left|\phi^u(x+he)+\phi^u(x-he)-2\phi^u(x)\right| \geq -Ch^2
\end{aligned}$$

for some constant $C > 0$ and for all sufficiently small $h > 0$. The desired (18.38) is consequence of the inequality above and the fact:

$$h^{-2}(V(x+he)+V(x-he)-2V(x)) \rightarrow \frac{\partial^2 V}{\partial e^2} \quad \text{as } h \rightarrow 0, \quad \text{in } D'(R^n).$$

Fact II: *Under (18.6a) and the non-degeneration hypothesis (18.6b), the function V is of class $W^{2,\infty}_{loc}(R^n \setminus \{0\})$.*

Indeed, Lemma 3.6 and (18.30) yield $L^u_{(1.1)}V \leq -\frac{1}{2}|x|^2$ in $D'(R^n)$, thus:

$$\left(L^u_{(1.1)}V\right) * \psi_\sigma \leq -\frac{1}{2}|x|^2 * \psi_\sigma \quad (\text{in } R^n)$$

The latter, by virtue of Lemma 3.5, implies that for any open bounded $Q \subset R^n \setminus \{0\}$ it holds that

$$L^u_{(1.1)}V_\sigma = L^u_{(1.1)}(V * \psi_\sigma) \leq C + \frac{1}{2}|x|^2 * \psi_\sigma \leq \overline{C}, \quad \forall\, x \in Q$$

for certain $\overline{C} > C > 0$. Then by taking into account (18.6b) and following the same procedure with this employed in [8, p. 128] it can be established that for any compact $Q \subset R^n \setminus \{0\}$ it holds that

$$D^2V \leq C' \quad \text{in } D'(Q)$$

for some $C' > 0$, which in conjunction with (18.38) implies

$$V \in W^{2,\infty}_{loc}(R^n \setminus \{0\}).$$

It was pointed out that the second statement of the key Lemma 2.1 in [8] asserts that $I_\sigma(x) \rightarrow 0$ as $\sigma \rightarrow 0$ for almost all x, provided that $V \in W^{1,\infty}_{loc}(R^n)$. Based on Fact II and following exactly the same analysis with this made in [8, pp. 139–141], the statement above is strengthened as follows:

Fact III: ("Strong" version of second statement of [Lemma 2.1, 8]): *Under the additional hypothesis (2.3a,b), for any compact $Q \subset R^n \setminus \{0\}$ we have:*

$$\sup_{x \in Q} I_\sigma(x) \to 0 \quad \text{as} \quad \sigma \to 0 \tag{18.39}$$

We are now in a position to establish the existence of a $C^\infty(R^n \setminus \{0\})$ Lyapunov function. Consider the family of mappings $V_\sigma(\cdot)$, $\sigma > 0$ as defined by (18.20). By taking into account (18.24), (18.27) and (18.28) it follows that each V_σ is $C^\infty(R^n)$ and for any compact $Q \subset R^n \setminus \{0\}$ and constant $\varepsilon > 0$ we have

$$|V(x) - V_\sigma(x)| \leq \varepsilon \tag{18.40a}$$

$$|DV(x) - DV_\sigma(x)| \leq \varepsilon \tag{18.40b}$$

$$\varepsilon + \frac{1}{2}k_1|x|^2 \leq V_\sigma(x) \leq \varepsilon + 2k_2|x|^2 \tag{18.40c}$$

$$|DV_\sigma(x)| \leq 2K|x| + \varepsilon \tag{18.40d}$$

for all $x \in Q$ and sufficiently small σ, where the constants k_1, k_2 and K are defined in (18.28) and (18.27), respectively, thus, they are independent of Q. Moreover, (18.39) in conjunction with Lemma (3.6) and (18.30) yield

$$L^u_{(1.1)}V_\sigma(x) \leq -\frac{1}{2}|x|^2 + \varepsilon, \quad \forall\, x \in Q, \quad u \in I \tag{18.40e}$$

for sufficiently small σ. Using (18.40) and repeating the same partition of unity approach in [7, 14] we can construct a $C^\infty(R^n \setminus \{0\}) \cap C^2(R^n)$ map V satisfying all desired properties (18.7) with $r = 2$.

Case II: $r \neq 2$

The case $r \neq 2$ is reduced to the previous one, as follows. Consider the map

$$\Theta(x) := \frac{x}{|x|^a}, \quad x \in R^n \setminus \{0\}, \quad a := \frac{2-r}{2} \tag{18.41}$$

Then, by using Ito's theorem for the process $Y(\cdot) = \Theta(X(\cdot))$, the original system (18.1) takes the equivalent form

$$dy = F(y,u)dt + G(y,u)dw \tag{18.42}$$

where

$$F := \begin{cases} \left.\left(D\Theta f + \dfrac{1}{2} \displaystyle\sum_{1 \leq i,j \leq n} \frac{\partial^2 \Theta}{\partial x_i \partial x_j}(gg')_{i,j} \right)\right|_{x=\Theta^{-1}(y)}, & y \neq 0 \\ 0, \quad \text{for } y = 0 \end{cases} \tag{18.43a}$$

$$G := \begin{cases} (D\Theta g)|_{x=\Theta^{-1}(y)}, & y \neq 0 \\ 0, & \text{for } y = 0 \end{cases} \tag{18.43b}$$

Notice that for all nonzero x we have:

$$|D\Theta(x)| \leq C_1|x|^{-a}, \quad \left|\sum \frac{\partial^2 \Theta}{\partial x_i \partial x_j}\right| \leq C_2|x|^{-a-1} \tag{18.44}$$

for some $C_1, C_2 > 0$. From (18.2), (18.6), (18.43) and (18.44) we can easily verify that the dynamics F and G have the same properties (18.2) and (18.6b) with those imposed for the original system and, since zero is r-expUGAS for (18.1), it follows from (18.41) that zero is 2-expUGAS for (18.42). For the case $r < 2$, however, the second derivatives of G with respect to y_i are in general unbounded closed to the origin; in particular, a constant $C > 0$ can be found such that

$$\left|\sum \frac{\partial^2 F}{\partial y_i \partial y_j}\right| + \left|\sum \frac{\partial^2 G}{\partial y_i \partial y_j}\right| \leq C|y|^{-1}, \quad \forall\ |y| \neq 0, \quad u \in I \tag{18.45}$$

Hence, in order to reduce the analysis the previous case $r = 2$, we apply change of time in (18.42) as follows: We first consider a smooth function $q : R^+ \to R^+$ which satisfies:

$$1 > q(s) > 0, \quad \forall\ s > 0 \tag{18.46}$$

and in such a way that $q(|y|)F(y,u)$ and $q^{1/2}(|y|)G(y,u)$ satisfy all properties (18.2) and (18.6b) and in addition their second derivatives with respect to y_i are bounded over $R^n \times I$. We note that existence of $q(\cdot)$ is guaranteed from boundedness of the first derivatives of F and G and (18.45). Consider an initial state y_0, time t_0 and input $u \in F(I)$ and apply in (18.42) the change of time:

$$t \to \tau_t := \inf\{s, \beta_s \geq t\} \tag{18.47a}$$

$$\beta_t := \int_0^t \frac{1}{q(|Y(s,t_0,y_0,u)|)} ds \tag{18.47b}$$

Notice that $\tau_0 = 0$, almost surely and, since zero is the unique equilibrium, for every $y_0 \neq 0$ we have $\beta_t < \infty$, almost surely. Moreover, due to (18.46), it holds that

$$\tau_t \leq t \tag{18.48}$$

According to well known results (see for instance, Theorems 7.24, 7.26, 7.30 and 7.31 in [9]) the process $t \to Y(\tau_t, t_0, y_0\tilde{u})$, $\tilde{u} := u(\tau_t)$, is a weak solution of the stochastic system

$$d\tilde{y} = q(|\tilde{y}|)F(\tilde{y},\tilde{u})dt + q^{1/2}(|\tilde{y}|)G(\tilde{y},\tilde{u})d\tilde{w} \tag{18.49}$$

where $\tilde{w}$ is the F_{τ_t}- adapted Wiener process

$$\tilde{w}(t) := \int_0^{\tau_t} q^{-1/2}(|Y(s,t_0,y_0,u)|)dw(s),$$

which in general depends on t_0, y_0 and u, and further

$$E|Y(t,t_0,y_0,u)|^2 = E|\tilde{Y}(\tau_t,t_0,y_0,\tilde{u})|^2 \tag{18.50}$$

Taking into account that zero is 2-expUGAS for (18.42), condition (18.50) implies that zero is also 2-expUGAS for (18.49) if we replace $\tilde{w}$ by any F_t-adapted Wiener process w (being independent of t_0, y_0 and u). This follows from the fact that the solution $\overline{Y}(t,t_0,y_0,\tilde{u})$ of (18.49) with w instead of $\tilde{w}$ is identical in law with the corresponding solution $\tilde{Y}(t,t_0,y_0,\tilde{u})$ of (18.49). Moreover, notice that each controller $\tilde{u} = u(\tau_t)$ is F_{τ_t}- adapted, hence, by virtue of (18.48), is F_t - adapted. It turns out that each solution $\overline{Y}(\cdot)$ of

$$dy = q(|y|)F(y,\tilde{u})dt + q^{1/2}(|y|)G(y,\tilde{u})dw \tag{18.51}$$

satisfies

$$E|\overline{Y}(t,t_0,y_0,\tilde{u})|^2 \le C|y_0|\exp(-\ell(t-t_0)), \quad \forall t \ge 0$$

for some $C > 0$, for all $y_0 \in R^n$ and $\tilde{u} \in F(I)$, and its dynamics satisfy (18.2) and (18.6). It is therefore possible to apply the same analysis with this employed for the previous case $r = 2$ to build a $C^2(R^n) \cap C^\infty(R^n \setminus \{0\})$ function $V(y)$ satisfying (18.7) with $r = 2$. Particularly, we have

$$L^u_{(3.42)}V(y) = q(|y|)L^u_{(3.33)}V(y) \le -k|y|^2, \quad \forall\, y \in R^n, \quad u \in I$$

for some $k > 0$, which by virtue of (18.46) yields

$$L^u_{(3.33)}V(y) \le -k|y|^2, \quad \forall\, y \in R^n, \quad u \in I \tag{18.52}$$

Let

$$\hat{V}(x) := V(\Theta(x)) \tag{18.53}$$

The above $\hat{V}$ is the desired Lyapunov function for the original system (18.1), namely, is $C^{[r]}(R^n) \cap C^\infty(R^n \setminus \{0\})$ and satisfies the desired properties (18.7) in the original coordinates. The claim is an immediate consequence of properties of V in the y-coordinates and the definition of $\Theta(\cdot)$. For reasons of completeness we note that, since

$$L^u_{(1.1)}V(\Theta(x))\Big|_{x=\Theta^{-1}(y)} = L^u_{(3.33)}V(y)$$

we get from (18.52), (18.53) and (18.41):

$$L^u_{(1.1)}\hat{V}(x) \le -k|\Theta(x)|^2 = -k|x|^{2-2a} = -k|x|^r, \quad \forall\, x \in R^n, \quad u \in I$$

and this proves (18.7c). Similarly, we can easily verify that (2.4a,b) hold as well.

As a consequence of Propositions 3.1, 3.2 and Theorem 3.4 we obtain:

Corollary 18.3.1. *Consider the system (18.1) and assume that (18.2) and (18.6) are fulfilled. Then, if zero is expUGAS, there exists a Lyapunov function $V \in C^\infty(R^n \setminus \{0\})$, which satisfies (18.7). It turns out that the following statements are equivalent, provided that (18.15) holds as well:*

(1) The system (18.1) admits a $C^\infty(R^n \setminus \{0\})$ Lyapunov function V satisfying (18.7);

(2) Zero is expUGAS;

(3) Zero is r-expUGAS for r sufficiently small.

Remark: If we restrict ourselves to stochastic differential equations, namely, to systems (18.1) where f and g are independent of u, the analysis made in proof of Theorem 3.4 is extremely simplified (in fact, it constitutes a modification of the approach employed in Khasminskii [Theorem 7.2 (p. 187), 3]) and the non–degeneration assumption is not required.

18.4 Lyapunov description of expISS

The following proposition is a consequence of Propositions 3.1, 3.2 and the converse stability theorem of previous section. Its proof is quite analogous to that given in [10, 11] for the deterministic case.

Proposition 18.4.1. *Consider the system (18.1) whose dynamics are C^2 and satisfy (18.8).*
(i) Suppose that there exists a $C^2(R^n \setminus \{0\})$ function V satisfying (2.4a,b) as well as (18.7c) for all $|u| \le \gamma(|x|)$, $\gamma : R^+ \to R^+$ being positive definite and $C^2(R^+)$ with bounded first derivative $\gamma^{(1)}$. Then (18.1) satisfies the r-expISS. Moreover, expISS is fulfilled if in addition we assume that either

$$g(x,u)g^T(x,u) \ge 0, \quad \forall\ |u| \le \gamma(|x|) \tag{18.54}$$

or

$$|g(x,u)| \le k|u|, \quad \forall\ (x,u) \in R^{n+1} \tag{18.55}$$

for some $k > 0$.
(ii) Conversely, existence of a Lyapunov function as above is guaranteed, if, in addition to (18.8), we assume that

- *for any compact* $I \subset R$

$$\frac{\partial^2 f}{\partial x_i \partial u}, \frac{\partial^2 g}{\partial x_i \partial u}, \quad i = 1, \ldots, n \quad \text{are bounded over } R^n \times I; \tag{18.56}$$

- *for any nonzero* x *there are constants* $k_x > 0$ *and* $u_0 \in R$ *with*

$$g(x, u_0) g^T(x, u_0) \geq k_x I_n \tag{18.57}$$

- *r-expISS holds for some gain function* $\gamma : R^+ \to R^+$ *being positive definite,* $C^2(R^+)$ *and such that*

$$\gamma^{(1)} \quad \text{and} \quad \gamma^{(2)} \quad \text{are bounded over} \quad R^+ \tag{18.58}$$

(iii) The notions of expISS and r-expISS are equivalent for r sufficiently small, provided that (18.56), (18.57) hold and further

$$g(x, u) g^T(x, u) \geq 0, \quad \forall (x, u) \in R^n \times R$$

Proof: All statements are direct consequences of the results of the previous section. We only prove that existence of a $C^2(R^n \setminus \{0\})$ function V satisfying (2.4a,b,c) with $|u| \leq \gamma(|x|)$ implies expISS provided that (18.55) holds. The rest part of proof is left to the reader. First, notice that V satisfies all properties (18.7) with respect to

$$dx = f(x, v\gamma(|x|))dt + g(x, v\gamma(|x|))dw, \quad v \in I := [-1, 1] \tag{18.59}$$

Moreover, expISS (r-expISS) is fulfilled for (18.1), if and only if zero is expUGAS (r-expUGAS), respectively, for (18.59). To establish expISS pick a positive definite C^2 function $\gamma^* < \gamma$ (we may select $\gamma^* = c\gamma$ for $c > 0$ small enough) in such a way that if we define

$$K := \sup_{x \neq 0, |v| \leq 1} \frac{|g(x, v\gamma^8(|x|))|}{|x|}$$

then (18.16) holds; (because of (18.55) and boundedeness assumption for $\gamma^{(1)}$ such selection is feasible). It follows by Proposition 3.2 that zero is expUGAS for (18.59) with $\gamma = \gamma^*$, or equivalently expISS holds for (18.1) with gain function $\gamma = \gamma^*$. Likewise, by virtue of Proposition 3.2 we can establish r-expISS with the same γ. The rest part of proof is also a consequence of Propositions 3.1, 3.2 and Theorem 3.4. For reasons of completeness we note that, under the additional assumptions made in statement (ii) for f, g and γ, the dynamics of the system (18.59) satisfy both (18.2) and (18.6). Details are left for the reader.

Applications of the stochastic ISS and the relative notion "noisy input - state stochastic stability" to feedback stabilization problems are found in [4, 12, 13].

18.5 Remarks for deterministic systems

For deterministic systems the notion of expUGAS is of course equivalent to r-expUGAS for arbitrary positive r, and the technique employed in proof of the converse Lyapunov theorem in Section 3 requires only the hypothesis (18.3). It turns out that for the deterministic case expISS is equivalent to the existence of a $C^2(R^n \setminus \{0\})$ function V satisfying (2.4a,b) and (18.7c) for $|u| \leq \gamma(|x|)$, provided that (18.8) holds. It should be pointed out here that using a more elegant procedure, specifically, by making appropriate modifications in proof of the general converse Lyapunov theorem in [10], we can establish that the converse Lyapunov theorems concerning exponential stability for deterministic systems are valid under the weaker hypethesis that the dynamics are localy Lipschitz. The clain follows

References

1. L. Arnold (1974) *Stochastic Differential Equations*, Theory and Applications, Wiley, New York.

2. H. Khalil (1996) *Nonlinear Systems*, Prentice Hall.

3. R. Z. Khasminskii (1980) *Stochastic stability of differential equations*, Sijthoff & Noordhaff,Alphen aan en Rijn.

4. M. Krstic and H. Deng (1998) *Stabilization of Nonlinear uncertain systems*, Springer–Verlag.

5. N. V. Krylov (1980) *Controlled Diffusion Processes*, Springer–Verlag.

6. J. Kushner (1967) *Stochastic stability and Control*, Academic Press, New York.

7. Y. Lin, E. D. Sontag and Y. Wang (1996) *A smooth converse Lyapunov theorem for robust stability*, SIAM J. Control and Opt., 34, 124–160.

8. P. L. Lions (1980) *Control of Diffusion Processes in R^n, Communications on Pure and Applied Mathematica*, Vol. XXXIV, 121–147.

9. B. Oksendal (1985) *Stochastic Differential Equations*, Springer–Verlag.

10. E. D. Sontag and Y. Wang (1995) *On characterizations of input-to-state stability properties*, Systems and Control Lett., 24, 226–231.

11. J. Tsinias (1997) *Input-to-state stability properties of nonlinear systems and applications to bounded feedback stabilization using saturation.* COCV–ESAIM, 2, 7.

12. J. Tsinias (1998) *Stochastic input–to–state stability and applications to global feedback stabilization*, Int. J. Control, special issue "Recent advances in the control of nonlinear systems" Ed. F. Lamnabhi - Lagarrigue, 71(5), 907–931.

13. J. Tsinias (to appear) *The concept of "exponential ISS" for stochastic systems and applications to feedback stabilization*, in Systems and Control Lett.

14. F. W. Wilson (1969) *Smoothing derivatives of functions and applications*, Transactions of the American Mathematical Society, 136, 413–428.

19. LMIs for robust stable neural model-based control

*Bart Wams**, *Miguel Ayala Botto***
*Ton van den Boom** *and José Sá da Costa***

*Delft University of Technology
Faculty of Information Technology and Systems
Department of Electrical Engineering, P.O. Box 5031
2600 GA Delft, The Netherlands
Phone: +31-15-2785114 Fax: +31-15-2786679
b.wams@its.tudelft.nl
vdboom@harding.et.tudelft.nl

**Instituto Superior Técnico, Technical University of Lisbon
Department of Mechanical Engineering, GCAR/IDMEC
Avenida Rovisco Pais
1096 Lisboa Codex, Portugal
Phone: +351-1-8419028 Fax: +351-1-8498097
migbotto@gcar.ist.utl.pt
msc@gcar.ist.utl.pt

Summary.

This paper presents a robust nonlinear controller design strategy based on approximate feedback linearization using neural network models. It is shown how model uncertainty due to model-plant mismatch propagates through the control loop such that the overall closed-loop uncertainty can be expressed in terms of the uncertainty in the weights of the neural network model. Further, a polytopic uncertainty description for the closed-loop is found which enables robust stability analysis to be translated into a Linear Matrix Inequality (LMI) problem, and thus be tackled with the use of computationally efficient techniques.

19.1 Introduction

An important research direction in modern control theory points toward finding more reliable solutions to tackle highly nonlinear problems. In this sense, model-based control has been widely accepted as a powerful tool for control engineering as proven by the growing number of reported successful industrial applications. One of the major aspects in model-based control is therefore the reliability of the model upon which the controller will be based. Over the last decade neural network models have become favorite candidates in the nonlinear systems identification field due to their excellence in representing multivariable nonlinear mappings, together with their capability of generalizing from fresh data, as well as their ability to keep learning during operation [1]. Furthermore, the benefits of using neural network models in nonlinear control applications have been widely reported in the literature [2, 3, 4, 5]. However, the usefulness of any model-based control approach relies upon the sufficient stability and modeling errors robustness conditions that can be formally established. In fact, while the primary objective is to guarantee the stability (and performance) of the resulting closed-loop system, the control design should also account for the effects of model mismatch in the overall control scheme. Therefore, a model uncertainty description providing quantitative information about the model mismatch should be obtained. Moreover, it is crucial to understand how such model uncertainty propagates through the control design and so an efficient tool to analyze the stability robustness properties of the resulting closed-loop system should be available as well. In this paper, these items are addressed in the context of *approximate* input-output feedback linearization using a discrete-time neural network model. Feedback linearization [6, 7] as been recognized as a powerful tool to tackle nonlinear control problems since it provides, under some mild assumptions, an exact linearization of the process over the complete operating range. More formally, for the general case of a square multivariable $(p \times p)$ discrete-time nonlinear system, and assuming without loss of generality unitary relative degrees for each of the p system outputs, a feedback linearizing control law can be obtained by solving the following equation in respect to the p system inputs, $\boldsymbol{u}_k$ [8, 9]

$$E(\boldsymbol{x}_k, \boldsymbol{u}_k) = CA\boldsymbol{x}_k + CB\boldsymbol{v}_k \tag{19.1}$$

where $E(\boldsymbol{x}_k, \boldsymbol{u}_k)$ represents the non-singular discrete decoupling $(p \times p)$ matrix of the system, with rank p around the equilibrium point, $\boldsymbol{x}_k$ represents the system state vector, while $\boldsymbol{v}_k$ is the newly created external signal. The resulting input-output decoupled linear system will then be described by

$$\boldsymbol{y}_{k+1} = CA\boldsymbol{x}_k + CB\boldsymbol{v}_k \tag{19.2}$$

with A, B and C being proper choices for the linear state-space matrices. However, finding a solution for $\boldsymbol{u}_k$ in (19.1) tend to be a hard numerical problem. In this paper, a solution is given by means of the approximate feedback linearization, which basically consists of performing the first-order Taylor's expansion of the input-output relation of the system, around a state trajectory, resulting in an affine in the input decoupling matrix relation [10]. In this case, provided that the inversion of the resulting jacobian matrix holds, $\boldsymbol{u}_k$ in (19.1) can be obtained through the solution of a simple linear numerical problem, therefore enabling a straightforward computation of the feedback linearizing control law.

In this paper, quantitative measures for the model uncertainty description are obtained by assuming that model mismatch is caused by deviation of the estimated network parameters from the *real* ones, *i.e.*, by assuming that the process is in the model set. Tools for selecting the model structure, and checking whether the process can be expected to be in the model set, can be found in [11, 1, 12, 13]. This idea is motivated by the fact that finding the best parameters for the neural network model (training), is essentially equivalent to a nonlinear parameter estimation problem. Therefore, since the training objective is to minimize the sum of squared errors between training data and neural network model output, a nonlinear least square estimation of the weights results. In this situation, it can be stated that the *real* weights are within an interval from the estimated weights, provided that certain conditions are met [14]. Then, the size of these intervals can be determined by employing statistical properties of the nonlinear least square estimation. Throughout this paper expressions will also be derived which express the resulting closed-loop uncertainty in terms of the uncertainty in the weights of the neural network model. Furthermore, a procedure which transforms the closed-loop uncertainty description in a form which directly allows for stability analysis is also presented. Moreover, it is outlined how the uncertain closed-loop description can be bounded by a polytopic system, which transforms the robust stability analysis problem in that of stability analysis of polytopic systems. The use of polytopic systems descriptions in robust control has become more popular over the last years [15, 16, 17]. The main reason is due to the overall stability problem could be reformulated as a set of Linear Matrix Inequalities (LMIs) which can be solved by using computationally efficient algorithms.

The structure of this paper is as follows: in section 19.2.1 the nominal case of approximate feedback linearization is outlined, while section 19.2.2 describes how the closed-loop system uncertainty can be expressed in terms of the model uncertainty by means of the approximate feedback linearization. Then section 19.2.3 presents the necessary ingredients for robustness analysis of the resulting closed-loop uncertainty description. Section 19.3 will formalize the previous developments for the case the system is modeled with a neural network model. Finally in section 19.4 some conclusions are drawn.

Notation. Vectors are notated by bold lower case letters, whereas scalars are notated with non-bold lower case letters. Matrices are notated with upper case letters where W^{ij} indicate the ij-th entry of a matrix W. The meaning of used subscripts, should be clear from the context.

19.2 Control setting

Approximate feedback linearization is outlined, based on a general nonlinear ARX description of the system under consideration. The feedback linearizing control law is presented in terms of the system description. In addition, approximate feedback linearization in the presence of model uncertainty is treated. Moreover, an uncertainty description for the closed-loop system in terms of the model uncertainty is derived, and a method to analyze the robustness properties of the closed-loop system is provided. These results are valid for any kind of nonlinear ARX system description, as no implementation of the system description is selected yet.

19.2.1 Nominal case

Consider the following square[1] input-output system, with p inputs and p outputs, given by the following state-space representation

$$\boldsymbol{y}_{k+1} = \boldsymbol{f}(\boldsymbol{x}_k, \boldsymbol{u}_k) \tag{19.3}$$

where

$$\boldsymbol{y}_{k+1} = [y^1_{k+1}, \dots, y^p_{k+1}]^T \tag{19.4}$$

$$\boldsymbol{x}_k = [y^1_k, \dots, y^1_{k-n1}, \dots, y^p_{k-np}, u^p_{k-1}, \dots, u^1_{k-m1}, \dots u^p_{k-mp}]^T \tag{19.5}$$

$$\boldsymbol{u}_k = [u^1_k, \dots u^p_k]^T \tag{19.6}$$

with ni and mj being the maximum delay of the i-th output and j-th and input, respectively, present in the regression vector $\boldsymbol{x}_k$. To allow a straightforward calculation of the approximate feedback linearizing control law, the first order of the Taylor's expansion of system (19.3) around the state trajectory $(\boldsymbol{x}_{k-1}, \boldsymbol{u}_{k-1})$ is performed, yielding

$$\boldsymbol{y}_{k+1} = \boldsymbol{f}(\boldsymbol{x}_{k-1}, \boldsymbol{u}_{k-1}) + F(\boldsymbol{x}_{k-1}, \boldsymbol{u}_{k-1})\Delta\boldsymbol{x}_k + E(\boldsymbol{x}_{k-1}, \boldsymbol{u}_{k-1})\Delta\boldsymbol{u}_k \tag{19.7}$$

where

$$\boldsymbol{f}(\boldsymbol{x}_{k-1}, \boldsymbol{u}_{k-1}) = \begin{bmatrix} f^1(\cdot) & \cdots & f^p(\cdot) \end{bmatrix}^T_{(\boldsymbol{x}_{k-1}, \boldsymbol{u}_{k-1})} \tag{19.8}$$

$$F(\boldsymbol{x}_{k-1}, \boldsymbol{u}_{k-1}) = \begin{bmatrix} \frac{\partial f^1(\cdot)}{\partial x} & \cdots & \frac{\partial f^p(\cdot)}{\partial x} \end{bmatrix}^T_{(\boldsymbol{x}_{k-1}, \boldsymbol{u}_{k-1})} \tag{19.9}$$

$$E(\boldsymbol{x}_{k-1}, \boldsymbol{u}_{k-1}) = \begin{bmatrix} \frac{\partial f^1(\cdot)}{\partial u} & \cdots & \frac{\partial f^p(\cdot)}{\partial u} \end{bmatrix}^T_{(\boldsymbol{x}_{k-1}, \boldsymbol{u}_{k-1})} \tag{19.10}$$

[1] Throughout this paper, only square systems are considered. However, the results can be extended to non-square systems, when the number of inputs exceeds the number of outputs, by replacing the matrix inversion by the left (pseudo) inverse.

and $\Delta = 1 - q^{-1}$, with q^{-1} being the delay operator, thus $\Delta\boldsymbol{x}_k = \boldsymbol{x}_k - \boldsymbol{x}_{k-1}$. Then, the approximate feedback linearizing control law can be derived as

$$\begin{aligned}\Delta\boldsymbol{u}_k = E(\boldsymbol{x}_{k-1}, \boldsymbol{u}_{k-1})^{-1}(-\boldsymbol{f}(\boldsymbol{x}_{k-1}, \boldsymbol{u}_{k-1}) - F(\boldsymbol{x}_{k-1}, \boldsymbol{u}_{k-1})\Delta\boldsymbol{x}_k + \\ +CA\boldsymbol{x}_k + CB\boldsymbol{v}_k)\end{aligned} \tag{19.11}$$

where A, B and C are user specified matrices of appropriate dimensions and $\boldsymbol{v}_k$ is the new linear imposed input signal. From expression (19.11) it can be seen that for a successful application of the approximate feedback linearization, matrix E must be invertible for all admissible pairs $(\boldsymbol{x}_{k-1}, \boldsymbol{u}_{k-1})$. However, as shown later in this paper, this is a reasonable assumption if considering a neural network model. Finally, the application of the feedback law (19.11) to the system (19.7) results in the following closed-loop system

$$\boldsymbol{y}_{k+1} = CA\boldsymbol{x}_k + CB\boldsymbol{v}_k \tag{19.12}$$

which can either be represented in the following state space form

$$\begin{aligned}\boldsymbol{x}_{k+1} &= A\boldsymbol{x}_k + B\boldsymbol{v}_k \\ \boldsymbol{y}_k &= C\boldsymbol{x}_k\end{aligned} \tag{19.13}$$

19.2.2 Uncertain case

In order to obtain an uncertainty description of the closed-loop system, which later on allows for a straightforward robust stability analysis procedure, a new notation must be introduced. In view of this, the first part of the right-hand side of (19.7) is rewritten as

$$\boldsymbol{f}(\boldsymbol{x}_{k-1}, \boldsymbol{u}_{k-1}) + F(\boldsymbol{x}_{k-1}, \boldsymbol{u}_{k-1})\Delta\boldsymbol{x}_k = \tilde{F}(\boldsymbol{x}_{k-1}, \boldsymbol{u}_{k-1})\tilde{\boldsymbol{x}}_k \tag{19.14}$$

where the new state vector $\tilde{\boldsymbol{x}}_k$ is now given by

$$\tilde{\boldsymbol{x}}_k = [y_{1,k}, \ldots, y_{1,k-n1-1}, \ldots, y_{p,k-np-1}, \Delta u_{1,k-1}, \ldots, \Delta u_{p,k-mp}]^T \tag{19.15}$$

Thus, the structure of $\tilde{F}(\boldsymbol{x}_{k-1}, \boldsymbol{u}_{k-1})$ in (19.14) can be generically determined by the following procedure. Let N_a be a matrix of dimension $(a \times (a+1))$ having the following structure

$$N_a = \begin{bmatrix} 1 & -1 & 0 & \ldots & 0 \\ 0 & 1 & -1\, 0 & \ldots & 0 \\ & & \vdots & \vdots & \\ & & & & \\ 0 & \ldots & & 1 & -1 \end{bmatrix} \tag{19.16}$$

and let $s_{(a,b)}$ be a vector of size $(1 \times (a+b+1))$ defined as

$$s_{a,b} = [a_0 \quad 1 \quad b_0] \tag{19.17}$$

where a_0 and b_0 are zero vectors of dimensions $(1 \times a)$ and $(1 \times b)$, respectively. With these definitions, $\tilde{F}(\boldsymbol{x}_{k-1}, \boldsymbol{u}_{k-1})$ can be constructed according to

$$\tilde{F}(\boldsymbol{x}_{k-1}, \boldsymbol{u}_{k-1}) = \begin{bmatrix} s_{(0,n_1+\cdots+n_p+2p-1)} \\ s_{(n_1+1,n_2+\cdots+n_p+2p-2)} \\ \vdots \\ s_{(n_1+\cdots+n_{p-1}+p-1,n_p+p)} \end{bmatrix} + F(\boldsymbol{x}_{k-1}, \boldsymbol{u}_{k-1}) \begin{bmatrix} N_{n_1} & 0 & & \dots & 0 \\ 0 & N_{n_2} & 0 & \dots & 0 \\ & & \ddots & & \\ & & & N_{n_p} & 0 \dots 0 \\ 0 & & \dots & & I \end{bmatrix} \tag{19.18}$$

where I is the $(p \times p)$ identity matrix. According to this new formulation, (19.7) is now given by

$$\boldsymbol{y}_{k+1} = \tilde{F}(\boldsymbol{x}_{k-1}, \boldsymbol{u}_{k-1})\tilde{\boldsymbol{x}}_k + E\Delta \boldsymbol{u}_k \tag{19.19}$$

In this case, the application of the following feedback linearizing control law

$$\Delta \boldsymbol{u}_k = E(\boldsymbol{x}_{k-1}, \boldsymbol{u}_{k-1})^{-1}(-\tilde{F}(\boldsymbol{x}_{k-1}, \boldsymbol{u}_{k-1})\tilde{\boldsymbol{x}}_k + CA\tilde{\boldsymbol{x}}_k + CB\boldsymbol{v}_k) \tag{19.20}$$

results in the closed-loop system described by

$$\boldsymbol{y}_{k+1} = CA\tilde{\boldsymbol{x}}_k + CB\boldsymbol{v}_k \tag{19.21}$$

which can be either represented in the following state-space form

$$\begin{aligned} \tilde{\boldsymbol{x}}_{k+1} &= A\tilde{\boldsymbol{x}}_k + B\boldsymbol{v}_k \\ \boldsymbol{y}_k &= C\tilde{\boldsymbol{x}}_k. \end{aligned} \tag{19.22}$$

For a more transparent notation, the argument $(\boldsymbol{x}_{k-1}, \boldsymbol{u}_{k-1})$ will be omitted from matrices $\tilde{F}$ and E on further developments. In order to analyze the influence of model uncertainty on the control strategy, an uncertainty description for the model will be adopted. It is assumed throughout that the *real* system can be described by

$$\boldsymbol{y}_{k+1} = (\tilde{F} + \delta\tilde{F})\tilde{\boldsymbol{x}}_k + (E + \delta E)\Delta \boldsymbol{u}_k \tag{19.23}$$

where $\delta\tilde{F}$ and δE are matrices of appropriate dimensions with unknown entries. In sections 19.3.2 and 19.3.3 it is outlined how (19.23) can be obtained in case the model is implemented with a single-layer feed-forward neural network. Further, a procedure is presented that provides quantitative measures for the intervals in which the entries of $\delta\tilde{F}$ and δE live. In the sequel, it will become obvious that this knowledge is sufficient for analyzing the overall closed-loop stability. In fact, the application of the control law (19.20) to the system (19.23), results in a closed-loop system which is described by

$$\begin{aligned} \boldsymbol{y}_{k+1} &= CA\tilde{\boldsymbol{x}}_k + CB\boldsymbol{v}_k + (\delta\tilde{F} - \delta E \cdot E^{-1}\tilde{F} + \delta E \cdot E^{-1}CA)\tilde{\boldsymbol{x}}_k + \\ &+\delta E \cdot E^{-1}CB\boldsymbol{v}_k \end{aligned} \tag{19.24}$$

Moreover, if C is chosen such that $C \cdot C^T = I$, the following state-space representation can be obtained

$$\begin{aligned} \tilde{\boldsymbol{x}}_{k+1} &= (A + C^T(\delta\tilde{F} - \delta E \cdot E^{-1}\tilde{F} + \delta E \cdot E^{-1}CA))\tilde{\boldsymbol{x}}_k + \\ &+(B + C^T\delta E \cdot E^{-1}CB)\boldsymbol{v}_k \\ \boldsymbol{y}_k &= C\tilde{\boldsymbol{x}}_k \end{aligned} \tag{19.25}$$

leading to the following description of the uncertain linear closed-loop system

$$\begin{aligned} \tilde{\boldsymbol{x}}_{k+1} &= (A + \delta A)\tilde{\boldsymbol{x}}_k + (B + \delta B)\boldsymbol{v}_k \\ \boldsymbol{y}_k &= C\tilde{\boldsymbol{x}}_k \end{aligned} \tag{19.26}$$

where the uncertainty terms δA and δB are now state dependent matrices, respectively given by

$$\begin{aligned} \delta A &= C^T(\delta\tilde{F} - \delta E \cdot E^{-1}\tilde{F} + \delta E \cdot E^{-1}CA) \\ \delta B &= C^T\delta E \cdot E^{-1}CB \end{aligned} \tag{19.27}$$

Clearly, the choice for the imposed linear dynamics A, B and C used in the feedback linearizing control law, will determine the characteristics of the resulting closed-loop system. While the condition $C \cdot C^T = I$ can always be satisfied by adopting the appropriate closed-loop canonical form, the choice for A and B should be guided by the properties of the system under control in account for the desired performance specifications of the overall control scheme.

19.2.3 Stability analysis

The analysis concerning model uncertainty which was performed in the previous section resulted in the uncertain closed-loop description given by (19.26). Assuming that

$$[A + \delta A(\cdot)\ B + \delta B(\cdot)] \in \mathbf{Co}\left([A_1\ B_1], [A_2\ B_2], \ldots, [A_L\ B_L]\right) \tag{19.28}$$

where **Co** stands for convex hull, *i.e.*, for some $\lambda_1, \lambda_2, \ldots, \lambda_L \geq 0$ summing to one

$$[A + \delta A(\cdot)\ B + \delta B(\cdot)] = \sum_{i=1}^{L} \lambda_i [A_i\ B_i] \tag{19.29}$$

Then, if B_i, with $i = 1, \ldots L$, is bounded, for showing closed-loop stability of the uncertain system it is sufficient to find a matrix P, such that

$$\begin{array}{r} A_i^T P A_i - P < 0 \\ P > 0 \end{array} \quad \forall i = 1, \ldots, L. \tag{19.30}$$

Moreover, expression (19.30) can be rewritten as a linear matrix inequality (LMI) by using $Q^{-1} = P$, resulting

$$\begin{array}{r} A_i^T Q^{-1} A_i - Q^{-1} < 0 \\ Q^{-1} > 0 \end{array} \quad \forall i = 1, \ldots, n \tag{19.31}$$

Then, by post- and pre-multiply (19.31) by Q, with $Q = Q^T$, leads to

$$\begin{array}{r} Q - (A_i Q)^T Q^{-1} A_i Q > 0 \\ Q > 0. \end{array} \tag{19.32}$$

According to the Schur's complement condition, this set of equations is equivalent to the following LMI

$$\begin{bmatrix} Q & (A_i Q)^T \\ A_i Q & Q \end{bmatrix} > 0 \tag{19.33}$$

Solving a set of LMIs can be easily made through the use of computationally efficient algorithms, which turn them into a practical tool for control engineering purposes [17]. Therefore, the key to the solution of the robust stability analysis problem is to build the necessary LMIs which enclose the closed-loop uncertainty description given by (19.26) in a polytope. These bounds can be found by determining the intervals within which the entries of matrices δA and δB in (19.27) can vary.

19.3 Implementation with a feed-forward neural network

This section applies the previously presented control strategy for the case system (19.3) describes a single-layer feed-forward neural network model. Further, a method for obtaining quantitative information concerning the neural network uncertainty will be outlined, enabling robust stability analysis of the closed-loop system to be performed. The development will be made according to the following steps

1. First, the expressions involving the nominal control design, (19.3, 19.7 and 19.11), are derived for the case the system under control is modeled with a single-layer feed-forward neural network.

2. Secondly, it is investigated how uncertainty in the neural network model parameters propagate through the control design. As a result, expressions for $\delta\tilde{F}$ and δE in (19.23) will be derived in terms of these model uncertainties. Then, based on (19.27), the uncertainty in the closed-loop system matrices A and B can be determined.
3. Finally, a polytopic description for the uncertain closed-loop system matrices is built. A procedure that finds the bounds for the entries of these state dependent uncertain matrices will be outlined.

The accomplishment of the three previous steps will provide the ingredients for designing a robust neural model-based control.

19.3.1 Nominal case

Assume that the $(p \times p)$ square MIMO system (19.3) describes a single layer feed-forward neural network with tangent hyperbolic activation function. In matrix notation, such neural network model can be described through the following notation

$$\boldsymbol{y}_{k+1} = W \tanh(V\boldsymbol{x}_k + G\boldsymbol{u}_k + \boldsymbol{b}) + \boldsymbol{d} \tag{19.34}$$

where $\boldsymbol{y}_{k+1}$, $\boldsymbol{x}_k$ and $\boldsymbol{u}_k$ are defined in (19.4), (19.5) and (19.6), respectively. Let nh be the number of neurons in the hidden layer, let nx be the length of the vector $\boldsymbol{x}$, then $W \in \mathbb{R}^{p\times nh}$, $V \in \mathbb{R}^{nh\times nx}$, $G \in \mathbb{R}^{nh\times p}$, $\boldsymbol{b} \in \mathbb{R}^{nh\times 1}$ and $\boldsymbol{d} \in \mathbb{R}^{p\times 1}$. The first order of Taylor's expansion around $(\boldsymbol{x}_{k-1}, \boldsymbol{u}_{k-1})$ produces:

$$\boldsymbol{y}_{k+1} = F_0 + F\Delta\boldsymbol{x}_k + E\Delta\boldsymbol{u}_k \tag{19.35}$$

where

$$F_0 = W \tanh(V\boldsymbol{x}_{k-1} + G\boldsymbol{u}_{k-1} + \boldsymbol{b}) + \boldsymbol{d} \tag{19.36}$$
$$F = W\Gamma_{k-1}V \tag{19.37}$$
$$E = W\Gamma_{k-1}G \tag{19.38}$$

with

$$\Gamma_{k-1} = I - \mathrm{diag}[\tanh^2(V\boldsymbol{x}_{k-1} + G\boldsymbol{u}_{k-1} + \boldsymbol{b})] \tag{19.39}$$

These expressions form the basis to proceed with the application of the feedback linearization procedure described in section (19.2.1).

19.3.2 Uncertain case

As stated before, the philosophy here considered assumes that model uncertainty is due to the uncertainty on the weights of the neural network model. This choice is motivated by the observation that, once the structure of the network (number of neurons, composition of the regression vector) has been chosen, training the neural network is nothing more than a parameter estimation problem. When the training objective of the network is chosen to be the minimization of the sum of squared errors between the training data and the outputs of the neural network, a nonlinear least-square estimation of the weights of the neural network results. As long as certain conditions are met, the statistical properties of a least-square estimation provides confidence regions for the estimated parameters [14]. In view of this, it is assumed that the *real* system can be described by the following model

$$\boldsymbol{y}_{k+1} = \hat{W} \tanh(\hat{V}\boldsymbol{x}_k + \hat{G}\boldsymbol{u}_k + \hat{\boldsymbol{b}}) + \hat{\boldsymbol{d}} \tag{19.40}$$

where, $\hat{W}=W+\delta W$, $\hat{V}=V+\delta V$, $\hat{G}=G+\delta G$, $\hat{\boldsymbol{b}}=\boldsymbol{b}+\delta\boldsymbol{b}$ and $\hat{\boldsymbol{d}}=\boldsymbol{d}+\delta\boldsymbol{d}$. Recall that the uncertainty description of the closed-loop system (19.26) was obtained by assuming that the uncertainty description (19.23) was available for the open-loop system. Such a description will now be derived for the system (19.40). In this way, the first order of Taylor's expansion around $(\boldsymbol{x}_{k-1}, \boldsymbol{u}_{k-1})$ results in

$$\boldsymbol{y}_{k+1} = \hat{F}_0 + \hat{F}\Delta\boldsymbol{x}_k + \hat{E}\Delta\boldsymbol{u}_k \tag{19.41}$$

with $\hat{F}_0$, $\hat{F}$ and $\hat{E}$, respectively given by

$$\hat{F}_0 = \hat{W} \tanh(\hat{V}\boldsymbol{x}_{k-1} + \hat{G}\boldsymbol{u}_{k-1} + \hat{\boldsymbol{b}}) + \hat{\boldsymbol{d}} \tag{19.42}$$

$$\hat{F} = \hat{W}\hat{\Gamma}_{k-1}\hat{V} \tag{19.43}$$

$$\hat{E} = \hat{W}\hat{\Gamma}_{k-1}\hat{G} \tag{19.44}$$

where

$$\hat{\Gamma}_{k-1} = I - \mathrm{diag}[\tanh^2(\hat{V}\boldsymbol{x}_{k-1} + \hat{G}\boldsymbol{u}_{k-1} + \hat{\boldsymbol{b}})] \tag{19.45}$$

A description is pursued in which the nominal part of (19.41) is described by (19.35), while the uncertain part is expressed in terms of the uncertainty weights δW, δV, δG, $\delta\boldsymbol{b}$ and $\delta\boldsymbol{d}$. Such a description can be obtained by performing the first order of Taylor's expansion of (19.41) around the *nominal* weights of the neural network model, yielding

$$\boldsymbol{y}_{k+1} = F_0 + (F + \delta F)\Delta\boldsymbol{x}_k + (E + \delta E)\Delta\boldsymbol{u}_k \tag{19.46}$$

where

$$\delta F = \delta W \Gamma_{k-1} V + W \Gamma_{k-1} \delta V + W \Gamma'_{k-1} \Phi^V_{k-1} V \tag{19.47}$$

$$\delta E = \delta W \Gamma_{k-1} G + W \Gamma_{k-1} \delta G + W \Gamma'_{k-1} \Phi^G_{k-1} G \tag{19.48}$$

with

$$\Gamma'_{k-1} = \mathrm{diag}[-2\tanh(V\boldsymbol{x}_{k-1} + G\boldsymbol{u}_{k-1} + \boldsymbol{b})] \tag{19.49}$$

$$\Phi^V_{k-1} = \mathrm{diag}[\delta V \boldsymbol{x}_{k-1} + \delta \boldsymbol{b}] \tag{19.50}$$

$$\Phi^G_{k-1} = \mathrm{diag}[\delta G \boldsymbol{u}_{k-1} + \delta \boldsymbol{b}] \tag{19.51}$$

and where F_0, F and E are defined in (19.36), (19.37) and (19.38), respectively. The final step would be to represent (19.46) in the form (19.23) through the procedure outlined in section 19.2.2. Then, an uncertainty description for the linear closed-loop system matrices A and B, defined by δA and δB, can be obtained through similar expressions as given in (19.27).

19.3.3 Towards uncertainty bounds

The remaining task to complete, in order to have all ingredients for robustness analysis available, is to bound the uncertainty interval of the entries of matrices A and B, such that a polytopic uncertainty description can be constructed. Considering the involved expressions, it is not likely that such bounds can be found analytically without introducing a great deal of conservatism.

Alternatively, a two-step approach can be adopted in which the appropriate bounds are determined by a combination of analytical and numerical analysis. In the first step an analytical expression for the bounds, $\delta \tilde{F}$ and δE in (19.23), *at a given operation point* $(\boldsymbol{x}_k, \boldsymbol{u}_k)$, have to be determined. This is straightforward giving a neural network model. Then, the second step consists of performing a numerical search over all admissible $(\boldsymbol{x}_k, \boldsymbol{u}_k)$ in order to find a maximum for each of these bounds. To ensure that only admissible $(\boldsymbol{x}_k, \boldsymbol{u}_k)$ are considered, a reference trajectory $\boldsymbol{v}_k$, as used in the feedback linearizing control law (19.20), will be designed and used to feedback linearize *the model.* As the model output travels along this trajectory, bounds on the uncertainty entries of matrices A and B can be computed, for each encountered operating point. Finally, the maximum bound encountered along that trajectory can be used to construct the polytopic description. Moreover, while this procedure is being performed, it can be automatically checked whether matrix E, as defined in (19.10), is invertible over all encountered operating points, a basic assumption for the success of the proposed control scheme.

19.4 Concluding remarks

The main contribution of this paper is to provide the necessary ingredients for a robust nonlinear control design based on approximate feedback linearization using neural network models. It was shown that parametric uncertainty in the weights of the neural network model can be transported through the proposed control scheme. Moreover, the overall closed-loop uncertainty could finally be expressed in terms of the uncertainty in the weights of the neural network. This result enabled to find a polytopic uncertainty description for the uncertain closed-loop. Based on this polytopic description, robust stability of the uncertain closed-loop could be verified by finding a proper Lyapunov function. Due to the special form of the closed-loop uncertainty description, finding the Lyapunov function could be translated into an LMI problem, and thus be tackled with the use of computationally efficient techniques.

References

1. J. Sjöberg. *Non-linear system identification with neural networks.* Phd thesis, Linköping University, Department of Electrical Engineering, Sweden, 1995.

2. K. S. Narendra and K. Parthasarathy. Identification and control of dynamical systems using neural networks. *IEEE Transactions on Neural Networks*, 1(1):4–27, March 1990.

3. K.J. Hunt and D. Sbarbaro. Neural networks for nonlinear internal model control. *IEE Proceedings-D*, 138(5):431–438, September 1991.

4. K.J. Hunt, D. Sbarbaro, R. Żbikowski, and P.J. Gawthrop. Neural networks for control systems – a survey. *Automatica*, 28(6):1083–1112, 1992.

5. J.A.K. Suykens, J.P.L. Vandewalle, and B.L.R. de Moor. *Artificial neural networks for modelling and control of non-linear systems.* Kluwer Academic Publishers, 1996.

6. J.-J. E. Slotine and Weiping Li. *Applied Nonlinear Control.* Prentice-Hall, New Jersey, USA, 1991.

7. H. Nijmeijer and A. J. van der Schaft. *Nonlinear Dynamical Control Systems.* Springer-Verlag, New York, 1990.

8. J. W. Grizzle. Local input-output decoupling of discrete-time non-linear systems. *International Journal of Control*, 43:1517–1530, 1986.

9. B. Jakubczyk. Feedback linearization of discrete-time systems. *Systems and Control Letters*, 9:411–416, 1987.

10. H.A.B. te Braake. *Neural control of biotechnological processes.* Phd thesis, Delft University of Technology, Control Laboratory, Department of Electrical Engineering, Delft, The Netherlands, 1997.

11. J. Sjöberg. Nonlinear black-box modeling in system identification: a unified overview. *Automatica*, 31(12):1691–1724, 1995.

12. S.A. Billings and W.S.F. Voon. Correlation based model validity test for nonlinear models. *International Journal of Control*, 4(1):235–244, 1986.

13. L. Ljung. *System Identification: Theory for the User.* Prentice-Hall, Englewood Cliffs, New Jersey, 1987.

14. G.A.F. Seber and C.J. Wild. *Nonlinear regression.* John Wiley and Sons, Inc., 1989.

15. M.V. Kothare. *Control of systems subject to constraints.* Phd thesis, California Institute of Technology, California, U.S.A., 1997.

16. M.V. Kothare, V. Balakrihnan, and M. Morari. Robust constrained predictive control using linear matrix inequalities. *Automatica*, 32(10):1361–1379, 1996.

17. S. Boyd, L. El Gahoui, E. Feron, and V. Balakrishnan. *Linear matrix inequalities in system and control.* Vol. 15 of Studies in applied mathematics, SIAM, 1994.

Lecture Notes in Control and Information Sciences

Edited by M. Thoma

1993–1999 Published Titles:

Vol. 186: Sreenath, N.
Systems Representation of Global Climate Change Models. Foundation for a Systems Science Approach.
288 pp. 1993 [3-540-19824-5]

Vol. 187: Morecki, A.; Bianchi, G.; Jaworeck, K. (Eds)
RoManSy 9: Proceedings of the Ninth CISM-IFToMM Symposium on Theory and Practice of Robots and Manipulators.
476 pp. 1993 [3-540-19834-2]

Vol. 188: Naidu, D. Subbaram
Aeroassisted Orbital Transfer: Guidance and Control Strategies
192 pp. 1993 [3-540-19819-9]

Vol. 189: Ilchmann, A.
Non-Identifier-Based High-Gain Adaptive Control
220 pp. 1993 [3-540-19845-8]

Vol. 190: Chatila, R.; Hirzinger, G. (Eds)
Experimental Robotics II: The 2nd International Symposium, Toulouse, France, June 25-27 1991
580 pp. 1993 [3-540-19851-2]

Vol. 191: Blondel, V.
Simultaneous Stabilization of Linear Systems
212 pp. 1993 [3-540-19862-8]

Vol. 192: Smith, R.S.; Dahleh, M. (Eds)
The Modeling of Uncertainty in Control Systems
412 pp. 1993 [3-540-19870-9]

Vol. 193: Zinober, A.S.I. (Ed.)
Variable Structure and Lyapunov Control
428 pp. 1993 [3-540-19869-5]

Vol. 194: Cao, Xi-Ren
Realization Probabilities: The Dynamics of Queuing Systems
336 pp. 1993 [3-540-19872-5]

Vol. 195: Liu, D.; Michel, A.N.
Dynamical Systems with Saturation Nonlinearities: Analysis and Design
212 pp. 1994 [3-540-19888-1]

Vol. 196: Battilotti, S.
Noninteracting Control with Stability for Nonlinear Systems
196 pp. 1994 [3-540-19891-1]

Vol. 197: Henry, J.; Yvon, J.P. (Eds)
System Modelling and Optimization
975 pp approx. 1994 [3-540-19893-8]

Vol. 198: Winter, H.; Nüßer, H.-G. (Eds)
Advanced Technologies for Air Traffic Flow Management
225 pp approx. 1994 [3-540-19895-4]

Vol. 199: Cohen, G.; Quadrat, J.-P. (Eds)
11th International Conference on Analysis and Optimization of Systems – Discrete Event Systems: Sophia-Antipolis, June 15–16–17, 1994
648 pp. 1994 [3-540-19896-2]

Vol. 200: Yoshikawa, T.; Miyazaki, F. (Eds)
Experimental Robotics III: The 3rd International Symposium, Kyoto, Japan, October 28-30, 1993
624 pp. 1994 [3-540-19905-5]

Vol. 201: Kogan, J.
Robust Stability and Convexity
192 pp. 1994 [3-540-19919-5]

Vol. 202: Francis, B.A.; Tannenbaum, A.R. (Eds)
Feedback Control, Nonlinear Systems, and Complexity
288 pp. 1995 [3-540-19943-8]

Vol. 203: Popkov, Y.S.
Macrosystems Theory and its Applications: Equilibrium Models
344 pp. 1995 [3-540-19955-1]

Vol. 204: Takahashi, S.; Takahara, Y.
Logical Approach to Systems Theory
192 pp. 1995 [3-540-19956-X]

Vol. 205: Kotta, U.
Inversion Method in the Discrete-time Nonlinear Control Systems Synthesis Problems
168 pp. 1995 [3-540-19966-7]

Vol. 206: Aganovic, Z.; Gajic, Z.
Linear Optimal Control of Bilinear Systems with Applications to Singular Perturbations and Weak Coupling
133 pp. 1995 [3-540-19976-4]

Vol. 207: Gabasov, R.; Kirillova, F.M.; Prischepova, S.V.
Optimal Feedback Control
224 pp. 1995 [3-540-19991-8]

Vol. 208: Khalil, H.K.; Chow, J.H.; Ioannou, P.A. (Eds)
Proceedings of Workshop on Advances inControl and its Applications
300 pp. 1995 [3-540-19993-4]

Vol. 209: Foias, C.; Özbay, H.; Tannenbaum, A.
Robust Control of Infinite Dimensional Systems: Frequency Domain Methods
230 pp. 1995 [3-540-19994-2]

Vol. 210: De Wilde, P.
Neural Network Models: An Analysis
164 pp. 1996 [3-540-19995-0]

Vol. 211: Gawronski, W.
Balanced Control of Flexible Structures
280 pp. 1996 [3-540-76017-2]

Vol. 212: Sanchez, A.
Formal Specification and Synthesis of Procedural Controllers for Process Systems
248 pp. 1996 [3-540-76021-0]

Vol. 213: Patra, A.; Rao, G.P.
General Hybrid Orthogonal Functions and their Applications in Systems and Control
144 pp. 1996 [3-540-76039-3]

Vol. 214: Yin, G.; Zhang, Q. (Eds)
Recent Advances in Control and Optimization of Manufacturing Systems
240 pp. 1996 [3-540-76055-5]

Vol. 215: Bonivento, C.; Marro, G.; Zanasi, R. (Eds)
Colloquium on Automatic Control
240 pp. 1996 [3-540-76060-1]

Vol. 216: Kulhavý, R.
Recursive Nonlinear Estimation: A Geometric Approach
244 pp. 1996 [3-540-76063-6]

Vol. 217: Garofalo, F.; Glielmo, L. (Eds)
Robust Control via Variable Structure and Lyapunov Techniques
336 pp. 1996 [3-540-76067-9]

Vol. 218: van der Schaft, A.
L_2 Gain and Passivity Techniques in Nonlinear Control
176 pp. 1996 [3-540-76074-1]

Vol. 219: Berger, M.-O.; Deriche, R.; Herlin, I.; Jaffré, J.; Morel, J.-M. (Eds)
ICAOS '96: 12th International Conference on Analysis and Optimization of Systems - Images, Wavelets and PDEs:
Paris, June 26-28 1996
378 pp. 1996 [3-540-76076-8]

Vol. 220: Brogliato, B.
Nonsmooth Impact Mechanics: Models, Dynamics and Control
420 pp. 1996 [3-540-76079-2]

Vol. 221: Kelkar, A.; Joshi, S.
Control of Nonlinear Multibody Flexible Space Structures
160 pp. 1996 [3-540-76093-8]

Vol. 222: Morse, A.S.
Control Using Logic-Based Switching
288 pp. 1997 [3-540-76097-0]

Vol. 223: Khatib, O.; Salisbury, J.K.
Experimental Robotics IV: The 4th International Symposium, Stanford, California, June 30 - July 2, 1995
596 pp. 1997 [3-540-76133-0]

Vol. 224: Magni, J.-F.; Bennani, S.; Terlouw, J. (Eds)
Robust Flight Control: A Design Challenge
664 pp. 1997 [3-540-76151-9]

Vol. 225: Poznyak, A.S.; Najim, K.
Learning Automata and Stochastic Optimization
219 pp. 1997 [3-540-76154-3]

Vol. 226: Cooperman, G.; Michler, G.; Vinck, H. (Eds)
Workshop on High Performance Computing and Gigabit Local Area Networks
248 pp. 1997 [3-540-76169-1]

Vol. 227: Tarbouriech, S.; Garcia, G. (Eds)
Control of Uncertain Systems with Bounded Inputs
203 pp. 1997 [3-540-76183-7]

Vol. 228: Dugard, L.; Verriest, E.I. (Eds)
Stability and Control of Time-delay Systems
344 pp. 1998 [3-540-76193-4]

Vol. 229: Laumond, J.-P. (Ed.)
Robot Motion Planning and Control
360 pp. 1998 [3-540-76219-1]

Vol. 230: Siciliano, B.; Valavanis, K.P. (Eds)
Control Problems in Robotics and Automation
328 pp. 1998 [3-540-76220-5]

Vol. 231: Emel'yanov, S.V.; Burovoi, I.A.; Levada, F.Yu.
Control of Indefinite Nonlinear Dynamic Systems
196 pp. 1998 [3-540-76245-0]

Vol. 232: Casals, A.; de Almeida, A.T. (Eds)
Experimental Robotics V: The Fifth International Symposium Barcelona, Catalonia, June 15-18, 1997
190 pp. 1998 [3-540-76218-3]

Vol. 233: Chiacchio, P.; Chiaverini, S. (Eds)
Complex Robotic Systems
189 pp. 1998 [3-540-76265-5]

Vol. 234: Arena, P.; Fortuna, L.; Muscato, G.; Xibilia, M.G.
Neural Networks in Multidimensional Domains: Fundamentals and New Trends in Modelling and Control
179 pp. 1998 [1-85233-006-6]

Vol. 235: Chen, B.M.
H∞ Control and Its Applications
361 pp. 1998 [1-85233-026-0]

Vol. 236: de Almeida, A.T.; Khatib, O. (Eds)
Autonomous Robotic Systems
283 pp. 1998 [1-85233-036-8]

Vol. 237: Kreigman, D.J.; Hagar, G.D.; Morse, A.S. (Eds)
The Confluence of Vision and Control
304 pp. 1998 [1-85233-025-2]

Vol. 238: Elia, N. ; Dahleh, M.A.
Computational Methods for Controller Design
200 pp. 1998 [1-85233-075-9]

Vol. 239: Wang, Q.G.; Lee, T.H.; Tan, K.K.
Finite Spectrum Assignment for Time-Delay Systems
200 pp. 1998 [1-85233-065-1]

Vol. 240: Lin, Z.
Low Gain Feedback
376 pp. 1999 [1-85233-081-3]

Vol. 241: Yamamoto, Y.; Hara S.
Learning, Control and Hybrid Systems
472 pp. 1999 [1-85233-076-7]

Vol. 242: Conte, G.; Moog, C.H.; Perdon A.M.
Nonlinear Control Systems
192 pp. 1999 [1-85233-151-8]

Vol. 243: Tzafestas, S.G.; Schmidt, G. (Eds)
Progress in Systems and Robot Analysis and Control Design
624 pp. 1999 [1-85233-123-2]

Vol. 244: Nijmeijer, H.; Fossen, T.I. (Eds)
New Directions in Nonlinear Observer Design
552pp: 1999 [1-85233-134-8]

Vol. 245: Garulli, A.; Tesi, A.; Vicino, A. (Eds)
Robustness in Identification and Control
448pp: 1999 [1-85233-179-8]